T0313319

carbon nanotubes

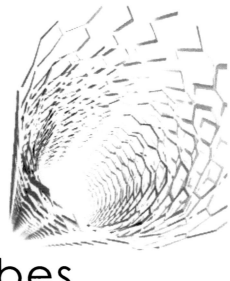

carbon
nanotubes

FROM BENCH CHEMISTRY TO PROMISING BIOMEDICAL APPLICATIONS

EDITOR
GIORGIA PASTORIN
National University of Singapore, Singapore

CRC Press
Taylor & Francis Group
Boca Raton London New York

CRC Press is an imprint of the
Taylor & Francis Group, an **informa** business

CARBON NANOTUBES: FROM BENCH CHEMISTRY TO PROMISING BIOMEDICAL APPLICATIONS

First published 2011 by Pan Stanford Publishing Pte. Ltd.

Published 2019 by CRC Press
Taylor & Francis Group
6000 Broken Sound Parkway NW, Suite 300
Boca Raton, FL 33487-2742

ISBN-13: 978-981-4241-68-7 (hbk)

This book contains information obtained from authentic and highly regarded sources. While all reasonable efforts have been made to publish reliable data and information, neither the author[s] nor the publisher can accept any legal responsibility or liability for any errors or omissions that may be made. The publishers wish to make clear that any views or opinions expressed in this book by individual editors, authors or contributors are personal to them and do not necessarily reflect the views/opinions of the publishers. The information or guidance contained in this book is intended for use by medical, scientific or health-care professionals and is provided strictly as a supplement to the medical or other professional's own judgement, their knowledge of the patient's medical history, relevant manufacturer's instructions and the appropriate best practice guidelines. Because of the rapid advances in medical science, any information or advice on dosages, procedures or diagnoses should be independently verified. The reader is strongly urged to consult the relevant national drug formulary and the drug companies' and device or material manufacturers' printed instructions, and their websites, before administering or utilizing any of the drugs, devices or materials mentioned in this book. This book does not indicate whether a particular treatment is appropriate or suitable for a particular individual. Ultimately it is the sole responsibility of the medical professional to make his or her own professional judgements, so as to advise and treat patients appropriately. The authors and publishers have also attempted to trace the copyright holders of all material reproduced in this publication and apologize to copyright holders if permission to publish in this form has not been obtained. If any copyright material has not been acknowledged please write and let us know so we may rectify in any future reprint.

Visit the Taylor & Francis Web site at
http://www.taylorandfrancis.com

and the CRC Press Web site at
http://www.crcpress.com

British Library Cataloguing-in-Publication Data
A catalogue record for this book is available from the British Library.

Contents

Contributors

 **Giorgia Pastorin** received her MSc in pharmaceutical chemistry and technology in 2000 and her PhD in 2004 from the University of Trieste (Italy), working on adenosine receptors' antagonists. She spent two years as a post-doc at CNRS in Strasbourg (France), where she acquired some skills in drug delivery. She joined the National University of Singapore in June 2006 as Assistant Professor in the Department of Pharmacy–Faculty of Science.

Dr Pastorin's research interests focus on both medicinal chemistry, through the synthesis of heterocyclic molecules as potent and selective antagonists towards different adenosine receptors' subtypes, and drug delivery, through the development of functionalised nanomaterials for a variety of potential therapeutic applications.

She is the editor of this book and co-author in many chapters.

 **Marisa van der Merwe** received a BPharm in 1998 and an MSc in pharmaceutics in 2000 from Potchefstroom University (South Africa). She additionally registered as a pharmacist in 2000 in South Africa. She was awarded a Nelson Mandela Scholarship by the University of Leiden (The Netherlands) to do most of her research for her PhD in pharmaceutics, which she obtained in 2003 from the University of Potchefstroom. Her research during both her MSc and PhD focused on the mucosal delivery of peptide drugs using *N*-trimethyl chitosan chloride as absorption enhancer. She spent a further 18 months as a post-doc at the North West University (South Africa) researching mucosal vaccine delivery for a pharmaceutical company. She joined the University of Portsmouth (England) in September 2004 and is a Senior Lecturer in Pharmaceutics in the School of Pharmacy and Biomedical Sciences. Her research interests include mucosal peptide, protein and vaccine delivery, as well as nanomaterials for drug delivery with a variety of potential therapeutic applications.

She is the main author of Chapter 1 on the functionalisation of carbon nanotubes.

 **Giampiero Spalluto** received his degree in chemistry and pharmaceutical technology in 1987 from the University of Ferrara. He obtained a PhD in organic chemistry from the University of Parma in 1992. Between 1995 and 1998 he was Assistant Professor of Medicinal Chemistry at the University of Ferrara. Since November1998, he has held the position of Associate Professor of Medicinal Chemistry at the University of Trieste and is a member of the Italian Chemical Society since 1989 (Medicinal Chemistry and Organic Chemistry divisions). Dr Spalluto's scientific interests have focused on the enantioselective synthesis of natural compounds and the structure activity relationships of ligands for adenosine receptor subtypes and antitumor agents. He has authored more than 150 articles published in international peer-reviewed journals.

He is the main author of Chapter 2 on carbon nanotubes for drug delivery.

 **Tatiana Da Ros** received her MSc in pharmaceutical chemistry and technology in 1995 and her PhD in medicinal chemistry in 1999.

She worked as post-doc at the Pharmaceutical Sciences' Department in Trieste and spent many periods abroad visiting Prof. Wudl's group at UCLA (USA) in 1999, Prof. Taylor's lab at Sussex University (UK) in 2000, the Biophysique lab at Museum National d'Histoire Naturelle (France) in 1999, 2000, 2001 and 2002, and Dr Murphy's group at the MRC in Cambridge (UK) in 2004. In 2002 she joined the Faculty of Pharmacy in Trieste as Assistant Professor.

Dr Da Ros's research is mainly focused on the study of fullerene and carbon nanotube derivatives' biological applications. She is the co-author of about 70 articles on peered international journals and of different book chapters. She is co-organiser of the annual symposium dedicated to the bioapplications of fullerenes, carbon nanotubes and nanostructures, in the Electrochemical Society Spring Meeting and co-editor of *Medicinal Chemistry and Pharmacological Potential of Fullerenes and Carbon Nanotubes* (Springer, 2008).

She is the main author of Chapter 3 on carbon nanotubes for cancer therapy.

Li Jian received his BSc in pharmacy in 2004 from Shanghai Jiao Tong University (China). He entered the National University of Singapore in January 2009 as a PhD candidate in the Department of Pharmacy–Faculty of Science. His research is focused on carbon nanotubes as drug delivery system.

He is the main author of Chapter 4 on carbon nanotubes for the delivery of vaccines and immunostimulants.

Venkata Sudheer Makam received his MSc in industrial chemistry from the Technical University of Munich (TUM) and National University of Singapore (NUS) in 2008, during which time he did his thesis, "Biocatalytical and Expression Studies of β-Aminopeptidases," at Swiss Federal Institute of Aquatic Science and Technology, Switzerland. Later, he started his career as a research assistant in the Biophysics laboratory at the National University of Singapore. In 2009, he joined Dr Giorgia Pastorin's group as research assistant in the Department of Pharmacy, NUS, where he focuses on lab-on-a-chip devices for cancer diagnostics. Makam is currently doing his PhD in the same group.

He is the main author of Chapter 5 on carbon nanotube–nucleic acid complexes.

Cécilia Ménard-Moyon received her MSc in organic chemistry in 2002 from the University of Pierre et Marie Curie in Paris. She obtained her PhD in 2005 at CEA/Saclay (France) working in the group of C. Mioskowski on carbon nanotubes and their applications for optical limitation, nanoelectronics, and the development of novel methods of functionalisation.

In 2006 she worked as a post-doc in the group of Richard J. K. Taylor on the total synthesis of a natural product ('upenamide) and on the development of novel methods of synthesis of heterocycles. She then joined, for 18 months, the R&D department of Nanocyl in Belgium, one of the main European producers of carbon nanotubes, and worked on the synthesis, dispersion and functionalisation of carbon nanotubes.

Since October 2008, Dr Ménard-Moyon holds the position of Researcher at CNRS in the group of A. Bianco in Strasbourg. Her research interests focus on the functionalisation of carbon nanotubes for the vectorisation of biologically active molecules.

She is the main author of Chapter 6 on the influence of carbon nanotubes in neuronal living networks and of the overview (Chapter 9) on the main research activities on carbon nanotubes in the world.

Yupeng Ren received his PhD in pharmaceutical sciences from the National University of Singapore (Singapore) in 2007, working on protein cages of plant viruses as potential anti-cancer drug delivery system. After finished his PhD, he worked as a research assistant at the Department of Pharmacy for one year and developed nano-drug delivery systems from carbon nanotubes. From November 2007 to January 2008, Dr Ren worked as an analyst for the Shanghai Institute for Food and Drug Institute. In February 2008, he joined the Shanghai Institute of Materia Medica, Chinese Academy of Sciences. As Associate Professor, his research is focused on the applications of nano-systems on drug delivery and analysis.

He is the main author of Chapter 7 on carbon nanotubes as biosensing and bio-interfacial materials.

Tapas Ranjan Nayak received his MTech in biochemical engineering and biotechnology in 2006 from the Indian Institute of Technology, Khargapur (India). He is currently continuing with his PhD at the National University of Singapore (Singapore). His research interests focus on toxicological studies and biomedical applications involving various nanomaterials such as carbon nanotubes, zinc oxide nanofibres and graphene.

He is the main author of Chapter 8 on the toxicity of carbon nanotubes.

Preface

Nanotechnology is a fast-emerging, sophisticated discipline that involves the study and manipulation of matter at atomic dimensions. It holds great promise to revolutionise and impact scientific research and industry, with opportunities for discovering new and exciting phenomena. This is largely due to nanotechnology being so different and counter-intuitive from previous technologies, resulting in past experience providing very little guidance about how to proceed. The fact that nanotechnology is *the technology of the 21st century* does not represent an exaggerated view of an ephemeral phenomenon, but instead echoes a real and immediate need for an extensive, "in-depth" investigation of what the synergy between Mother Nature and human ingenuity has to offer. Scientists, as is usual to their nature, have risen to the challenge with great gusto. This has led, among other things, to the realisation of advanced and extremely precise instruments that capitalise on the fact that material in the nanoscale dimensions allows integrated and compact systems to be fabricated. Nanotechnology includes not only great challenges such as the use of nanomaterials in novel scientific applications but also the understanding and manipulation of biological specimen at its fundamental levels. Carbon-based materials, among which carbon nanotubes (CNTs) represent a fascinating example, have shown extraordinary effects. CNTs represent interesting materials not only because they have high mechanical stability and nanoscale dimensions, but also because, depending on how the constitutive graphene sheets are rolled up, they share electronic properties of both metals and semiconductors. In addition, differently from spherical nanoparticles, they present a large inner volume that could be filled with several biomolecules ranging from small derivatives to proteins. This offers the advantage to load the inside of CNTs with a drug, while imparting chemical properties through the functionalisation of the external walls and thus rendering these tubes water-soluble and biocompatible.

However, there also exist cautious, almost mistrustful, but justified, opinions on nanotechnology and its consequences. A good reason is the effect on personal health or environmental pollution, because nanoparticles might escape the normal phagocytic defences in the body or might fluctuate and accumulate in the atmosphere. The reason behind such scepticism is that there is the general consciousness that the laws of physics and chemistry are pretty different when particles get down to the nanoscale. As a consequence, even substances that are normally innocuous can trigger intense chemical reactions and biological anomalies as nanospecies.

This has led to the stimulation of attitudes for and against this new science. This book addresses both these aspects by offering a general overview of the main factors that render CNTs unique for further promising applications, as well as the potentially risky aspects associated with these still-unknown carbon-based nanomaterials. It is particularly suitable for young scientists who have been involved in nanotechnology recently, or those who are simply curious about one of the most debated topics of their generation. The main authors of the present volume have been specifically picked from the pool of expert researchers and professors involved in nanotechnologies, but who are younger than 50, with the intention of providing dynamic visions and fresh perspectives of the actual "state of the art" of CNTs. To reiterate, the common undeniable opinion is that, although it is too early to say whether these "nano-structures" will wean the world from its current limitations, or monumentally backfire to cause harm, a superficial understanding might provide good ideas, but a deep knowledge favours great discoveries, even at the nanoscale.

Giorgia Pastorin

Chapter 1

STABILISATION OF CARBON NANOTUBE SUSPENSIONS

Dimitrios G. Fatouros,[a] Marta Roldo[b] and Susanna M. van der Merwe[b]

[a] *Department of Pharmaceutical Technology, School of Pharmacy,*
Aristotle University of Thessaloniki, 54124 Thessaloniki, Greece
[b] *School of Pharmacy and Biomedical Sciences, St Michael's Building,*
White Swan Road, Portsmouth PO1 2DT, UK
marisa.vdmerwe@port.ac.uk

1.1 INTRODUCTION

Carbon nanotubes (CNTs) are allotropes of carbon that are composed entirely of carbon atoms arranged into a series of condensed benzene rings, known as graphene sheets, "rolled" into a cylindrical structure. They belong to the family of fullerenes, the third allotrope of carbon after graphite and diamond.[1-3]

CNTs can be classified into two categories according to their structure: (i) single-walled carbon nanotubes (SWCNTs), comprising a single cylindrical graphene layer capped at both ends in a hemispherical arrangement of carbon networks, and (ii) multi-walled carbon nanotubes (MWCNTs), consisting of numerous concentric cylinders of graphene sheets. SWCNTs have outer diameters ranging from 1.0 to 3.0 nm and inner diameters ranging from 0.6 to 2.4 nm, whereas for MWCNTs, outer diameters range from 2.5 to 100 nm and inner diameters range from 1.5 to 15 nm. MWCNTs can consist of varying amounts of concentric SWCNT layers, which are separated by a distance of approximately 0.34 nm (Fig.1.1).[4,5]

CNTs are highly versatile because of their physicochemical features. They possess ordered structures with a high ratio of length to diameter (aspect ratio) and are ultra-light-weight.

Carbon Nanotubes: From Bench Chemistry to Promising Biomedical Applications
Edited by Giorgia Pastorin
Copyright © 2011 Pan Stanford Publishing Pte. Ltd.
www.panstanford.com

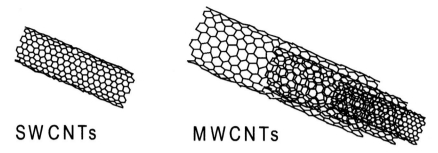

SWCNTs   MWCNTs

Figure 1.1 Schematic representation of CNTs in the form or either single-walled (SWCNTs) or multi-walled (MWCNTs) tubes.

They have high mechanical and tensile strength and high electrical and thermal conductivity. They exhibit both semi-metallic and metallic behaviour and have large surface areas.[3] They possess outstanding chemical and thermal stability.[6] The interaction between cells and CNTs, and hence their internalisation into cells, needs to be clarified to ascertain their future potential as drug delivery systems.[7] Numerous studies have been conducted using biocompatible CNTs whereby CNTs have undergone covalent or non-covalent functionalisation rendering them soluble in aqueous media, and hence biologically compatible.[8] Overall, the general consensus is that CNTs are capable of crossing many types of cell membranes; however, the mechanism by which this occurs is not clearly understood, and there are discrepancies between different authors. CNTs have been shown to be internalised within cells by using a simple tracking process of CNTs labelled with a fluorescing agent.[9]

It was initially observed that CNTs were capable of penetrating the cell membrane primarily via a passive and endocytosis process. This was confirmed to occur depending on the cell type and CNT characteristics, such as surface charge or the nature of the functional groups attached to the CNTs.[7] A hypothesis of functionalised carbon nanotubes (*f*-CNTs) acting as "nanoneedles" (Fig. 1.2) was proposed on the basis of images obtained from high-resolution transmission electron microscopy (TEM), which showed CNT interaction with mammalian cells where the CNTs adopted a perpendicular orientation towards the plasma membrane of the cells during the process of internalisation within the cells. It has been shown that CNTs can passively traverse numerous types of cell membranes via a translocation mechanism termed the *nanoneedle mechanism*.[5,7] These nanoneedles are the tiniest of needles that have the potential to channel therapeutic agents into tumour cells.

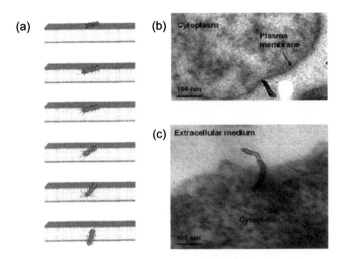

Figure 1.2 CNTs acting as nanoneedles. (a) A schematic of a CNT crossing the plasma membrane; (b) a TEM image of MWNT-NH$_3^+$ interacting with the plasma membrane of A549 cells; (c) a TEM image of MWNT-NH$_3^+$ crossing the plasma membrane of HeLa cells. Reproduced from Lacerda *et al.*[7] with permission. See also Colour Insert.

Administration of free drugs has numerous limitations: limited solubility, poor biodistribution, lack of selectivity, unfavourable pharmacokinetics, as well as the propensity to cause collateral damage to healthy tissue. A drug delivery system allows for the enhancement of the pharmacological and therapeutic profiles of free drugs.

Advances in nanotechnology have resulted in CNTs being used as pharmaceutical excipients and as building blocks for delivery systems. CNTs have been shown to exhibit properties that are desirable for efficient drug delivery systems, such as the ability to achieve controlled and targeted delivery. The interaction between CNTs and pharmaceutically active compounds can occur in three ways. First, the CNT can act as a porous matrix which entraps active compounds within the CNT mesh or bundle (Fig. 1.3a). Second, the compound can attach itself to the exterior surface of the CNT (Fig. 1.3b). The final mechanism of interaction involves the interior channel of CNTs acting as a "nanocatheter" or "nanocontainer" (Fig. 1.3c).[10]

The purpose of targeted drug delivery is to enhance the efficiency, while diminishing the noxious effects, of the therapeutic agent. CNTs can chemically undergo surface modification to achieve targeted delivery by attachment of ligands to the functional groups on the CNT surface. These ligands, which are specific to certain receptors, can carry CNTs directly to the specific site without affecting non-target sites. On the other hand, diagnostic moieties like

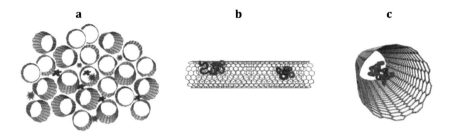

Figure 1.3 A schematic representation of how drugs can interact with CNTs. (a) A bundle of CNTs can act as a porous matrix encapsulating drug molecules between the grooves of individual CNTs. (b) Drug moieties can be attached to the exterior of a CNT either by covalent bonding to the CNT wall or by hydrophobic interaction. (c) The drug can be encapsulated within the internal nanochannel of a CNT. Reproduced from Foldvari and Bagonluri[10] with permission. See also Colour Insert.

fluorescein isothiocyanate (FITC) can also be attached to CNTs for probing their way to the cell nucleus. CNTs can also act as controlled-release systems for drugs by releasing the loaded drugs for a long period of time. In this way CNTs can be used multifunctionally for drug delivery and targeting.

1.2 FUNCTIONALISED CNTS FOR DRUG DELIVERY

From a pharmaceutical perspective, solubility of CNTs in a biological milieu is essential for biocompatibility, and therefore CNTs must be dispersed before they are incorporated in therapeutic formulations. CNT dispersions should also be uniform and stable to ensure that accurate data can be obtained *in vivo*.

The main obstacle in the application of CNTs in drug delivery is that pristine CNTs (non-functionalised) are inherently hydrophobic and hence have poor solubility in most solvents compatible with the aqueous-based biological milieu. CNTs also have a tendency to aggregate to form large bundles which also contribute to their inability to form stable suspensions in aqueous solutions.[11] The aggregation of the CNTs is a result of van der Waals (VDW) attractive forces, hydrophobic interactions and π stacking between individual CNTs.[12] VDW attractions supersede any existing electrostatic or steric repulsive forces that may render these suspensions thermodynamically unstable.[13]

To overcome this barrier and render CNTs more hydrophilic, CNTs can be structurally modified by functionalisation with different functional groups through adsorption, electrostatic interaction or covalent bonding of different molecules.[2] The two main approaches adopted for CNT modification are the covalent and non-covalent modification. Covalent modification is when the sidewalls or defect sites can be modified by various grafting reactions or covalent binding of hydrophilic moieties to the CNT surface, which enhances their solubility and biocompatibility profiles. Non-covalent adsorption or wrapping of various functional molecules is used to form supramolecular complexes. Typical examples of molecules that can be adsorbed onto the hydrophobic surface of CNTs to form stable suspensions are surface-active agents, which include surfactants, synthetic molecules and biopolymers.[2]

The stability of non-covalently functionalised CNT dispersions depends on the efficiency of the physical wrapping of molecular units around CNTs. This "physical wrapping" involves forces that are relatively weaker than those involved in covalent functionalisation, and hence the latter is expected to produce the most stable dispersion. However, covalent functionalisation alters the electronic structure of CNTs and hence potentially also affects their physical properties.[2] Non-covalent chemical modification of CNTs is particularly attractive as it offers the facility of associating functional groups to the CNT surface without modifying the π system (conjugation) of the graphene lattice and thereby not modifying their electrical or physical properties.[14] This indicates non-covalent modification of CNTs to be the preferred approach, and in the following section we will focus on surfactant adsorption onto the CNT surface in order to obtain stable and homogeneous aqueous dispersions.

1.3 SURFACE-ACTIVE AGENTS IN STABILISING CNT SUSPENSIONS

Surface-active agents have a tendency to accumulate at the boundary between two phases because of their amphiphilic nature, whereby they exhibit both hydrophilic and lipophilic properties. Surfactant molecules possess a hydrophobic "tail" and a hydrophilic "head", which have been shown to lower the interfacial tension between insoluble particles and the suspending medium through adsorption onto the insoluble particles. This process enables particles to be dispersed in the form of a suspension. The hydrophobic regions usually consist of saturated or unsaturated hydrocarbon chains, rarely heterocyclic or aromatic systems. The hydrophilic regions can be anionic (negatively charged), cationic (positively charged) or non-ionic (no charge). Surfactants are usually classified by the charge and nature of the hydrophilic portion.[15]

The surface tension of a surfactant solution reduces as the concentration of the surfactant increases where an increasing number of molecules enter the interfacial layer. At a particular concentration termed the *critical micelle concentration* (CMC), this layer becomes saturated and the surfactant molecules adopt a supramolecular micellar structure in which the hydrophobic regions of the surfactant molecules orient themselves in the core of the almost spherical aggregates, termed *micelles*. This shields the hydrophobic components of the surfactant molecules from the aqueous environment and positions the hydrophilic regions towards the outer surface of the micelles. The outermost part of the micelle is hence composed of these hydrophilic groups which maintain the solubility of the aggregates in an aqueous environment.[15]

Surfactants have the ability to suspend individual CNTs by distributing the charges over the graphitic surface and by modifying the particle-suspending medium interface, which prevents their re-aggregation over longer periods of time.[16] They provide an additional repulsive force (electrostatic and steric) which reduces the surface energy and alters the rheological surface properties, which in turn contribute to enhancing suspension stability.[13]

Micelles are increasingly being employed as solubilising and stabilising agents for nanoparticles, such as CNTs, for two reasons. Firstly, they act to stabilise and hence disperse the inherently hydrophobic CNTs, but they also reduce their high toxicity.[17] Sodium dodecyl sulphate (SDS) is an example of a traditional surfactant, one of the most widely used and extensively studied surfactants; however, it only produces stable CNT suspensions at very high concentrations, and SDS itself has raised concern regarding toxicity issues.[18]

Phospholipids are natural amphiphiles that occur in the cell membrane. They are therefore biocompatible and pose significantly less risk than other non-biocompatible surfactants. It has been found that lysophospholipids (Fig. 1.3), or single-tailed phospholipids, can form supramolecular complexes with SWCNTs and offer unsurpassed solubility for SWCNTs compared with other surfactants such as SDS.[19] A comparison of SWCNT solubility with four different pure phospholipids – lysophosphatidylcholine (LPC), dimyristoyl phosphatidylcholine (PC), 1,2-dioleoylphosphatidylglycerol (PG) and 1,2-dipalmitoylphosphatidylethanolamine (PE) – in a phosphate buffered saline (PBS) solution showed complete solubilisation of CNTs by LPC following one hour of bath sonication.[11] In the same paper, by Wu *et al.*,[11] a comparison of SWCNT solubility in LPC, lysophosphatidylglycerol (LPG) and SDS solutions revealed LPC to show superiority over the other two lipid agents in dispersing CNTs in PBS. The authors attributed this to LPC's possessing a bulkier head group for interaction with water and a longer acyl chain for binding to SWCNTs. Furthermore, the experimental data in this article revealed that LPC exhibited enhanced binding affinity for SWCNTs compared with LPG and that single-chain phospholipids showed exceptional solubilisation of SWCNTs while double-chained phospholipids were ineffective.[11] It has been recently shown that the binding of lysophospholipids onto CNTs is dependent on the charge and geometry of the lipids and the pH of the solvent and is not affected by the temperature of the solvent.[12] Additionally, it has been demonstrated that solubilising SWCNTs with lysophospholipids is more effective than solubilising them with nucleic acids, including both single-stranded (ss)

DNA and RNA and proteins such as bovine serum albumin (BSA).[11] L-α-Lysophosphatidylcholine (LPC), depicted in Fig. 1.4, is a major component of oxidised low-density lipoproteins (LDLs). It is a signalling molecule that occurs naturally in cell membranes and thus promotes even greater biocompatibility of SWCNTs when associated with it.

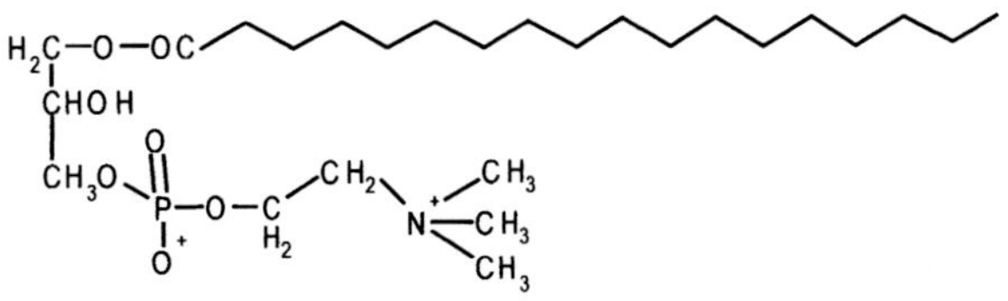

LPC 18:0

Figure 1.4 Structure of LPC 18:0. The numbers "18" and "0" in LPC 18:0, respectively, denote the number of carbon atoms and double bonds in the fatty acyl chain.

As described previously, there are three main models that illustrate the adsorption of surfactants onto CNTs.[19] Table 1.1 displays the schematic representations of these models by which surfactants disperse SWCNTs.

Table 1.1 Schematic representations of the existing models illustrating surfactant interaction with SWCNTs when forming stable dispersions[a]

1.	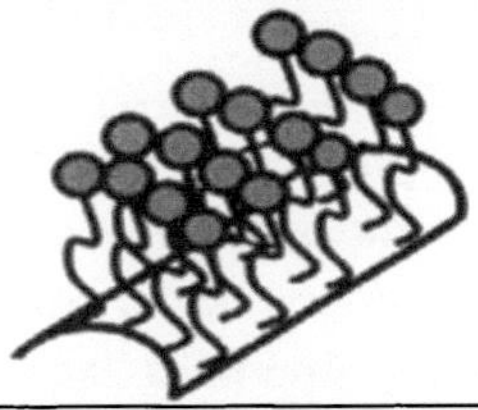	The SWCNT is encapsulated in a cylindrical surfactant micelle. In this diagram, only a portion of the CNT is shown as a curved surface.
2.	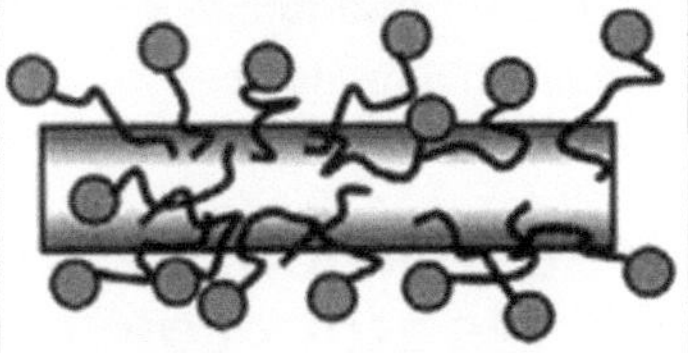	The surfactant molecules randomly adsorb onto the CNT surface. The CNT is represented by the grey cylinder.
3.	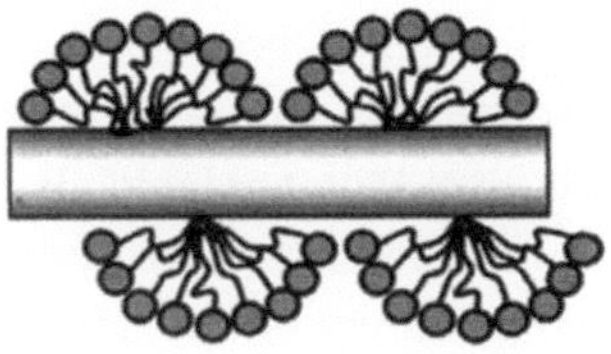	Hemi-micelles of surfactant molecules adsorb onto the CNT surface.

[a] Reproduced from Ke[12] with permission.

To assess the stacking motif of LPC micelles onto the CNT sidewalls, atomic force microscopy (AFM) imaging was utilised.[20] An uneven distribution of micelles over SWCNT surfaces was observed. The calculated height value for the uncoated part of the nanostructure was 1.4 ± 0.1 nm, typical of individual single-walled nanotubes. In sharp contrast, the height value for the coated region of the nanostructure was *ca.* 7.4 ± 0.4 nm (*n* = 10). The increased height values can be attributed to the presence of lipid moieties coating the graphitic surface.

In a recent study, single bilayer membranes of 1-palmitoyl-2-oleoyl-*sn*-glycero-3-phosphocholine (POPC) were fused onto a network of hydrophobic CNTs. By doping the nanotubes to enhance hydrophilicity, it was possible to create a structure that may act as a nanoporous support for a single lipid bilayer.[21] Such systems might be used for biomaterials or biosensors.

In another approach, phosphatidylserine (PS)-coated SWCNTs were used for targeted delivery into macrophages to control their functions, including inflammatory responses to SWCNTs themselves. More specifically, PS-coated SWCNTs were able to successfully deliver cytochrome c (cyt c), a pro-apoptotic death signal, and cause apoptosis in macrophages, emphasizing that non-covalent modification of SWCNTs with specific phospholipid molecules can be employed for targeted delivery and regulation of phagocytes.[22]

Lipid vesicles composed of 1-stearoyl-2-oleoyl-*sn*-glycero-3-[phospho-L-serine] sodium salt (SOPS), 1-palmitoyl-2-oleoyl-*sn*-glycero-3-phosphocholine (POPC), and 2-(4,4-difluoro-5-methyl-4-bora-3a,4a-diaza-s-indacene-3-do-decanoyl)-1-hexadecanoyl-*sn*-glycero-3-phosphocholine (BODIPY-PC) in the ratio 75:23:2 (SOPS:POPC:BODIPY-PC) were incubated with polymer-coated CNTs to produce self-assembled phospholipids into tubular one-dimensional geometry. Given that lipid membranes can support a large number of membrane proteins and receptors, *f*-CNTs with lipid bilayers could further be employed for utilising membrane proteins in nanodevices.[23]

1.4 STABILISATION OF AQUEOUS SUSPENSIONS OF CARBON NANOTUBES BY SELF-ASSEMBLING BLOCK COPOLYMERS

Self-assembling polymers offer an ideal system for the non-covalent stabilisation of hydrophobic molecules in aqueous solutions. They tend to form micellar systems with a hydrophobic core that can be loaded with the water-insoluble molecule and a hydrophilic corona that interacts with the surrounding water. Owing to the hydrophobic nature of CNTs, these micellar systems are very effective for aqueous stabilisation.

Pluronics, or poloxamers, are commercially available block copolymers that are extensively used in drug delivery. These consist of three polymeric blocks, poly(ethylene oxide)$_x$-poly(propylene oxide)$_y$-poly(ethylene oxide)$_z$ (PEO-PPO-PEO), with the central block being the hydrophobic one, forming the core of the micelle, and the PEO being the hydrophilic block, forming a hydrated micellar corona responsible for the biocompatibility of the polymers and the prolonged *in vivo* circulation time of these systems.[24] Efficient use of pluronics in the preparation of stable suspensions of individual SWCNTs and MWCNTs was shown by TEM images. This was found to be true also for polymer solutions of concentrations lower than the critical micellar concentration (CMC) and at temperatures below the critical micellar temperature (CMT); in fact it was found that the presence of SWCNTs affected the value of the polymer CMT, suggesting that a new type of hybrid between CNTs and pluronics is formed.[25] Differential scanning calorimetry (DSC) studies revealed that SWCNTs form larger aggregates compared with those formed by the polymer alone, while MWCNTs form smaller aggregates than both SWCNTs and the polymer alone, because the small diameter of the SWCNTs does not induce perturbation of the dynamics of polymeric assembly. Therefore, the structure of the system is very similar to that of the original micelle but with an elongated form. While the bigger diameter of MWCNTs does not allow formation of the micelle as the core diameter, it is smaller than the diameter of CNTs, and therefore the polymer adsorbs to the MWCNTs forming a different type of aggregate.[25] These findings are further confirmed by spin probe electron paramagnetic resonance (EPR) data, suggesting that the formation of micelles in the presence of an SWCNT-polymer hybrid is suppressed and the composite nanostructure dominates the system.[26]

A similar approach to the stabilisation of CNTs in an aqueous environment was applied by Wang *et al.*,[24] who used the triblock copolymer poly(ethylene glycol)-poly(acrylic acid)-poly(styrene) (PEG-PAA-PS) (Fig. 1.5a). The rationale behind the choice of this polymer is based on the fact that the PS end can interact with the hydrophobic sidewalls of the nanotubes, while the PEG end will stabilise the complex in the aqueous environment. The PAA core has been introduced to allow cross-linking of the polymer once the CNT-polymer complex is formed; this was identified as a way to improve *in vivo* stability, as previously prepared CNT-pluronic complexes were found to undergo polymer displacement by blood proteins.[27] The so-called SWCNT PEG-eggs showed efficient water dispersion and improved *in vivo* stability, and at the same time the cross-linked coating did not diminish the CNTs' intrinsic near-infrared (NIR) fluorescence, which could be exploited in *in vivo* imaging, and the complex did not show acute cytotoxicity.[27,28]

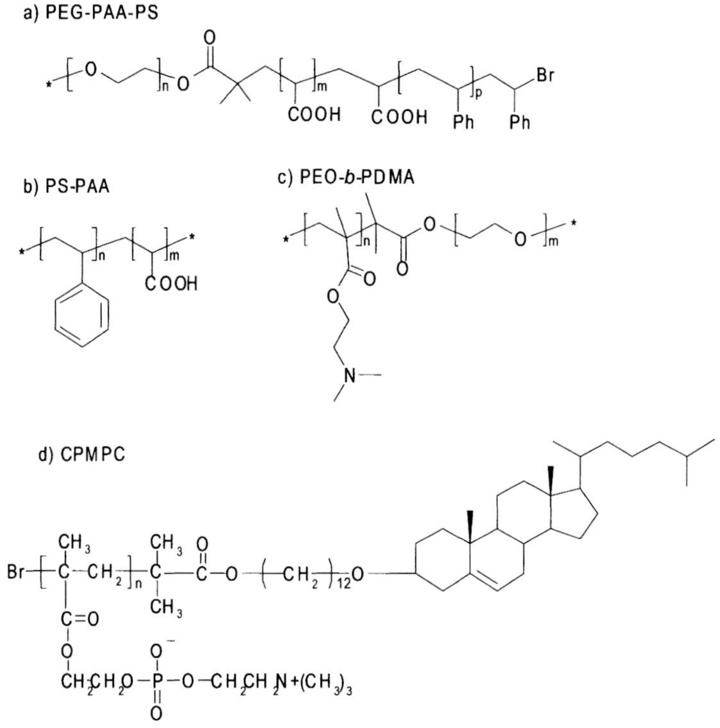

a) PEG-PAA-PS

b) PS-PAA

c) PEO-*b*-PDMA

d) CPMPC

Figure 1.5 Chemical structure of block copolymers: (a) PEG-PAA-PS, (b) PS-PAA, (c) PEO-*b*-PDMA, and (d) CPMPC.

This cross-linking method was previously successfully attempted with poly(styrene)-block-poly(acrylic acid) (PS-PAA) (Fig. 1.5b) and produced complexes soluble and stable for weeks in both hydrophilic and hydrophobic solvents.[29] In both cases, the encapsulation of the SWCNTs preceded the cross-linking process; the SWCNTs and the polymer were initially mixed in a solvent that solvates all the polymer blocks but that does not induce micellar formation, and water was then added dropwise to the CNT-polymer mixture to induce the stepwise formation of the hybrid, as shown in Fig. 1.6.[24,29]

Non-covalent modification of CNTs is also achieved by zwitterionic interaction between the carboxylic groups present on the surface of oxidised nanotubes and a polycationic polymer; this type of interaction is pH dependent, and such a characteristic could have important applications in drug delivery and CNT purification.[30]

Zwitterionic interactions between the double-hydrophilic block copolymer poly(ethylene oxide)-*b*-poly[3-(*N,N*-dimethylamino-ethyl) methacrylate] (PEO-*b*-PDMA) and oxidised SWCNTs were confirmed by ^{1}H-NMR, where the

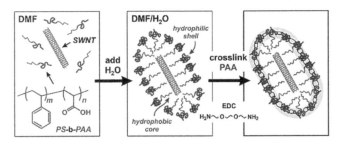

Figure 1.6 Schematic representation of the mechanism of encapsulation of SWCNTs into block copolymers. Reproduced from Kang *et al.*[29] with permission. See also Colour Insert.

peaks of the protons next to the amino groups are shifted downfield. Furthermore, thermogravimetric analysis (TGA) data showed that 26% wt of the complex was formed by the polymer; it was also found that the grafting procedure reached saturation when the polymer was employed at a concentration of ~10 mg/mL. In saturation condition, the complex presented an excess of free amino groups ($NH_2/COOH = 1.4$).[30]

Xu *et al.*[31] created a novel biocompatible block copolymer, cholesterol-end-capped-poly(2-methacryloyloxyethyl phosphorylcholine) (CPMPC) that formed complexes with MWCNTs by simple mixing and brief sonication (30 s); this polymer showed great efficacy in individually suspending MWCNTs in water up to concentrations of 3.307 mg/mL (Fig. 1.7).[17]

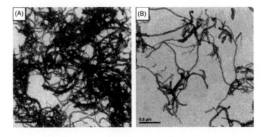

Figure 1.7 TEM images of (a) pristine CNTs and (b) CPMPC-coated CNTs. The images show the effective isolation of individual nanotubes by the formation of CNT-block copolymer complexes. Reproduced from Xu *et al.*[17] with permission.

The use of block copolymers in the stabilisation of aqueous suspensions of CNTs has so far been demonstrated to be a very promising approach to the preparation of CNT-polymer complexes with stability and biocompatibility characteristics that will allow their use *in vivo*. This is a very promising advance towards the development of novel systems for drug delivery, gene transfection, *in vivo* imaging and targeted thermoablation.[24] The major advantage offered by the use of self-assembling block copolymers is that covalent modification of CNTs' sidewalls, which introduces defects, reduces the strength of the nanotubes and induces a perturbation of their electronic structure, is not required.[32]

1.5 STABILISATION OF AQUEOUS SUSPENSIONS OF CARBON NANOTUBES BY CHITOSAN AND ITS DERIVATIVES

Covalent and non-covalent stabilisation of aqueous suspensions of CNTs has recently been attempted by several research groups by using chitosan and its derivatives. Chitosan is a biocompatible, safe, stable and biodegradable semi-synthetic polymer obtained by alkaline deacetylation of chitin (Fig. 1.8), the most abundant natural polysaccharide after cellulose.[33–35] Chitosan is a linear, semi-rigid polysaccharide that presents a rigid crystalline structure through inter- and intramolecular hydrogen bonding.[36] It is insoluble in water and most organic solvents but soluble in aqueous diluted acids as the amine groups of the polymer become protonated and result in a soluble, positively charged polysaccharide that has a high charge density.[37]

Chitosan exhibits a wide variety of activities; its immuno-stimulatory effect has been investigated for applications in wound healing and regenerative medicine,[38] and its antibacterial effect has also been studied for biomedical and food industry applications.[39] Furthermore, chitosan has been found to be a versatile carrier for biologically active species and drugs because of its

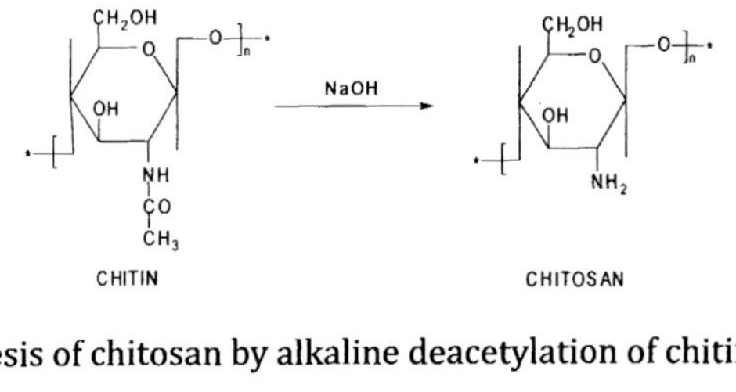

Figure 1.8 Synthesis of chitosan by alkaline deacetylation of chitin.

properties as an absorption enhancer and mucoadhesive polymer,[40,41] and it has also been extensively studied as a vaccine and gene therapy carrier.[42,43] The presence of functional groups such as primary and secondary hydroxyl groups and amino groups along the polymeric backbone have allowed for easy modification of the polysaccharide and led to the synthesis of derivatives with new and improved properties.[44–48] Owing to the variety of properties and possible applications of chitosan and its derivatives, a lot of interest has been raised in their use for the preparation of polymer-CNT composites.

The preparation of stable aqueous suspensions of CNTs by non-covalent hybridisation with chitosan is relatively simple. Most authors report mixing pristine CNTs with 0.5–1% wt/wt acidic solutions of chitosan, followed by sonication for 1 to 5 h to help dispersion and centrifugation to eliminate any large bundle still present in the suspension.[49-54] Similar methods have been employed with oxidised CNTs (CNT-COOH)[55,56] or chitosan derivatives.[57] However, Long *et al.*[58] reported on a different approach: they ground the CNTs and the polymer together in a mortar and washed the granular solid obtained with water prior to its resuspension in the same solvent. The suspensions obtained with both methods were observed to be stable at room temperature for long periods of time; this stabilisation has been thermodynamically explained as the disruption of intertubular attractions and reduced hydrophobicity of the carbon surface in contact with water.[59]

To obtain a permanent coating, Liu *et al.*[60] used controlled chitosan surface deposition followed by glutaraldehyde cross-linking, which resulted in CNT surfaces decorated with a non-destroyable coating (Fig. 1.9).

Covalent binding of the polysaccharide on the surface of CNTs has also been successfully attempted following different methods.

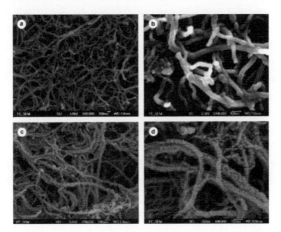

Figure 1.9 SEM images of pristine MWCNTs (a and b), and the corresponding chitosan surface-decorated MWCNTs (c and d). Reproduced with permission from Liu *et al.*[60] with permission.

Treatment of carboxylated CNTs with acidic solutions of chitosan at 98°C for 24 hours produced stable chitosan-grafted CNT derivatives, in which the chitosan content was about 25% wt of the composite, that resulted in improved water stability.[61] Other groups applied a similar method of modification to MWCNTs, but these were initially shortened to improve their dispersivity in organic solvents and their reactivity.[62] The shortened MWCNTs were then subjected to oxidation, which led to the formation of carboxylated CNTs; these were further modified to obtain reactive acyl chloride groups, used for the functionalisation with chitosan of 9 kDa[62] and 100 kDa.[63] Successful covalent bonding between the polysaccharide and the fullerene pipes was demonstrated by Fourier transform infrared spectroscopy (FTIR), X-ray photoelectron spectroscopy (XPS) and nuclear magnetic resonance (NMR).[62] The data collected showed that both the chitosan amino group in C2 and the hydroxyl group in C6 reacted with the carboxylic groups on the CNTs' surface when chitosan with a degree of deacetylation of 84.7% was used,[62] while only the amino groups reacted when a completely deacetylated chitosan was employed.[63] Furthermore, H-bonding between the free amino groups on the polymer and the free carboxylic groups on the nanotubes' sidewalls has been shown to increase the stability of the binding as well as the crystallinity of the polymer. TGA data showed that 58% wt of the novel composite[62] was formed by the polysaccharidic polymer as opposed to the 11–17% wt yield obtained by non-covalent complexation.[58]

From the studies carried out so far, it is evident that covalent modification of CNTs with chitosan can lead to higher modification degrees. However, further studies would need to be undertaken to verify whether the extra time and cost involved in the covalent binding process is justified by a real advantage in the characteristics of the final product, as non-covalent binding has shown to provide very stable suspensions by employing a much simpler method. On the other hand, it is also worth investigating new improved and more convenient methods for the covalent modification of CNTs – for example, the method described by Yu *et al.*[64] that employs microwaves. Carboxylated MWCNTs were obtained by a 1 h procedure involving 30 min of sonication and 30 min of reaction under microwaves of a suspension of MWCNTs in 70% HNO_3; the so-obtained MWCNTs-COOH were then reacted with a solution of chitosan for 20 min in a microwave oven affording a composite with a chitosan content higher than 25% wt.

Besides the chosen method of preparation, the molecular weight of the polymer used affects the characteristics of the composite. In literature we find evidence of the fact that increasing the molecular weight of chitosan from 20 to 200 kDa can improve the suspension efficiency from ~36% to ~47%, according to UV absorption measurements.[51] Long *et al.*[58] reported that when using a water-soluble chitosan derivative, the carboxymethylated chitosan, of increasing molecular weight, they obtained MWCNTs with increasing thickness of coating, respectively 1–2 nm for a 7 kDa polymer and 3 nm for a 17 kDa polymer. These data are in good agreement with those obtained for the thickness of coating using other chitosan derivatives, namely trimethyl chitosan chloride (TMC)[57] and *N*-octyl-*O*-sulphate-chitosan (NOSC)

(Fig. 1.10).[65] Both TMC and NOSC polymers were synthesised from low-viscous chitosan, which has an average molecular weight of 200 kDa. The

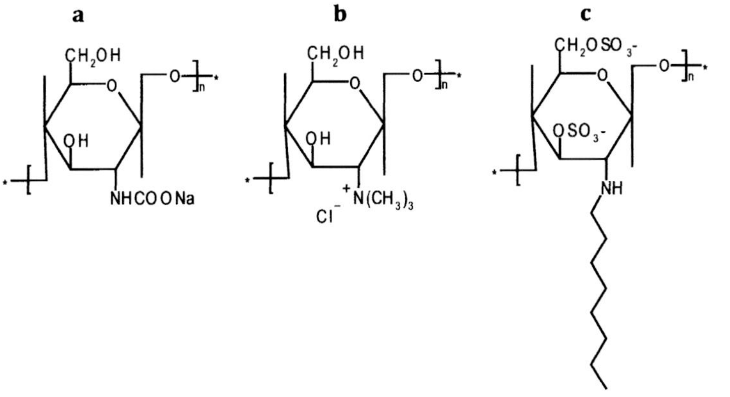

a   **b**   **c**

Figure 1.10 Chemical structure of some water-soluble derivatives of chitosan: (a) carboxymethylated chitosan, (b) thrimethylchitosan chloride (TMC) and (c) *N*-octyl-*O*-sulphate-chitosan (NOSC).

TMC and NOSC polymers present comparable thickness of coating, respectively 3.7–4.4 nm and 4.7 nm, as observed by AFM imaging (Fig. 1.11). Further studies should be carried out to identify the actual effect of coating thickness on the stability and characteristics of the CNT-polymer composite.

Figure 1.11 An AFM image of SWNT treated with octyl chitosan (0.5 mg/mL) after centrifugation. A = coated region (height = 6.6 ± 1.2 nm, n = 10) and B = uncoated region (height = 1.9 ± 0.4 nm, n = 10). See also Colour Insert.

The use of chitosan as a suspending agent also provides a way of purifying CNTs. Zhang *et al.*[53] reported that when chitosan-MWCNT composites were analysed by TGA, the peak at 470°C, assigned to the degradation of amorphous carbon impurities, disappeared. Chitosan has been shown not only to separate CNTs from carbonaceous impurities but also to specifically segregate CNTs according to their diameter and chirality. Yang *et al.*[52] analysed the supernatant and the precipitate of a SWCNT suspension obtained by sonication in the presence of high molecular weight chitosan, by Raman spectroscopy. They found that the supernatant was richer in small-diameter (0.91 and 0.82 nm) semiconducting CNTs as shown by the presence of only two radial breathing mode (RBM) bands at 258 and 284 cm^{-1} as opposed to the increased intensity of bands below 240 cm^{-1}, typical of large-diameter conducting CNTs, observed in the precipitate. Similar results were previously obtained by Takahashi *et al.*[49] Furthermore, the chitosan derivative TMC has shown size segregation capacity; in fact, SWCNTs suspended in TMC solutions presented a Raman spectrum with a reduction in lower-frequency RBM bands (<220 cm^{-1}) corresponding to larger-diameter tubes. These data were also verified by AFM studies.[57]

The complexation of CNTs with chitosan produces composites that present the mechanical strength, electrical conductivity and thermal stability of nanotubes, combined with the biocompatibility and pH sensitivity of the polysaccharide. These properties are promising for several nanotechnology and biotechnology applications.[66] One application that has attracted considerable attention is the use of CNT-chitosan complexes in the construction of biosensors.[61,67] Recent literature reports their use in the preparation of amperometric sensors, such as oxygen peroxide biosensors,[68] sulfite sensors,[69] immunosensors for α-fetoprotein,[70] glucose oxidase sensors[71] and biosensors for the detection of deep DNA damage,[72] to mention only a few. CNT-chitosan complexes have also shown enhanced DNA condensation properties compared with chitosan alone; this could be useful in the development of gene delivery systems.[73] Furthermore, the metal-binding properties of chitosan, combined with absorption properties of CNTs, give the scope for further investigation into environmentally friendly nanocomposites.[62]

Unmodified CNTs are very hydrophobic; they readily aggregate and therefore find it difficult to interface with biological materials. Various systems have been investigated to stabilise CNT suspensions in water yielding unbundled, individual CNTs to increase biocompatibility. However, more efforts are needed to achieve stable suspensions with high concentrations of individual CNTs, which could be further used as biomaterials in the field of pharmaceutical nanotechnology.

References

1. Bianco, A., Kostarelos, K., Partidos, C. D., and Prato, M. (2005) Biomedical applications of functionalised carbon nanotubes, *Chem. Commun.*, 571–577.

2. Foldvari, M., and Bagonluri, M. (2008) Carbon nanotubes as functional excipients for nanomedicines: I. pharmaceutical properties, *Nanomed. Nanotechnol. Biol. Med.*, **4**, 173–182.

3. Lacerda, L., Bianco, A., Prato, M., and Kostarelos, K. (2006) Carbon nanotubes as nanomedicines: from toxicology to pharmacology, *Adv. Drug Delivery Rev.*, **58**, 1460–1470.

4. Bianco, A., Kostarelos, K., and Prato, M. (2005) Applications of carbon nanotubes in drug delivery, *Curr. Opin. Chem. Biol.*, **9**, 674–679.

5. Prato, M., Kostarelos, K., and Bianco, A. (2007) Functionalized carbon nanotubes in drug design and discovery, *Acc. Chem.Res.*, **41**, 60–68.

6. Smart, S. K., Cassady, A. I., Lu, G. Q., and Martin, D. J. (2006) The biocompatibility of carbon nanotubes, *Carbon*, **44**, 1034–1047.

7. Lacerda, L., Raffa, S., Prato, M., Bianco, A., and Kostarelos, K. (2007) Cell-penetrating CNTs for delivery of therapeutics, *Nano Today*, **2**, 38–43.

8. Kostarelos, K., Lacerda, L., Pastorin, G., Wu, W., Wieckowski, S., Luangsivilay, J., Godefroy, S., Pantarotto, D., Briand, J.-P., Muller, S., Prato, M., and Bianco, A. (2007) Cellular uptake of functionalized carbon nanotubes is independent of functional group and cell type, *Nat. Nanotechnol.*, **2**, 108–113.

9. Feazell, R. P., Nakayama-Ratchford, N., Dai, H., and Lippard, S. J. (2007) Soluble single-walled carbon nanotubes as longboat delivery systems for platinum(IV) anticancer drug design, *J. Am. Chem.Soc.*, **129**, 8438–8439.

10. Foldvari, M., and Bagonluri, M. (2008) Carbon nanotubes as functional excipients for nanomedicines: II. drug delivery and biocompatibility issues, *Nanomed. Nanotechnol. Biol. Med.*, **4**, 183–200.

11. Wu, Y., Hudson, J. S., Lu, Q., Moore, J. M., Mount, A. S., Rao, A. M., Alexov, E., and Ke, P. C. (2006) Coating single-walled carbon nanotubes with phospholipids, *J. Phys. Chem. B*, **110**, 2475–2478.

12. Ke, P. C. (2007) Fiddling the string of carbon nanotubes with amphiphiles, *Phys. Chem. Chem. Phys.*, **9**, 439–447.

13. Bonard, J., Stora, T., Salvetat, J., Maier, F., Stöckli, T., Duschl, C., Forró, L., de Heer, W., and Châtelain, A. (1997) Purification and size-selection of carbon nanotubes, *Adv. Mater.*, **9**, 827–831.

14. Panchakarla, L., and Govindaraj, A. (2008) Covalent and non-covalent functionalization and solubilization of double-walled carbon nanotubes in nonpolar and aqueous media, *J. Chem. Sci.*, **120**, 607–611.

15. Florence, A. T., and Attwood, D. (2006) *Physicochemical principles of pharmacy*, 4th ed., Pharmaceutical Press, London.

16. Douroumis, D., Fatouros, D., Bouropoulos, N., Papagelis, K., and Tasis, D. (2007) Colloidal stability of carbon nanotubes in an aqueous dispersion of phospholipid, *Int. J. Nanomed.*, **2**, 761–766.

17. Xu, F.-M., Xu, J.-P., Ji, J., and Shen, J.-C. (2008) A novel biomimetic polymer as amphiphilic surfactant for soluble and biocompatible carbon nanotubes (CNTs), *Colloids Surf. B*, **67**, 67–72.

18. Xu, F., Xu, J., Ji, J., and Shen, J. (2008) A cell membrane biomimetic polymer for surface modification of carbon nanotubes, *Acta Polym. Sin.*, **8**, 1006–1009.

19. Qiao, R., and Ke, P. C. (2006) Lipid-carbon nanotube self-assembly in aqueous solution, *J. Am. Chem. Soc.*, **128**, 13656–13657.

20. Tasis, D., Papagelis, K., Douroumis, D., Smith, J. R., Bouropoulos, N., and Fatouros, D. G. (2008) Diameter-selective solubilization of carbon nanotubes by lipid micelles, *J. Nanosci. Nanotechnol.*, **8**, 420–423.

21. Gagner, J., Johnson, H., Watkins, E., Li, Q., Terrones, M., and Majewski, J. (2006) Carbon nanotube supported single phospholipid bilayer, *Langmuir*, **22**, 10909–10911.

22. Konduru, N. V., Tyurina, Y. Y., Feng, W., Basova, L. V., Belikova, N. A., Bayir, H., Clark, K., Rubin, M., Stolz, D., Vallhov, H., Scheynius, A., Witasp, E., Fadeel, B., Kichambare, P. D., Star, A., Kisin, E. R., Murray, A. R., Shvedova, A. A., and Kagan, V. E. (2009) Phosphatidylserine targets single-walled carbon nanotubes to professional phagocytes *in vitro* and *in vivo*, *PLoS ONE*, **4**, e4398.

23. Artyukhin, A. B., Shestakov, A., Harper, J., Bakajin, O., Stroeve, P., and Noy, A. (2005) Functional one-dimensional lipid bilayers on carbon nanotube templates, *J. Am. Chem. Soc.*, **127**, 7538–7542.

24. Wang, R., Cherukuri, P., Duque, J. G., Leeuw, T. K., Lackey, M. K., Moran, C. H., Moore, V. C., Conyers, J. L., Smalley, R. E., Schmidt, H. K., Weisman, R. B., and Engel, P. S. (2007) SWCNT PEG-eggs: single-walled carbon nanotubes in biocompatible shell-crosslinked micelles, *Carbon*, **45**, 2388–2393.

25. Shvartzman-Cohen, R., Florent, M., Goldfarb, D., Szleifer, I., and Yerushalmi-Rozen, R. (2008) Aggregation and self-assembly of amphiphilic block copolymers in aqueous dispersions of carbon nanotubes, *Langmuir*, **24**, 4625–4632.

26. Florent, M., Shvartzman-Cohen, R., Goldfarb, D., and Yerushalmi-Rozen, R. (2008) Self-assembly of pluronic block copolymers in aqueous dispersions of single-wall carbon nanotubes as observed by spin probe EPR, *Langmuir*, **24**, 3773–3779.

27. Cherukuri, P., Gannon, C., Leeuw, T., Schmidt, H., Smalley, R., Curley, S., and Weisman, R. (2006) Mammalian pharmacokinetics of carbon nanotubes using intrinsic near-infrared fluorescence, *Proc. Natl. Acad. Sci. U.S.A.*, **103**, 18882–18886.

28. Cherukuri, P., Bachilo, S. M., Litovsky, S. H., and Weisman, R. B. (2004) Near-infrared fluorescence microscopy of single-walled carbon nanotubes in phagocytic cells, *J. Am. Chem. Soc.*, **126**, 15638–15639.

29. Kang, Y., and Taton, T. A. (2003) Micelle-encapsulated carbon nanotubes: a route to nanotube composites, *J. Am. Chem. Soc.*, **125**, 5650–5651.

30. Wang, Z., Liu, Q., Zhu, H., Liu, H., Chen, Y., and Yang, M. (2007) Dispersing multi-walled carbon nanotubes with water-soluble block copolymers and their use as supports for metal nanoparticles, *Carbon*, **45**, 285–292.

31. Xu, J.-P., Ji, J., Chen, W.-D., and Shen, J.-C. (2005) Novel biomimetic polymersomes as polymer therapeutics for drug delivery, *J. Controlled Release*, **107**, 502–512.

32. Xie, L., Xu, F., Qiu, F., Lu, H., and Yang, Y. (2007) Single-walled carbon nanotubes functionalized with high bonding density of polymer layers and enhanced mechanical properties of composites, *Macromolecules*, **40**, 3296–3305.

33. Kumar, M. N. V. R., Muzzarelli, R. A. A., Muzzarelli, C., Sashiwa, H., and Domb, A. J. (2004) Chitosan chemistry and pharmaceutical perspectives, *Chem. Rev.*, **104**, 6017–6084.

34. Rinaudo, M. (2008) Main properties and current applications of some polysaccharides as biomaterials, *Polym. Int.*, **57**, 397–430.

35. Hirano, S., Seino, H., Akiyama, Y., and Nonaka, I. (1988) Bio-compatibility of chitosan by oral and intravenous administrations, *Proc. ACS Div. Polym. Mater. Sci. Eng.*, **59**, 897–901.

36. Wu, Y., Seo, T., Sasaki, T., Irie, S., and Sakurai, K. (2006) Layered structures of hydrophobically modified chitosan derivatives, *Carbohydr. Polym.*, **63**, 493–499.

37. Hejazi, R., and Amiji, M. (2003) Chitosan-based gastrointestinal delivery systems, *J. Controlled Release*, **89**, 151–165.

38. Shi, C., Zhu, Y., Ran, X., Wang, M., Su, Y., and Cheng, T. (2006) Therapeutic potential of chitosan and its derivatives in regenerative medicine, *J. Surg. Res.*, **133**, 185–192.

39. Fernandez-Saiz, P., Lagaron, J. M., and Ocio, M. J. (2009) Optimization of the biocide properties of chitosan for its application in the design of active films of interest in the food area, *Food Hydrocolloids*, **23**, 913–921.

40. Thanou, M., Verhoef, J. C., and Junginger, H. E. (2001) Chitosan and its derivatives as intestinal absorption enhancers, *Adv. Drug Delivery Rev.*, **50**, S91–S101.

41. Thanou, M., Verhoef, J. C., and Junginger, H. E. (2001) Oral drug absorption enhancement by chitosan and its derivatives, *Adv. Drug Delivery Rev.*, **52**, 117–126.

42. Borchard, G. (2001) Chitosans for gene delivery, *Adv. Drug Delivery Rev.*, **52**, 145–150.

43. van der Lubben, I. M., Verhoef, J. C., Borchard, G., and Junginger, H. E. (2001) Chitosan for mucosal vaccination, *Adv. Drug Delivery Rev.*, **52**, 139–144.

44. Alves, N. M., and Mano, J. F. (2008) Chitosan derivatives obtained by chemical modifications for biomedical and environmental applications, *Int. J. Biol. Macromol.*, **43**, 401–414.

45. van der Merwe, S. M., Verhoef, J. C., Verheijden, J. H. M., Kotzé, A. F., and Junginger, H. E. (2004) Trimethylated chitosan as polymeric absorption enhancer for improved peroral delivery of peptide drugs, *Eur. J. Pharm. Biopharm.*, **58**, 225–235.

46. Roldo, M., Hornof, M., Caliceti, P., and Bernkop-Schnurch, A. (2004) Mucoadhesive thiolated chitosans as platforms for oral controlled drug delivery: synthesis and in vitro evaluation, *Eur. J. Pharm. Biopharm.*, **57**, 115–121.

47. Muzzarelli, R. A. A. (1988) Carboxymethylated chitins and chitosans, *Carbohydr. Polym.*, **8**, 1–21.

48. Jayakumar, R., Nwe, N., Tokura, S., and Tamura, H. (2007) Sulfated chitin and chitosan as novel biomaterials, *Int. J. Biol. Macromol.*, **40**, 175–181.

49. Takahashi, T., Luculescu, C., Uchida, K., Ishii, T., and Yajima, H. (2005) Dispersion behaviour and spectroscopic properties of single-walled carbon nanotubes in chitosan acidic aqueous solutions, *Chem. Lett.*, **34**, 1516–1517.

50. Wang, S., Shen, L., Zhang, W., and Tong, Y. (2005) Preparation and mechanical properties of chitosan/carbon nanotubes composites, *Biomacromolecules*, **6**, 3067–3072.

51. Haggenmueller, R., Rahatekar, S., Fagan, J., Chun, J., Becker, M., Naik, R., Krauss, T., Carlson, L., Kadla, J., Trulove, P., Fox, D., DeLong, H., Fang, Z., Kelley, S., and Gilman, J. (2008) Comparison of the quality of aqueous dispersions of single wall carbon nanotubes using surfactants and biomolecules, *Langmuir*, **24**, 5070–5078.

52. Yang, H., Wang, S., Mercier, P., and Akins, D. (2006) Diameter-selective dispersion of single-walled carbon nanotubes using a water-soluble, biocompatible polymer, *Chem. Comm.*, 1425–1427.

53. Zhang, M., Smith, A., and Gorski, W. (2004) Carbon nanotube-chitosan system for electrochemical sensing based on dehydrogenase enzymes, *Anal. Chem.*, **76**, 5045–5050.

54. Spinks, G. M., Shin, S. R., Wallace, G. G., Whitten, P. G., Kim, S. I., and Kim, S. J. (2006) Mechanical properties of chitosan/CNT microfibers obtained with improved dispersion, *Sens. Actuators, B.*, **115**, 678–684.

55. Kang, B., Yu, D., Chang, S., Chen, D., Dai, Y., and Ding, Y. (2008) Intracellular uptake, trafficking and subcellular distribution of folate conjugated single walled carbon nanotubes within living cells, *Nanotechnology*, **19**, 1–8.

56. Ozarkar, S., Jassal, M., and Agrawal, A. (2008) Improved dispersion of carbon nanotubes in chitosan, *Fibers Polym.*, **9**, 410–415.

57. Wise, A., Smith, J., Bouropoulos, N., Yannopoulos, S., van der Merwe, S. M., and Fatouros, D. (2008) Single-walled carbon nanotube dispersions stabilised with N-trimethyl-chitosan, *J. Biomed. Nanotechnol.*, **4**, 67–72.

58. Long, D., Wu, G., and Zhu, G. (2008) Noncovalently modified carbon nanotubes with carboxymethylated chitosan: a controllable donor-acceptor nanohybrid, *Int. J. Mol. Sci.*, **9**, 120–130.

59. O'Connell, M. J., Boul, P., Ericson, L. M., Huffman, C., Wang, Y., Haroz, E., Kuper, C., Tour, J., Ausman, K. D., and Smalley, R. E. (2001) Reversible water-solubilization of single-walled carbon nanotubes by polymer wrapping, *Chem. Phys. Lett.*, **342**, 265–271.

60. Liu, Y., Tang, J., Chen, X., and Xin, J. H. (2005) Decoration of carbon nanotubes with chitosan, *Carbon*, **43**, 3178–3180.

61. Shieh, Y.-T., and Yang, Y.-F. (2006) Significant improvements in mechanical property and water stability of chitosan by carbon nanotubes, *Eur. Polym. J.*, **42**, 3162–3170.

62. Ke, G., Guan, W., Tang, C., Guan, W., Zeng, D., and Deng, F. (2007) Covalent functionalization of multiwalled carbon nanotubes with a low molecular weight chitosan, *Biomacromolecules*, **8**, 322–326.

63. Wu, Z., Feng, W., Feng, Y., Liu, Q., Xu, X., Sekino, T., Fujii, A., and Ozaki, M. (2007) Preparation and characterization of chitosan-grafted multiwalled carbon nanotubes and their electrochemical properties, *Carbon*, **45**, 1212–1218.

64. Yu, J.-G., Huang, K.-L., Tang, J.-c., Yang, Q., and Huang, D.-S. (2009) Rapid microwave synthesis of chitosan modified carbon nanotube composites, *Int. J. Biol. Macromol.*, **44**, 316–319.

65. Roldo, M., Power, K., Smith, J., Cox, P., Papagelis, K., Bouropoulos, N., and Fatouros, D. (2009) N-octyl-O-sulfate chitosan stabilises single wall carbon nanotubes in aqueous media and bestows biocompatibility, *Nanoscale*, **1**, 1–9.

66. Li, A. Z. F., Luo, A. G. H., Zhou, A. W. P., F, W. A., Xiang, A. R., and Liu, A. Y. P. (2006) The quantitative characterization of the concentration and dispersion of multi-walled carbon nanotubes in suspension by spectrophotometry, *Nanotechnology*, **17**, 3692–3698.

67. Lau, C., Cooney, M. J., and Atanassov, P. (2008) Conductive macroporous composite chitosan/carbon nanotube scaffolds, *Langmuir*, **24**, 7004–7010.

68. Qian, L., and Yang, X. (2006) Composite film of carbon nanotubes and chitosan for preparation of amperometric hydrogen peroxide biosensor, *Talanta*, **68**, 721–727.

69. Zhou, H., Yang, W., and Sun, C. (2008) Amperometric sulfite sensor based on multiwalled carbon nanotubes/ferrocene-branched chitosan composites, *Talanta*, **77**, 366–371.

70. Lin, J., He, C., Zhang, L., and Zhang, S. (2009) Sensitive amperometric immunosensor for alpha-fetoprotein based on carbon nanotube/gold nanoparticle doped chitosan film, *Anal. Biochem.*, **384**, 130–135.

71. Zhou, Y., Yang, H., and Chen, H. (2008) Direct electrochemistry and reagentless biosensing of glucose oxidase immobilized on chitosan wrapped single-walled carbon nanotubes, *Talanta*, **76**, 419–423.

72. Galandova, J., Ziyatdinova, G., and Labuda, J. (2008) Disposable electrochemical biosensor with multiwalled carbon nanotubes – chitosan composite layer for the detection of deep DNA damage, *Anal. Sci.*, **24**, 711–716.

73. Liu, Y., Yu, Z.-L., Zhang, Y.-M., Guo, D.-S., and Liu, Y.-P. (2008) Supramolecular architectures of beta-cyclodextrin-modified chitosan and pyrene derivatives mediated by carbon nanotubes and their DNA condensation, *J. Am. Chem. Soc.*, **130**, 10431–10439.

Chapter 2

BIOMEDICAL APPLICATIONS I: DELIVERY OF DRUGS

Giampiero Spalluto,[a] Stephanie Federico,[a] Barbara Cacciari,[b] Alberto Bianco,[c] Siew Lee Cheong[d] and Maurizio Prato[a]

[a] *Dipartimento di Scienze Farmaceutiche, Università di Trieste, Trieste 34127, Italy*

[b] *Department of Pharmaceutical Sciences, University of Ferrara, 44100 Ferrara, Italy*

[c] *CNRS, Institut de Biologie Moléculaire et Cellulaire, UPR 9021 Immunologie et Chimie Thérapeutiques, 67000 Strasbourg, France*

[d] *Department of Pharmacy, National University of Singapore, Singapore 117543*
spalluto@univ.trieste.it

2.1 INTRODUCTION

In the last few years several nanosystems, derived from simple or more complex materials, have been strongly investigated as drug delivery systems (DDS), considering their ability to transverse several physiological barriers, which represent a challenging obstacle for drug targeting. In addition, DDS could be considered fundamental to avoid several limitations presented by various therapeutic agents such as poor solubility in biological fluids, rapid deactivation, unfavourable pharmacokinetics, limited biodistribution and unwanted side effects intrinsically associated with systemic administration.[1]

Considering all these aspects, an ideal DDS should release a therapeutic agent to the target site without collateral adverse damage, protect the molecule from deactivation, improve the pharmacokinetic profile and enhance intracellular penetration and biodistribution.[2]

In general, nano-size DDS show a reduced size ranging from 1 to 100 nm, and this aspect permits a suitable manipulation at the molecular level.

Carbon Nanotubes: From Bench Chemistry to Promising Biomedical Applications
Edited by Giorgia Pastorin
Copyright © 2011 Pan Stanford Publishing Pte. Ltd.
www.panstanford.com

Several DDS have been investigated, such as liposomes, dendrimers and smart polymers, iron and gold nanoparticles, fullerenes and nanohorns. Nevertheless, none of the above-mentioned systems can be considered "ideal" for drug delivery. For example, liposomes (the most studied DDS) show excellent biocompatibility and low toxicity but show also big dimensions and significant instability in solution.[3-5] Dendrimers and smart polymers possess a high controllable size and surface functionalization but display also a slow release rate and cytotoxicity (up to 200 nM).[6-12] A quite significant toxicity was also observed for gold and iron nanoparticles,[13-25] while fullerenes showed accumulation in the liver[26-30] and nanohorns were found to be poorly soluble in aqueous media and resulted in self-assembly (potential toxicity) as a consequence.[31-33]

Taking into account all these experimental observations, the use of carbon nanotubes (CNTs) as potential carrier for DDS appeared to be a promising option.[34]

In fact, they show several interesting properties, such as high aspect ratio, ultra-light weight, great strength, high thermal conductivity and remarkable electronic properties similar to those of metallic to semiconductors.[35-39] Produced carbon nanotubes (pCNTs) are made of a series of condensed benzene rings and wrapped in a tubular form and can contain either one (SWCNTs) or multiple (MWCNTs) graphene sheets.

At present it is not possible to demonstrate which of the two systems is more advantageous. In fact, while SWCNTs offer the additional photoluminescence property that can be considered promising in diagnostics, MWCNTs present a wider surface and internal volume that can facilitate encapsulation and external functionalisation with active molecules.[40-44]

One of the major problems related to the use of CNTs as DDS is the lack of solubility (both in organic solvents and aqueous solutions), which significantly compromises their biocompatibility and immunogenicity.[45] However, these observations can be considered appropriate for pCNTs only, indicating that further modifications can lead to functionalised CNTs (f-CNTs) with such desirable properties that may enable them to be considered as good biomaterials for drug delivery. In fact, introduction of multiple functions on their surface can render them dispersible in aqueous media, thereby overcoming the major disadvantage of pCNTs.[46,47]

In particular, the application of f-CNTs as new nanovectors for drug delivery became feasible soon after the demonstration of cellular uptake of this new material. In fact, it has been clearly demonstrated that f-CNTs can be internalised into the cells regardless the type of functionalisation on the surface of carbon nanotube, indicating that different chemical procedures can be adopted to introduce several groups and functionalities.[34,48-55]

In Scheme 2.1 are briefly reported a few approaches to functionalise CNTs which can be performed in a covalent or non-covalent manner.[56-71]

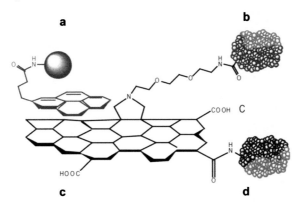

Scheme 2.1 Examples of functionalisations on CNTs' sidewalls and tips for drug delivery purposes: (a) non-covalent, (b) covalent (1,3-dipolar cycloaddition), (c) "defect" and (d) covalent (via oxidation). See also Colour Insert.

CNTs can be also considered good DDS after the experimental observation that SWCNTs are able to encapsulate small molecules. This can be a new approach in DDS; in fact encapsulation of drugs into the nanotubes can prevent their inactivation or degradation and increase their half time[72-74] (Scheme 2.2).

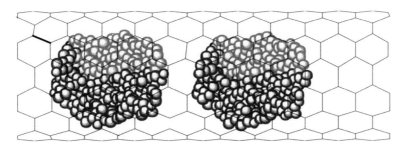

Scheme 2.2 Encapsulation of bioactive molecules in the inner cavity of CNTs. See also Colour Insert.

However, even though a lot of diversified CNT functionalisations have been successfully achieved, only a few examples of delivery of small molecules (antibacterial, antiviral and anticancer agents) using *f*-CNTs are currently reported in the literature (Table 2.1).[55,56,66,75-89]

Table 2.1 Functionalisations of carbon nanotubes (CNTs) and their use as drug delivery systems (DDS)[a]

	CNTs	Diam.	Functionalisation	Therapeutic agents incorporated	Applications	Ref.
Non-covalent physical adsorption on CNTs' external walls	pSWCNTs	~1 nm	1-pyrenebutanoic acid, succinimidyl ester onto CNTs	Proteins (ferritin and streptavidin)	Cancer therapy	(56)
			Amphiphilic (PEG)-based copolymer	Doxorubicin (DOX)	Cancer therapy	(75)
			Copolymer Pluronic F127	Doxorubicin (DOX)	Cancer therapy	(76)
			PEG-8 caprylic/capric glycerides	Erythropoietin (EPO)	Anaemia	(77)
			Phospholipid (PL)–folic acid (FA) copolymer + NIR	—	Carcer therapy	(78)
			—	DNA	Gene therapy	(79)
			PEG-platinum(IV) construct	Prodrug Pt(IV)	Cancer therapy	(80)
	SWCNTs-ox	~1 nm, shorter tubes	SWCNTs-CONH-$C_6H_{12}NH_3^+$	siRNA	Cancer therapy	(81)
	pMWCNTs	10–50 nm	Copolymer Pluronic F127	Doxorubicin (DOX)	Cancer therapy	(76)
	SWCNTs-ox	~1 nm, shorter tubes	Oxidation + EDC + Biotin/ Streptavidin	Streptavidin	Cancer therapy	(121)
			Oxidation + carbodiimide (EDAC)	BSA	Immunohistochemistry, enzyme stabilisation	(82)
			Oxidation + EDC	DNA	Gene therapy	(83, 84)
	MWCNTs-ox	10–50, shorter tubes	Oxidation + carbodiimide (EDAC)	BSA	Immunohistochemistry, enzyme stabilisation	(83)
			Oxidation + EDC + NHS	Gonadotropin-releasing hormone	Cancer therapy	(85)

	SWCNTs	~1 nm	Nitrene cycloaddition- substituted C_2B_9 carborane units	Boron	Boron capture neutron therapy (BNCT)	(66)
	MWCNTs	10–50 nm	1,3-dipolar cycloaddition of azomethine ylides	Methotrexate (MTX)	Cancer therapy	(86)
			Oxidation and 1,3-dipolar cycloaddition of azomethine ylides	Amphotericin B (AmB)	Fungal infection	(87)
Encapsulation	SWCNTs	~1 nm	Annealing at 350º	β-Carotene	Photonic tech- nology	(74)
	SWCNTs + DWCNTs	~2 nm	Annealing at 550º + nano- extraction	Altretamine (HMM)	Cancer therapy	(88)
	MWCNTs	10–50 nm	Thermal treatment + Oxidation	Carboplatin	Cancer therapy	(139)

[a] Reproduced with permission from G. Pastorin, *Pharmaceutical Research* **2009**, 26(4), 746–769."

Another aspect should be also considered for the use of CNTs in drug delivery. In fact it has been recently demonstrated that they behave like asbestos,[90] showing carcinogenic effects. Anyway, it should be underlined that the toxicity of CNTs is still uncertain. All the studies performed till now are contradictory and not uniform, indicating that several factors related to both cell lines and materials are involved in the toxicity of CNTs. In fact the toxicity of CNTs seems to be related to the dimension of the tubes and, most important, to their functionalisation.[45,90–94] It has been recently observed that *f*-CNTs are not toxic, but more studies should be performed to better clarify this fundamental parameter for considering CNTs an ideal candidate for drug delivery.

In this chapter we wish to briefly summarise the possible approaches for use of CNTs as DDS and to analyse in more detail the few examples reported in the literature.

2.2 NON-COVALENT FUNCTIONALISATION ON THE EXTERNAL WALLS

Functionalisation of CNTs for their application in biomedicine is focused on the chemical strategies that can render this material biocompatible as well as functional.

The simplest procedure includes the attachment, by physical adsorption, of pCNTs to several molecules such as pyrene, naphthalene derivatives,[95] sulfonated polyaniline,[96] poly(acrylic acid),[97] proteins, DNA[98-100] and gold nanoparticles.[101] The nanotube–adsorbate conjugation consists of π–π stacking interactions between the aromatic part of the adsorbate and the graphitic sidewall of nanotubes. Therefore, this does not affect CNTs' whole integrity. An interesting approach is summarised in Scheme 2.3, which on one side enabled the irreversible adsorption onto the surface of SWCNTs through π–π interactions, while its additional succinimidyl ester group allowed the covalent attachment of various molecules via the nucleophilic attack of primary or secondary amines (such as in ferritin, streptavidin or biotin-polyethyleneoxyde-amine).[56]

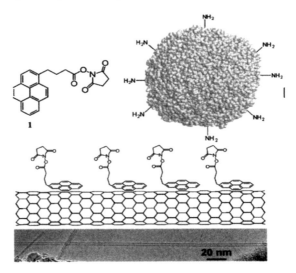

Scheme 2.3 Amine groups on a protein react with the anchored succinimidyl ester to form amide bonds for protein immobilisation. Reproduced from J. Chen *et al.*[56] with permission.

The significant increase in CNT dispersibilty[59,102-105] is another beneficial effect of this interaction, and it has been utilised for purifications of CNTs from contaminations represented mainly by amorphous carbon.[106-108]

With similar methodologies, debundling CNTs may be obtained using surfactants[56,57,109-113] or solubilising polymers.[59,114] Several surfactants (either anionic, nonionic or cationic) are able not only to suspend CNTs in aqueous solution,[109-112] but also to prevent re-aggregation of the tubes by Coulomb repulsion between surfactant-coated CNTs.[114] In addition, if the surfactant presents aromatic groups in its hydrophobic part, additional π–π

stacking interactions take place with the graphitic sidewalls of CNTs, while hydrophilic groups are exposed to the aqueous solution. Surfactants such as sodium dodecyl sulfate (SDS) were found to be very effective in dispersing SWCNTs.[115] Similarly, water-soluble polymers such as polyvinylpyrrolidone (PVP) wrapped around the surface of SWCNTs facilitated the nanotubes' dissolution in aqueous phases.[114]

Even though water dispersibility of CNTs is extremely important and re-aggregation of the tubes should be minimised, the above-mentioned procedures do not always allow an effective, stable incorporation of additional bioactive molecules. Therefore, such non-covalent functionalisation does not guarantee efficient DDS. An exception in this sense is represented by a recent work by Park and collaborators,[75] who designed a "trivalent" amphiphilic polymer by preparing a polymer bearing both hydrophobic and hydrophilic residues and thus able to disperse CNTs in water, and by subsequently integrating a polyethylene glycol (PEG)-based copolymer useful as a solubilising agent and as protection against protein adsorption (the so-called anti-biofouling effect, which is often encountered during *in vivo* studies). Doxorubicin, an antitumour agent was then condensed to nanotubes. The positively charged amino group of the drug interacted with the carboxylic functions of the polymer, while its aromatic rings stabilised the π–π stacking interactions at the surface of CNTs. Non-covalent functionalisation with polyethylene glycol[116] or the block copolymer Pluronic F127[76] was also used to adsorb the same drug (DOX) onto SWCNTs and MWCNTs, respectively, but noticeably different results might induce us to think that further mechanistic evaluation should disclose whether the cell internalisation of CNTs is also affected by drug–nanotube complexes. With a similar procedure, it was observed that short fibres of CNTs were able to increase both delivery and absorption of erythropoietin (EPO), thereby identifying key factors to improve oral delivery of drug proteins.[77]

2.3. "DEFECT" FUNCTIONALISATION AT THE TIPS AND SIDEWALLS

Besides non-covalent procedures, CNTs can also be cut and functionalised simultaneously by simply treating them with oxidising agents such as HNO_3, $KMnO_4/H_2SO_4$, O_2, $K_2Cr_2O_7/H_2SO_4$ or OsO_4, thereby making them soluble in polar organic solvents, acids and water without the aid of sonication, surfactants or any other means.[59–61] After this treatment, it is possible to use the carboxylic acid groups and the carboxylated fractions introduced by oxidisation treatment, to further functionalise the nanotubes via amidation, esterification or the zwitterionic $COO^-NH_3^+$ formation.[117] This oxidising procedure is usually known as "defect functionalisation", since it takes place at the ends or at specific positions of pre-existing defects of CNTs. If not exaggerated, it preserves the macroscopic features of CNTs so that their electronic and mechanical properties are not lost.[118-120] In addition, it has been employed to link biological molecules to CNTs via stable covalent bonds. For example, bovine serum albumin (BSA) has been incorporated on *f*-CNTs (both single- and multi-walled) via diimide-activated amidation, demonstrating that the protein, once bound to nanotubes, retains its activity.[82] Similarly, streptavidin (a protein with potential clinical applications in anticancer therapy)[121] and DNA have been also bound to CNTs via amide linkage.[83,84]

In conclusion, this process permits the introduction of carboxylic groups and carboxylated fractions that enable further manipulation of both the activity of incorporated biomolecules and the spectroscopic properties of *f*-CNTs. However, it can also introduce an excess of defects or generate ultra-short *f*-CNTs.[122]

2.4 COVALENT FUNCTIONALISATION ON THE EXTERNAL SIDEWALLS

Among the most powerful methodologies aimed to functionalise CNTs, the use of 1,3-dipolar cycloaddition represents a fascinating example of covalent bonding because it is extremely versatile and easy to execute. It requires only an α-amino acid (or a correspondent ester) reacting with an aldehyde or a ketone, to generate *in situ* azomethine ylides, which are very reactive species, and affording the formation of pyrrolidine rings on the sidewalls of CNTs (Scheme 2.4).

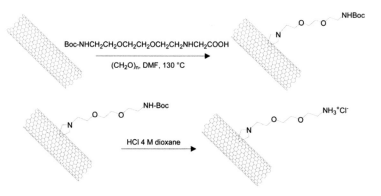

Scheme 2.4 1,3-Dipolar cycloaddition of azomethine ylides.

In general, the covalent functionlisation of nanotubes is more robust and better controllable compared with procedures based on non-covalent methods, and it offers the possibility of introducing multiple functionalities. In particular, some strategies to integrate multiple groups on the tubes' sidewalls have been developed. One of them consists in the application of 1,3-dipolar cycloaddition to introduce *N*-functionalised pyrrolidine rings on the external walls of the tubes. Two orthogonally protected α-amino acids were introduced in the presence of paraformaldehyde. Selective elimination of the phthalimidic group (Pht) in ethanolic hydrazine allowed the introduction of the fluorescent molecule (FITC), while the acidic conditions removed the *tert*-butyloxycarbonyl (Boc) group and permitted the coupling of the generated free amino function with the activated α or γ carboxylic group of the anticancer drug methotrexate (MTX) (Scheme 2.5).[86]

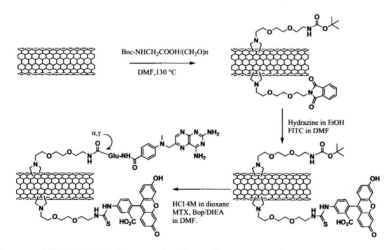

Scheme 2.5 MTX-FITC-functionalized carbon nanotubes. Reagents: (a) R-NHCH$_2$COOH/(CH$_2$O)$_n$ in DMF, 130 °C; (b) hydrazine in EtOH; (c) FITC in DMF; (d) HCl 4 M in dioxane; (e) MTX, Bop/DIEA in DMF.

MTX is a drug that is widely used against cancer, but it displays toxic side effects and a reduced cellular uptake.[123] The limited capacity of MTX to cross the cell membrane was overcome by conjugating it to *f*-CNTs, which are capable to enhance cell uptake of linked moieties. The presence of fluorescent tubes inside the cells around the nuclear membrane and the time and dose dependence of the internalization process have been demonstrated.

In a slightly different approach, MWCNTs have been functionalised with amphotericin B (AmB), which is a potent antimycotic drug normally used for the treatment of chronic fungal infections.[87] However, AmB displays also a remarkable toxicity towards mammalian cells,[124] presumably because of its low water solubility and its tendency to form aggregates[125] (Scheme 2.6).

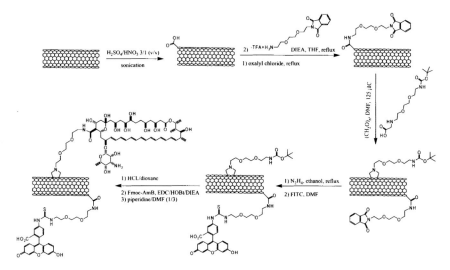

Scheme 2.6 AmB-FITC-functionalised carbon nanotubes.

In the study with AmB-functionalised CNTs it was demonstrated that the free drug at a dose of 10 µg/mL, which corresponds to the amount of the drug covalently bound to 40 µg/mL of MWCNTs, determined more than 40% of cell death, while all cells remained alive following the treatment with the AmB-CNT conjugate. Very interestingly, once linked to the nanotubes, AmB preserved its high antifungal activity. To verify this characteristic, different types of pathogens comprising *Candida albicans*, *Candida paropsilosis* and *Cryptococcus neoformans* were treated with AmB-CNT conjugates, and the results were in some cases superior than those obtained using the drug alone. Although the reason for such improvement in activity are still unclear, it might be that the increase in the solubility of the drug, together with its favourable multi-presentation to the fungal membrane by CNTs, determined the enhancement of its therapeutic effect by decreasing mammalian toxicity and increasing antifungal activity (Scheme 2.7).

In addition, it was also observed that AmB-CNT conjugates were rapidly internalised into mammalian (Jurkat) cells in a dose-dependent manner and with a mechanism that excluded endocytosis. About this aspect, there are two versions regarding the uptake mechanism. The first corresponds to nanopenetration, passive insertion/diffusion experimentally demonstrated by several groups,[49,50,87,126–128] while the second is based on a phagocytosis/endocytosis internalisation processes proposed by other researchers.[53,68,79,129,130] The great divergence seems to be attributable to significant differences both in the nanotube materials and in the type of functionalisation.

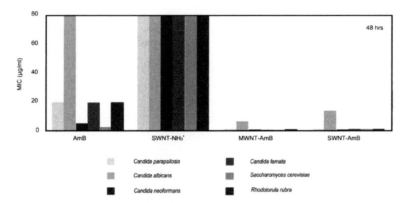

Scheme 2.7 MIC of AmB-FITC-functionalised carbon nanotubes on fungi and yeast after 48 hours of incubation. AmB alone (**AmB**), CNTs alone (**SWNT-NH₃⁺**) and CNTs with AmB covalently bound (**MWNT-AmB** and **SWNT-AmB**). See also Colour Insert.

Poor or non-existent cellular uptake was observed for large proteins (e.g., human immunoglobulin), presumably due to the large size of the cargo or to inefficient endocytosis of big conjugates.[129]

2.5 ENCAPSULATION INSIDE CNTS

As previously mentioned, the use of CNTs to encapsulate molecules has rendered these nanosystems particularly suitable for additional applications such as material storage,[72] compound synthesis[73] and drug delivery.[74,88] Successful encapsulation of organic molecules inside SWCNTs has already been reported (Scheme 2.8).[88,131–136] The advantage of this methodology lies in the ability of CNTs to provide protection and to control the release of loaded molecules, thus prolonging the effect of eventual drugs.

An interesting example is the incorporation, inside CNTs, of a natural pigment, β-carotene,[74] whose application has been hampered by its fast degradation, under light exposure.[137,138]

From the results obtained in these experiments, the crucial role of carbon nanotubes in the protection of a natural substance from easy degradation was obvious, but no details about its delivery were provided.

An interesting encapsulation of a drug inside CNTs has been obtained by Hampel *et al.*, who investigated the influence of CNTs on tumour cell growth.[139] They relied on two main aspects: (i) the incorporation of carboplatin and (ii) the use of MWCNTs presenting a wider inner diameter. Hence, these types of tubes determined a higher drug loading that was further increased by controlled heat. The subsequent cell viability assays revealed that treatment with free carboplatin resulted in a concentration-dependent decrease in cell number and an increase in cell apoptosis, with about 50% of cells alive at a concentration of just 20 µg/mL. Interestingly, empty CNTs did not affect cell viability. On the contrary, the addition of carboplatin to empty tubes determined a synergistic effect, probably because MWCNTs altered the integrity of the cell membrane and increased the uptake of the drug. These results suggested that, even though the long-term influence of CNTs on cells should be deeply investigated, CNTs seem to be promising carriers with remarkable mechanical and chemical stability, albeit with still unclear immunogenic effects.

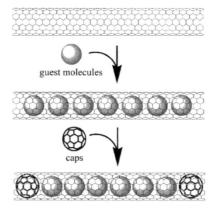

Scheme 2.8 Anticancer drug hexamethylmelamine (HMM) encapsulated inside SWCNTs and capped with fullerenes (C_{60}). Reproduced from Ren and Pastorin[88] with permission. See also Colour Insert.

2.6 CONCLUSIONS AND PERSPECTIVES

Considering the various drug delivery systems that are currently utilised at the nanoscale, together with the intrinsic properties of CNTs and the various examples of their applications, we can conclude that CNTs are promising materials, especially for potential drug delivery and for multimodality cancer therapy and imaging.

Functionalised CNTs permit incorporating simultaneously several drugs, targeting agents and even metals (e.g., iron) able to induce hyperthermia and thus improve therapeutic activities. The great advantage of CNTs is that they present a reactive external surface, which can be chemically modified to improve their role as DDS, and at the same time, their natural huge aspect ratio allows them to behave like nanoneedles that do not disrupt the integrity of external membranes during their cellular uptake.

In addition, their extraordinary strength has shown to preserve their structure, as has been clearly demonstrated by their excretion as intact tubes after intravenous administration in mice. However, we wish to underline that many scientists are using nano-objects to reduce the toxicity and undesirable effects of drugs, but till now they have not realised that carrier systems themselves may impose relevant risks.

Therefore, more studies are indispensable for demonstrating the real properties of CNTs as potential DDS – in particular, studies regarding their toxicity and, most importantly, about the delivery of drugs covalently or non-covalently linked to the tubes.

References

1. Vasir, J. K., Maram, M. K., and Labhasetwar, V. D. (2005) Nano-systems in drug targeting: opportunities and challenges, *Curr. Nanosci.*, **1**, 47–67.

2. Gabizon, A., Shmeeda, H., and Barenholz, Y. (2003) Pharmacokinetics of 0pegylated liposomal Doxorubicin: review of animal and human studies, *Clin. Pharmacokinet.*, **42**, 419–436.

3. Gabizon, A., Isacson, R., Libson, E., Kaufman, B., Uziely, B., Catane, R., Bendor, C. G., Rabello, E., Cass, Y., Peretz, T., Sulkes, A., Chisin, R., and Barenholz, Y. (1994) Clinical studies of liposome-encapsulated doxorubicin, *Acta Oncol.*, **33**, 779–786.

4. Gordon, K. B., Tajuddin, A., Guitart, J., Kuzel, T. M., Eramo, L. R., and Vonroenn, J. (2006) Hand-foot syndrome associated with liposome-encapsulated doxorubicin therapy, *Cancer*, **75**, 2169–2173.

5. Lyass, O., Uziely, B., Ben, R. Yosef, Tzemach, D., Heshing, N. I., Lotem, M., Brufman, G., and Gabizon, A. (2000) Correlation of toxicity with pharmacokinetics of pegylated liposomal doxorubicin (Doxil) in metastatic breast carcinoma, *Cancer*, **89**, 1037–1047.

6. Majors, I. J., Myc, A., Thomas, T., Menhta, C. B., and Baker, J. R, Jr. (2006) PAMAM dendrimer-based multifunctional conjugate for cancer therapy: synthesis, characterization, and functionality, *Biomacromolecules*, **7**, 572–579.

7. Gebhart, C. L., and Kabanov, A. V. (2001) Evaluation of polyplexes as gene transfer agents, *Control, J. Rel.*, **73**, 401–416.

8. T. D. McCarthy, P. Karellas, Henderson, S. A., Giannis, M., O, D. F.'Keefe, Heery, G., Paull, J. R., Matthews, B. R., and Holan, G. (2005) Dendrimers as drugs: discovery and preclinical and clinical development of dendrimer-based microbicides for HIV and STI prevention. *Mol. Pharm.*, **2**, 312–318.

9. Ghadiri, M. R., Granja, J. R., Milligan, R. A., McRee, D., and Khazanovich, N. (1993) Self-assembled organic nanotubes based on a cyclic peptide, *Nature*, **366**, 324–327.

10. Khazanovich, N., Granja, J. R., McRee, D., Milligan, R. A., and Ghadiri, M. R. (1994) Nanoscale tubular ensembles with specific internal diameters. Design of self-assembled nanotube with a 13 Å pore, *J. Am. Chem. Soc.*, **116**, 6011–6012.

11. Fernadanez, S.-Lopez, Kim, H. S., Choi, E. C., Delgado, M., Granja, J. R., Khasanov, A., Kraehenbuehl, K., Long, G., Weinberger, D. A., Wilcoxen, K. M., and Ghadiri, M. R. (2001) Antibacterial agents based on the cyclic D-L-α-peptide architecture, *Nature*, **412**, 452–455.

12. Ghadiri, M. R., Granja, J. R., and Buehler, L. K. (1994) Artificial transmembrane ion channels from self-assembling peptide nanotubes, *Nature*, **369**, 301–304.

13. P. Couvreur, and Vauthier, C. (2006) Nanotechnology: intelligent design to treat complex disease. *Pharm. Res.*, **23**, 1417–1450.

14. Tanaka, K., Kitamura, N., and Chujo, Y. (2008) Properties of superparamagnetic iron oxide nanoparticles assembled on nucleic acids. *Nucleic Acids Symp. Ser.*, **52**, 693–694.

15. Alexiou, C., Arnold, W., Klein, R. J., Parak, F. G., Hulin, P., Bergemann, C., Erhardt, W., Wagenpfeil, S., and Lubbe, A. S. (2000) Locoregional cancer treatment with magnetic drug targeting, *Cancer Res.*, **60**, 6641–6648.

16. T. K. Jain, Richey, J., Strand, M., Leslie, D. L.-Pelecky, Flask, C. A., and Labhasetwar, V. (2008) Magnetic nanoparticles with dual functional properties: drug delivery and magnetic resonance imaging, *Biomaterials*, **29**, 4012–4021.

17. M. Babincov, Altanerov, V., Altaner, C., Bergemann, C., and Babinec, P. (2008) In vitro analysis of cisplatin functionalized magnetic nanoparticles in combined cancer chemotherapy and electromagnetic hyperthermia, *IEEE Trans Nanobioscience.*, **7**, 15–19.

18. Arias, J. L., Linares, F.-Molinero, Gallardo, V., and Delgado, A. V. (2008) Study of carbonyl iron/poly (butylcyanoacrylate) (core/shell) particles as anticancer drug delivery systems Loading and release properties, *Eur. Pharm., J. Sci.,* **233**, 252–261.

19. Chen, S., Zhang, X. Z., Cheng, S. X., Zhuo, R. X., and Gu, Z. W. (2008) Functionalized amphiphilic hyperbranched polymers for targeted drug delivery, *Biomacromolecules*, **9**, 2578–2585.

20. Hillyer, J. F., and Albrecht, R. M. (1998) Correlative instrumental neutron activation analysis, light microscopy, transmission electron microscopy, and X-ray microanalysis for qualitative and quantitative detection of colloidal gold spheres in biological specimens, *Microsc. Microanal.*, **4**, 481–490.

21. Petersen, J., and Bendtzen, K. (1983) Immunosuppressive actions of gold salts, *Scand. J. Rheumatol. Suppl.*, **51**, 28–35.

22. Finkelstein, A. E., Walz, D. T., Batista, V., Mizraji, M., Roisman, F., and Misher, A. (1976) Auranofin. New oral gold compound for treatment of rheumatoid arthritis, *Ann. Rheum. Dis.*, **35**, 251–257.

23. Mottram, P. L. (2003) Past, present and future drug treatment for rheumatoid arthritis and systemic lupus erythematosus, *Immunol. Cell Biol.*, **81**, 350–353.

24. Patra, C. R., Bhattacharya, R., Wang, E., Katarya, A., Lau, J. S., Dutta, S., Murders, M., Wang, S., Buhrow, S. A., Safgren, S. L., Yaszemski, M. J., Reid, J. M., Ames, M. M., Mukherjee, P., and Mukhopadhyay, D. (2008) Targeted delivery of gemcitabine to pancreatic adenocarcinoma using cetuximab as a targeting agent, *Cancer Res.*, **68**, 1970–1978.

25. Portney, N. G., and Ozkan, M. (2006) Nano-oncology: drug delivery, imaging, and sensing, *Anal. Bioanal. Chem.*, **384**, 620–630.

26. Zakharian, T. Y., Seryshev, A., Sitharaman, B., Gilbert, B. E., Knight, V., and Wilson, L. J. (2005) A fullerene-paclitaxel chemotherapeutic: synthesis, characterization, and study of biological activity in tissue culture, *J. Am. Chem. Soc.*, **127**, 12508–12509.

27. Bakry, R., Vallant, R. M., Najam-ul-Haq, M., Rainer, M., Szabo, Z., Huck, C. W., and Bonn, G. K. (2007) Medicinal applications of fullerenes, *Int. Nanomedicine, J.*, **2**, 639–649.

28. Gharbi, N., Pressac, M., Hadchouel, M., Szwarc, H., Wilson, S. R., and Moussa, F. (2005) [60] Fullerene is a powerful antioxidant in vivo with no acute or subacute toxicity, *Nano Lett.*, **5**, 2578–2585.

29. Taglietti, M., Hawkins, C. N., and Rao, J. (2008) Novel topical drug delivery systems and their potential use in acne vulgaris, *Skin Therapy Lett.*, **13**, 6–8.

30. Mashino, T., Nishikawa, D., Takahashi, K., Usui, N., Yamori, T., Seki, M., Endo, T., and Mochizuki, M. (2003) Antibacterial and antiproliferative activity of cationic fullerene derivatives, *Bioorg. Med. Chem. Lett.*, **13**, 4395–4397.

31. Murakami, T., Ajima, K., Miyawaki, J., Yudasaka, M., Iijima, S., and Shiba, K. (2004) Drug-loaded carbon nanohorns: adsorption and release of dexamethasone in vitro, *Mol. Pharm.*, **1**, 399–405.

32. Ajima, K., Yudasaka, M., Murakami, T., Maigne, A., Shiba, K., and Iijima, S. (2005) Carbon nanohorns as anticancer drug carriers, *Mol. Pharm.*, **2**, 475–480.

33. Bianco, A., Kostarelos, K., and Prato, M. (2008) Opportunities and challenges of carbon-based nanomaterials for cancer therapy, *Expert Opin. Drug Deliv.*, **5**, 331–342.

34. Martin, C. R., and Kohli, P. (2003) The emerging field of nanotube biotechnology, *Nat. Rev. Drug Discov.*, **2**, 29–37.

35. Yu, M.-F., Files, B. S., Arepalli, S., and Ruoff, R. S. (2000) Tensile loading of ropes of single wall carbon nanotubes and their mechanical properties, *Phys. Rev. Lett.*, **84**, 5552–5555.

36. Saito, R., Dresselhaus, G., and Dresselhaus, M. S. (1998) *Physical Properties of Carbon Nanotubes*, Imperial College Press, London.

37. Saito, R., Fujita, M., Dresselhaus, G., and Dresselhaus, M. S. (1992) Electronic structure of chiral graphene tubules, *Appl. Phys. Lett.*, **60**, 2204–2206.

38. Saito, R., Fujita, M., Dresselhaus, G., and Dresselhaus, M. S. (1992) Electronic structure of graphene tubules based on C60, *Phys. Rev. B*, **46**, 1804–1811.

39. Kaiser, A. B., Düsberg, G., and Roth, S. (1998) Heterogeneous model for conduction in carbon. Nanotubes, *Phys. Rev. B.*, **57**, 1418–1421.

40. Bachilo, S. M., Strano, M. S., Kittrell, C., Hauge, R. H., Smalley, R. E., and Weisman, R. B. (2002) Structure-assigned optical spectra of single-walled carbon nanotubes, *Science*, **298**, 2361–2366.

41. Agüí, L., Yáñez-Sedeño, P., and Pingarrón, J. M. (2008) Role of carbon nanotubes in electroanalytical chemistry: a review, *Anal. Chim. Acta*, **622**, 11–47.

42. Star, A., Gabriel, J. P., Bradley, K., and Gruner, G. (2003) Electronic detection of specific protein binding using nanotube FET devices, *Nano Lett.*, **3**, 459–463.

43. Avouris, P., Chen, Z., and Perebeinos, V. (2007) Carbon-based electronics, *Nat. Nanotechnol.*, **2**, 605–615.

44. Yoshimoto, S., Murata, Y., Kubo, K., Tomita, K., Motoyoshi, K., Kimura, T., Okino, H., Hobara, R., Matsuda, I., Honda, S., Katayama, M., and Hasegawa, S. (2007) Four-point probe resistance measurements using PtIr-coated carbon nanotube tips, *Nano Lett.*, **7**, 956–959.

45. Singh, R., Pantarotto, D., Lacerda, L., Pastorin, G., Klumpp, C., Prato, M., Bianco, A., and Kostarelos, K. (2006) Tissue biodistribution and blood clearance rates of intravenously administered carbon nanotube radiotracers, *Proc. Natl. Acad. Sci. USA*, **103**, 3357–3362.

46. Hudson, J. L., Casavant, M. J., and Tour, J. M. (2004) Water-Soluble, Exfoliated, Nonroping Single-Wall Carbon Nanotubes, *J. Am. Chem. Soc.*, **126**, 11158–11159.

47. Hu, H., Ni, Y., Montana, V., Haddon, R. C., and Parpura, V. (2004) Chemically functionalized carbon nanotubes as substrates for neuronal growth, *Nano Lett.*, **4**, 507–511.

48. Pouton, C. W., and Seymour, L. W. (2001) Key issues in non-viral gene delivery, *Adv. Drug. Deliv. Rev.*, **46**, 187–203.

49. Pantarotto, D., Briand, J.-P., Prato, M., and Bianco, A. (2004) Translocation of bioactive peptides across cell membranes by carbon nanotubes, *Chem. Commun.*, 16–17.

50. Pantarotto, D., Singh, R., McCarthy, D., Erhardt, M., Briand, J. P., Prato, M., Kostarelos, K., and Bianco, A. (2004) Functionalized carbon nanotubes for plasmid DNA gene delivery, *Angew. Chem. Int. Ed.*, **43**, 5242–5236.

51. Cato, M. H., D'Annibale, F., Mills, D. M., Cerignoli, F., Dawson, M. I., Bergamaschi, E., Bottini, N., Magrini, A., Bergamaschi, A., Rosato, N., Rickert, R. C., Mustelin, T., and Bottini, M. (2008) Cell-type specific and cytoplasmic targeting of PEGylated carbon nanotube-based nanoassemblies, *J. Nanosci. Nanotechnol.*, **8**, 2259–2269.

52. Gannon, C. J., Cherukuri, P., Yakobson, B. I., Cognet, L., Kanzius, J. S., Kittrell, C., Weisman, R. B., Pasquali, M., Schmidt, H. K., Smalley, R. E., and Curley, S. A. (2007) Carbon nanotube-enhanced thermal destruction of cancer cells in a noninvasive radiofrequency field, *Cancer*, **110**, 2654–2665.

53. Cherukuri, P., Bachilo, S. M., Litovsky, S. H., and Weisman, R. B. (2004) Near-infrared fluorescence microscopy of single-walled carbon nanotubes in phagocytic cells, *J. Am. Chem. Soc.*, **126**, 15638–15639.

54. Dumortier, H., Lacotte, S., Pastorin, G., Marega, R., Wu, W., Bonifazi, D., J. -Briand, P., Muller, S., Prato, M., and Bianco, A. (2006) Functionalized carbon nanotubes are non toxic and preserve the functionality of primary immune cells, *Nano Lett.*, **6**, 1522–1528.

55. Kostarelos, K., Lacerda, L., Pastorin, G., Wu, W., Wieckowski, S., Luangsivilay, J., Godefroy, S., Pantarotto, D., Briand, J.-P., Muller, S., Prato, M., and Bianco, A. (2007) Cellular uptake of functionalized carbon nanotubes is independent of functional group and cell type, *Nat. Nanotechnol.*, **2**, 108–113.

56. Chen, R. J., Zhang, Y., Wang, D., and Dai, H. (2001) Noncovalent sidewall functionalization of single-walled carbon nanotubes for protein immobilization, *J. Am. Chem. Soc.*, **123**, 3838–3839.

57. Jin, Z., Huang, L., Goh, S. H., Xu, G., and Ji, W. (2000) Characterization and nonlinear optical properties of a poly (acrylic acid)–surfactant–multi-walled carbon nanotube complex, *Chem. Phys. Lett.*, **332**, 461–466.

58. Vigolo, B., Penicaud, A., Coulon, C., Sauder, C., Pailler, R., Journet, C. Bernier, P., and Poulin, P. (2000) Macroscopic fibers and ribbons of oriented carbon nanotubes, *Science*, **290**, 1331–1334.

59. Star, A., Stoddart, J. F., Steuerman, D., Diehl, M., Boukai, A., Wong, E. W., Yang, X., Chung, S. W., Choi, H., and Heath, J. R. (2001) Preparation and properties of polymer-wrapped single-walled carbon nanotubes, *Angew. Chem. Int. Ed.*, **40**, 1721–1725.

60. Rinzler, A., Liu, J., Dai, H., Nikolaev, P., Huffman, C., Rodriguez, F.-Macias, Boul, P., Lu, A., Heymann, D., Colbert, D. T., Lee, R. S., Fischer, J., Rao, A., Eklund, P. C., and Smalley, R. E. (1998) Large-scale purification of single-wall carbon nanotubes: process, product, and characterization, *Appl. Phys. A*, **67**, 29–37.

61. Duesberg, G. S., Muster, J., Krstic, V., Burghard, M., and Roth, S. (1998) Chromatographic size separation of single-wall carbon nanotubes, *Appl. Phys. A*, **67**, 117–119.

62. Holzinger, M., Hirsh, A., Bernier, P., Duesberg, G. S., and Burghard, M. (2000) A new purification method for single-wall carbon nanotubes (SWNTs), *Appl. Phys. A*, **70**, 599–602.

63. Pagona, G., and Tagmatarchis, N. (2006) carbon nanotubes: materials for medicinal chemistry and biotechnological applications, *Curr. Med. Chem.*, **13**, 1789–1798.

64. Holzinger, M., Vostrowsky, O., Hirsh, A., Hennrich, F., Kappes, M., Weiss, R., and Jellen, F. (2001) Sidewall functionalization of carbon nanotubes, *Angew. Chem. Int. Ed.*, **40**, 4002–4005.

65. Holzinger, M., Abraham, J., Whelan, P., Graupner, R., Ley, L., Hennrich, F., Kappes, M., and Hirsh, A. (2003) Functionalization of single-walled carbon nanotubes with (R-)oxycarbonyl nitrenes, *J. Am. Chem. Soc.*, **125**, 8566–8580.

66. Yinghuai, Z., Peng, A. T., Carpenter, K., Maguire, J. A., Hosmane, N. S., and Takagaki, M. (2005) Substituted carborane-appended water-soluble single-wall carbon nanotubes: new approach to boron neutron capture therapy drug delivery, *J. Am. Chem. Soc.*, **127**, 9875–9880.

67. Holzinger, M., Steinmetz, J., Samaille, D., Glerup, M., Paillet, M., Bernier, P., Ley, L., and Graupner, R. (2004) [2+1] Cycloaddition for cross-linking SWNTs, *Carbon*, **42**, 941–947.

68. Bahr, J. L., and Tour, J. M. (2001) Highly functionalized carbon nanotubes using in situ generated diazonium compounds, *Chem. Mater.*, **13**, 3823–3824.

69. Bahr, J. L., Yang, J., Kosynkin, D. V., Bronikowski, M. J., Smalley, R. E., and Tour, J. M. (2001) Functionalization of carbon nanotubes by electrochemical reduction of aryl diazonium salts: a bucky paper electrode, *J. Am. Chem. Soc.*, **123**, 6536–6542.

70. Dyke, C. A., and Tour, J. M. (2003) Solvent-free functionalization of carbon nanotubes, *J. Am. Chem. Soc.*, **125**, 1156–1157.

71. Hudson, J. L., Caavant, M. J., and Tour, J. M. (2004) Water-soluble, exfoliated, nonroping single-wall carbon nanotubes, *J. Am. Chem. Soc.*, **126**, 11158–11159.

72. Ohtsuki, T., Yuki, H., Muto, M., Kasagi, J., and Ohno, K. (2004) Enhanced electron-capture decay rate of 7Be encapsulated in C60 cages, *Phys. Rev. Lett.*, **93**, 112501.

73. Bandow, S., Takizawa, M., Hirahara, K., Yudasaka, M., and Iijima, S. (2001) Raman scattering study of double-wall carbon nanotubes derived from the chains of fullerenes in single-wall carbon nanotubes, *Chem. Phys. Lett.*, **337**, 48–54.

74. Yanagi, K., Miyata, Y., and Kataura, H. (2006) Highly stabilized β-carotene in carbon nanotubes, *Adv. Mater.*, **18**, 437–441.

75. Park, S., Yang, H. S., Kim, D., Jo, K., and Jon, S. (2008) Rational design of amphiphilic polymers to make carbon nanotubes water-dispersible, anti-biofouling, and functionalizable, *Chem. Commun.*, 2876–2878.

76. Ali, H.-Boucetta, Al, K.-Jamal, McCarthy, D., Prato, M., Bianco, A., and Kostarelos, K. (2008) Multiwalled carbon nanotube-doxorubicin supramolecular complexes for cancer therapeutics, *Chem. Commun.*, 459–461.

77. Ito, Y., Venkatesan, N., Hirako, N., Sugioka, N., and Takada, K. (2007) Effect of fiber length of carbon nanotubes on the absorption of erythropoietin from rat small intestine, *Int. J. Pharm.*, **337**, 357–360.

78. Kam, S, N. W., M. O'Connell, Wisdom, J. A., and Dai, H. (2005) Carbon nanotubes as multifunctional biological transporters and near-infrared agents for selective cancer cell destruction, *Proc. Natl. Acad. Sci. USA*, **102**, 11600–11605.

79. Wong, N. Shi Kam, Liu, Z., and Dai, H. (2006) Carbon nanotubes as intracellular transporters for proteins and DNA: an investigation of the uptake mechanism and pathway, *Angew. Chem. Int. Ed.*, **45**, 577–581.

80. Feazell, R. P., Nakayama, N.-Ratchford, Dai, H., and Lippard, S. J. (2007) Soluble single-walled carbon nanotubes as longboat delivery systems for platinum (IV) anticancer drug design, *J. Am. Chem. Soc.*, **129**, 8438–8439.

81. Zhang, Z., Yang, X., Zhang, Y., Zeng, B., Wang, S., Zhu, T., Roden, S, R. B., Chen, Y., and Yang, R. (2006) Delivery of telomerase reverse transcriptase small interfering RNA in complex with positively charged single-walled carbon nanotubes suppresses tumour growth, *Clin. Cancer Res.*, **12**, 4933–4939.

82. Huang, W., Taylor, S., Fu, K., Lin, Y., Zhang, D., Hanks, T. W., Rao, A. M., and Sun, Y.-P. (2002) Attaching proteins to carbon nanotubes via diimide-activated amidation, *Nano Lett.*, **2**, 311–314.

83. Baker, S. E., Cai, W., Lasseter, T. L., Weidkamp, K. P., and Hamers, R. J. (2002) Covalently bonded adducts of deoxyribonucleic acid (DNA) oligonucleotides with single-wall carbon nanotubes: synthesis and hybridization, *Nano Lett.*, **2**, 1413–1417.

84. Hazani, M., Naaman, R., Hennrich, F., and Kappes, M. M. (2003) Confocal fluorescence imaging of DNA-functionalized carbon nanotubes, *Nano Lett.*, **3**, 153–155.

85. Yu, B. Z., Yang, J. S., and Li, W. X. (2007) In vitro capability of multi-walled carbon nanotube modified with gonadotrophin releasing hormone on killing cancer cells, *Carbon*, **45**, 1921–1927.

86. Pastorin, G., Wu, W., Wieckowski, S., Briand, J.-P., Kostarelos, K., Prato, M., and Bianco, A. (2006) Double functionalisation of carbon nanotubes for multimodal drug delivery, *Chem. Commun.*, 1182–1184.

87. Wu, W., Wieckowski, S., Pastorin, G., Benincasa, M., Klumpp, C., Briand, J.-P., Gennaro, R., Prato, M., and Bianco, A. (2005) Targeted delivery of amphotericin B to cells using functionalised carbon nanotubes, *Angew. Chem. Int. Ed.*, **44**, 6358–6362.

88. Ren, Y., and Pastorin, G. (2008) Incorporation of hexamethylmelamine inside capped carbon nanotubes, *Adv. Mater.*, **20**, 2031–2036.

89. Hampel, S., Kunze, D., Haase, D., Krämer, K., Rauschenbach, M., Ritschel, M., Leonhardt, A., Thomas, J., Oswald, S., Hoffman, V., and Büchner, B. (2008) Carbon nanotubes filled with a chemotherapeutic agent: a nanocarrier mediates inhibition of tumor cell growth, *Nanomedicine*, **3**, 175–182.

90. Poland, C. A., Duffin, R., Kinloch, I., Maynard, A., Wallace, H, W. A., Seaton, A., Stone, V., Brown, S., Mnee, W., and Donaldson, K. (2008) Carbon nanotubes introduced into the abdominal cavity of mice show asbestos-like pathogenicity in a pilot study, *Nat. Nanotechnol.*, **3**, 423–428.

91. Lacerda, L., Ali, H.-Bouchetta, M. A, Herrero, Pastorin, G., Bianco, A., Prato, M., and Kostarelos, K. (2008) Tissue histology and physiology following intravenous administration of different types of functionalized multiwalled carbon nanotubes, *Nanomedicine*, **3**, 149–161.

92. Sato, Y., Yokoyama, A., Shibata, K., Akimoto, Y., Ogino, S., Nodasaka, Y., Kohgo, T., Tamura, K., Akasaka, T., Uo, M., Motomiya, K., Jeyadevan, B., Ishiguro, M., Hatakeyama, R., Watari, F., and Tohji, K. (2005) Influence of length on cytotoxicity of multi-walled carbon nanotubes against human acute monocytic leukemia cell line THP-1 in vitro and subcutaneous tissue of rats in vivo, *Mol. Biosyst.*, **1**, 176–182.

93. Schipper, M. L., Nakayama, N.-Ratchford, Davis, C. R., Wong, N. Shi Kam, Chu, P., Liu, Z., Sun, X., Dai, H., and Gambhir, S. S. (2008) A pilot toxicology study of single-walled carbon nanotubes in a small sample of mice, *Nat. Nanotechnol.*, **3**, 216–221.

94. Wang, H. F., Wang, J., Deng, X. Y., Sun, H. F., Shi, Z. J., Gu, Z. N., Liu, Y. F., and Zhao, Y. L. (2004) Biodistribution of carbon single-wall carbon nanotubes in mice, *J. Nanosci. Nanotechnol.*, **4**, 1019–1024.

95. Paloniemi, H., Aäritalo, T., Laiho, T., Liuke, H., Kocharova, N., Haapakka, K., Terzi, F., Seeber, R., and Lukkari, J. (2005) Water-soluble full-length single-wall carbon nanotube polyelectrolytes: preparation and characterization, *J. Phys. Chem. B*, **109**, 8634–8642.

96. Zhang, H., Li, H. X., and Cheng, H. M. (2006) Water-soluble multiwalled carbon nanotubes functionalized with sulfonated polyaniline, *J. Phys. Chem. B*, **110**, 9095–9099.

97. Liu, A., Watanabe, T., Honma, I., Wang, J., and Zhou, H. (2006) Effect of solution pH and ionic strength on the stability of poly(acrylic acid)-encapsulated multiwalled carbon nanotubes aqueous dispersion and its application for NADH sensor, *Biosens. Bioelectron.*, **22**, 694–699.

98. Besteman, K., J. -Lee, O., Wiertz, M, F. G., Heering, H. A., and Dekker, C. (2003) Enzyme-coated carbon nanotubes as single-molecule biosensors, *Nano Lett.*, **3**, 727–730.

99. Xin, H., and Woolley, A. T. (2003) DNA-templated nanotube localization, *J. Am. Chem. Soc.*, **125**, 8710–8711.

100. Taft, B. J., Lazareck, A. D., Withey, G. D., Yin, A., Xu, J. M., and Kelley, S. O. (2004) Site-specific assembly of DNA and appended cargo on arrayed carbon nanotubes, *J. Am. Chem. Soc.*, **126**, 12750–12751.

101. Liu, L., Wang, T., Li, J., Guo, Z., Dai, L., Zhang, D., and Zhu, D. (2003) Self-assembly of gold nanoparticles to carbon nanotubes using a thiol-terminated pyrene as interlinker, *Chem. Phys. Lett.*, **367**, 747–752.

102. Dalton, A. B., Stephan, C., Coleman, J. N., McCarthy, B., Ajayan, P. M., Lefrant, S., Bernier, P., Blau, W. J., and Byrne, H. J. (2000) Selective Interaction of a semiconjugated organic polymer with single-wall nanotubes, *J. Phys. Chem. B*, **104**, 10012–10016.

103. Steuerman, D. W., Star, A., Narizzano, R., Choi, H., Ries, R. S., Nicolini, C., Stoddart, J. F., and Heath, J. R. (2002) Interactions between conjugated polymers and single-walled carbon nanotubes, *J. Phys. Chem. B*, **106**, 3124–3130.

104. Star, A., and Stoddart, J. F. (2002) Dispersion and solubilization of single-walled carbon nanotubes with a hyperbranched polymer, *Macromolecules*, **35**, 7516–7520.

105. Mitchell, C. A., Bahr, J. L., Arepalli, S., Tour, J. M., and Krishnamoorti, R. (2002) Dispersion of functionalized carbon nanotubes in polystyrene, *Macromolecules*, **35**, 8825–8830.

106. Coleman, J. N., Dalton, A. B., Curran, S., Rubio, A., Davey, A. P., Drury, A., McCarthy, B., Lahr, B., Ajayan, P. M., Roth, S., Barklie, R. C., and Blau, W. J. (2000) Phase separation of carbon nanotubes and turbostratic graphite using a functional organic polymer, *Adv. Mater.*, **12**, 213–216.

107. Murphy, R., Coleman, J. N., Cadek, M., McCarthy, B., Bent, M., Drury, A., Barklie, R. C., and Blau, W. J (2002) High-yield, nondestructive purification and quantification method for multiwalled carbon nanotubes, *J. Phys. Chem. B*, **106**, 3087–3091.

108. Coleman, J. N., O'Brien, D. F., Dalton, A. B., McCarthy, B., Lahr, B., Barklie, R. C., and Blau, W. J. (2000) Electron paramagnetic resonance as a quantitative tool for the study of multiwalled carbon nanotubes, *J. Phys. Chem.*, **113**, 9788–9793.

109. Islam, M. F., Rojas, E., Bergey, D. M., Johnson, A. T., and Yodh, A. G. (2003) High weight fraction surfactant solubilization of single-wall carbon nanotubes in water, *Nano Lett.*, **3**, 269–273.

110. Richard, C., Balavoine, F., Schultz, P., Ebbesen, T. W., and Mioskowski, C. (2003) Supramolecular self-assembly of lipid derivatives on carbon nanotubes, *Science*, **300**, 775–778.

111. O, M. J.'Connell, Bachilo, S. M., Huffman, C. B., Moore, V. C., Strano, M. S., Haroz, E. H., Rialon, K. L., Boul, P. J., Noon, W. H., Kittrell, C., Ma, J., Hauge, R. H., Weisman, R. B., and Smalley, R. E. (2002) Band gap fluorescence from individual single-walled carbon nanotubes, *Science*, **297**, 593–596.

112. Moore, V. C., Strano, M. S., Haroz, E. H., Hauge, R. H., and Smalley, R. E. (2003) Individually suspended single-walled carbon nanotubes in various surfactants, *Nano Lett.*, **3**, 1379–1382.

113. Wenseleers, W., Vlasov, I. I., Goovaerts, E., Obraztsova, E. D., Lobach, A. S., and Bouwen, A. (2004) Efficient isolation and solubilization of pristine single-walled nanotubes in bile salt micelles, *Adv. Funct. Mater.*, **14**, 1105–1112.

114. O'Connell, M. J., Boul, P., Ericson, L. M., Huffman, C., Wang, Y., Haroz, E., Kuper, C., Tour, J., Ausman, K. D., and Smalley, R. E. (2001) Reversible water-solubilization of single-walled carbon nanotubes by polymer wrapping, *Chem. Phys. Lett.*, **342**, 265–271.

115. Singh, I., Bhatnagar, P. K., Mathur, P. C., and Bharadwaj, L. M. (2008) Optical absorption spectrum of single-walled carbon nanotubes dispersed in sodium cholate and sodium dodecyl sulphate, *J. Mater. Res.*, **23**, 632–636.

116. Liu, Z., Sun, X., Nakayama, N.-Ratchford, and Dai, H. (2007) Supramolecular chemistry on water-soluble carbon nanotubes for drug loading and delivery, *ACS Nano*, **1**, 50–56.

117. Salzmann, C. G., Llewellyn, S. A., Tobias, G., Ward, M. A. H, Huh, Y., and Green, M. L H. (2007) The role of carboxylated carbonaceous fragments in the functionalization and spectroscopy of a single-walled carbon-nanotube material, *Adv. Mater.*, **19**, 883–887.

118. Hamon, M. A., Chen, J., Hu, H., Chen, Y., Itkis, M. E., Rao, A. M., Eklund, P. C., and Haddon, R. C. (1999) Dissolution of single-walled carbon nanotubes, *Adv. Mater.*, **11**, 834–840.

119. Kukovecz, A., Kramberger, C., Holzinger, M., Kuzmany, H., Schalko, J., Mannsberger, M., and Hirsch, A. (2002) On the stacking behavior of functionalized single-wall carbon nanotubes, *J. Phys. Chem. B*, **106**, 6374–6380.

120. Chen, J., Rao, A. M., Lyuksyutov, S., Itkis, M. E., Hamon, M. A., Hu, H., Cohn, R. W., Eklund, P. C., Colbert, D. T., Smalley, R. E., and Haddon, R. C. (2001) Dissolution of full-length single-walled carbon nanotubes, *J. Phys. Chem. B*, **105**, 2525–2528.

121. Kam, N. W. S, Jessop, T. C., Wender, P. A., and Dai, H. (2004) Nanotube molecular transporters: internalization of carbon nanotube-protein conjugates into mammalian cells, *J. Am. Chem. Soc.*, **126**, 6850–6851.

122. Chen, Z., Kobashi, K., Rauwald, U., Booker, R., Fan, H., Hwang, W. F., and Tour, J. M. (2006) Soluble ultra-short single-walled carbon nanotubes, *J. Am. Chem. Soc.*, **128**, 10568–10571.

123. Pignatello, R., Guccione, S., Forte, S., Di, C. Giacomo, Sorrenti, V., Vicari, L., Uccello, G. Barretta, Balzano, F., and Puglisi, G. (2004) Lipophilic conjugates of methotrexate with short-chain alkylamino acids as DHFR inhibitors. Synthesis, biological evaluation, and molecular modelling, *Bioorg. Med. Chem.*, **12**, 2951–2964.

124. Zotchev, S. B. (2003) Polyene macrolide antibiotics and their applications in human therapy, *Curr. Med. Chem.*, **10**, 211–223.

125. Szlinder-Richert, J., Cybulska, B., Grzybowska, J., Bolard, J., and Borowski, E. (2004) Interaction of amphotericin B and its low toxic derivative, N-methyl-N-D-fructosyl amphotericin B methyl ester, with fungal, mammalian and bacterial cells measured by the energy transfer method, *Farmaco*, **59**, 289–296.

126. Pantarotto, D., Partidos, C. D., Hoebeke, J., Brown, F., Kramer, E., Briand, J.-P., Muller, S., Prato, M., and Bianco, A. (2003) Immunization with peptide-functionalized carbon nanotubes enhances virus-specific neutralizing antibody responses, *Chem. Biol.*, **10**, 961–966.

127. Cai, D., Mataraza, J. M., Zheng-Hong. Q., Huang, Z., Huang, J., Chiles, T. C., Carnahan, D., Kempa, K., and Ren, Z. (2005) Highly efficient molecular delivery into mammalian cells using carbon nanotube spearing, *Nat. Methods.*, **2**, 449–454.

128. Lu, Q., Moore, J. M., Huang, G., Mount, A. S., Rao, A. M., Larcom, L. L., and Ke, P. C. (2004) RNA Polymer translocation with single-walled carbon nanotubes, *Nano Lett.*, **4**, 2473–2477.

129. Kam, N. W. S., and Dai, H. (2005) Carbon nanotubes as intracellular protein transporters: generality and biological functionality, *J. Am. Chem. Soc.*, **127**, 6021–6026.

130. Liu, Y., Wu, D.-C., Zhang, W.-D., Jiang, X., He, C.-B., Chung, T. S., Goh, S. H., and Leong, K. W. (2005) Polyethylenimine-grafted multiwalled carbon nanotubes for secure noncovalent immobilization and efficient delivery of DNA, *Angew. Chem. Int. Ed.*, **44**, 4782–4785.

131. Kataura, H., Maniwa, Y., Kodama, T., Kikuchi, K., Hirahara, K., Suenaga, K., Iijima, S., Suzuki, S., Achiba, Y., and Krätschmer, W. (2001) High-yield fullerene encapsulation in single-wall carbon nanotubes, *Synth. Met.*, **121**, 1195–1196.

132. Li, L. J., Khlobystov, N., Wiltshire, J. G., Briggs, D, G. A., and Nicholas, R. J. (2005) Diameter-selective encapsulation of metallocenes in single-walled carbon nanotubes, *Nat. Mater.*, **4**, 481–485.

133. Smith, B. W., Mothioux, M., and Luzzi, D. E. (1998) Encapsulated C_{60} in carbon nanotubes, *Nature*, **396**, 323–324.

134. Takenobu, T., Takano, T., Shiraishi, M., Murakami, Y., Ata, M., Kataura, H., Achiba, Y., and Iwasa, Y. (2003) Stable and controlled amphoteric doping by encapsulation of organic molecules inside carbon nanotubes, *Nat. Mater.*, **2**, 683–688.

135. Simon, F., Kuzmany, H., Rauf, H., Pichler, T., Bernardi, J., Peterlik, H., Korecz, L., Fülöp, F., and Jánossy, A. (2004)Low temperature fullerene encapsulation in single wall carbon nanotubes: synthesis of N@C60@SWCNT, *Chem. Phys. Lett.*, **383**, 362–367.

136. Shao, L., T. -Lin, W., Tobias, G., and Green, M. L H. (2008) A simple method for the containment and purification of filled open-ended single wall carbon nanotubes using C_{60} molecules, *Chem. Commun.*, 2164–2166.

137. Krinsky, N. I., and Yeum, K. J. (2003) Carotenoid-radical interactions, *Biochem. Biophys. Res. Commun.*, **305**, 754–760.

138. Chen, B. H., and Huang, J. H. (1998) Degradation and isomerization of chlorophyll a and β-carotene as affected by various heating and illumination treatments, *Food Chem.*, **62**, 299–307.

139. Hampel, S., Kunze, D., Haase, D., Krämer, K., Rauschenbach, M., Ritschel, M., Leonhardt, A., Thomas, J., Oswald, S., Hoffman, V., and B. Büchner. (2008) Carbon nanotubes filled with a chemotherapeutic agent: a nanocarrier mediates inhibition of tumor cell growth, *Nanomedicine*, **3**, 175–182.

Chapter 3

BIOMEDICAL APPLICATIONS II: INFLUENCE OF CARBON NANOTUBES IN CANCER THERAPY

Chiara Fabbro,* Francesca Maria Toma* and Tatiana Da Ros
Dipartimento di Scienze Farmaceutiche, Università di Trieste, Trieste 34127, Italy
daros@units.it

After cardiovascular pathologies, cancer is one of the most important diseases and affects a large portion of the western world. The aetiology of this disease, which is comprehensive and consists of a number of different typologies, is related to genetic instability and/or multiple DNA alterations. One of the main problems is to selectively distinguish between healthy and tumour cells. Considering this aspect, nanotechnology's recent achievements are promising major breakthroughs in patient care than the current approaches in cancer treatment. As the U.S. National Cancer Institute states,[1] nanotechnology offers the unprecedented opportunity of studying normal and cancer cells in real time, at molecular and cellular scales, and during the earliest stages of the cancer process. Nanoscale molecular structures are similar in size to large biomolecules, and under the right conditions they can easily enter cells or cross blood vessel walls.

3.1 IMPORTANCE OF NANOTECHNOLOGY IN CANCER THERAPY

The biomedical nanotechnology approach (namely, nanomedicine) can be really useful in cancer therapy at three different stages of the tumoural process: early detection, imaging and therapeutic treatment of tumours.[2] For

*These authors contributed equally to this work.

instance, delivery systems make it possible to target therapeutic or contrast agents directly to the desired organ or tissue using nanovectors. They typically consist of a core decorated on the surface with targeting molecules and/ or solubilising agents and bearing inside therapeutic or imaging payloads (Fig. 3.1).

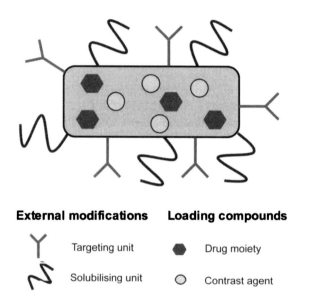

External modifications   **Loading compounds**

    Targeting unit   Drug moiety

    Solubilising unit   Contrast agent

Figure 3.1 Schematic representation of nanovectors. See also Colour Insert.

They are, in fact, engineered to release the loaded units after having reached the target and to carry out their activity with decreased side effects. The recognition units are generally molecules, which are recognised by receptors expressed (or in a favourable case overexpressed) on the cell surface in tumour tissues. The targeting can be driven by an active recognition, due to the presence of a special unit (active targeting), but can also happen passively, because natural cellular uptake is, in this case, strictly related to enhanced permeability and retention effect (EPR). EPR concerns tumour angiogenesis, where blood vessels bear more numerous and larger fenestrations than in normal tissues, because of the demands of a rapid growth. The endothelium of cancer blood vessels is, in fact, characterised by the presence of few pericytes and smooth muscle cells, with a consequent different morphology of the vessel walls.[2] Moreover, in tumour tissue there is usually a lack of lymphatic vessels, and these two characteristics, together with the cellular expression of receptors for growth factors and cell adhesion, allow the use of multiple targeting strategies that undergo extravasation and protracted lodging.[3] In this scenario, the reticuloendothelial system (RES), as a part of the immune system and consisting mostly of macrophages and monocytes localised in the lymph nodes, spleen and Kupffer cells, plays a determinant role, and nanovectors can be designed to target or escape RES cells by simply tuning vector size.

The approval of Abraxane™ by FDA has been a major success for cancer nanotechnology in the last years. It consists of paclitaxel coupled with albumin as delivery agent. This formulation has been successfully adopted in metastatic breast cancer therapy.[4] Generally, the advantages of using nanovectors are the possibility of overcoming poor solubility of some drugs, which can be protected from metabolic and immune system attacks, altering the biodistribution by tuning the nanovector size (exploiting the EPR effect) and controlling the release profile by coupling nanovectors with enzyme-sensitive linkers.

Nanotechnology can also be used to induce cellular damage to a specific target by exploiting the characteristics of the nanoparticles themselves. A good example is the induction of hyperthermia, which involves the use of external energy to promote a localised cytotoxic effect.[5] This non-invasive therapy could be useful when surgery is not possible. Some nanoparticles can act like antennas: when irradiated, they absorb energy and then release high doses of heat. The consequent increase in the temperature of the cells where they have been loaded promotes tumour tissue necrosis, so nanoparticles can be used to both target and destroy cells specifically. Metallic nanoparticles such as Au particles or Au nanoshells, oxides such as superparamagnetic nanoparticles, and organic nanoparticles such as carbon nanotubes (CNTs) are the most important examples in this approach (see Section 3.5).[2] CNTs play a major role because they can exploit the absorption of near-infrared (NIR) radiation, combined with their excellent thermal conductivity. It is very well known that NIR radiation can penetrate human tissues to many centimetres without causing damage, because cellular chromophores do not absorb NIR wavelengths.[6]

Another important aspect in which nanotechnology can play a crucial role is the enhancement of tumour imaging efficiency and resolution. Applications of optical imaging are limited because of the low penetration depth into tissue and lack of anatomic resolution and spatial information.[4] Although NIR wavelengths can be used to increase the penetration depth and three-dimensional (3D) fluorescence tomography could give spatial resolution, magnetic resonance imaging (MRI) is the most used visualisation technique because it gives better images than do tomography and 3D imaging. The development of new "dual-modality" contrast agents, capable of being monitored through their magnetic properties, can solve the problem of good imaging systems for both optical imaging and MRI.[4] In this context, nanoparticles are really fundamental. In fact optical and electronic properties are size dependent and can be tuned by changing particle size. *In vivo* imaging with superparamagnetic oxide nanoparticles have already been used as contrast agents in MRI,[5] because they change the spin-spin relaxation times of closed water molecules.[7] Quantum dots (QDs) are also promising candidates in the imaging field and in the monitoring of cellular events because of their unique tunable fluorescence emission, their great photostability and signal intensity and their large absorption coefficient,[7] thereby achieving high sensitivity in the detection of small tumour masses.[4]

Early detection of tumour is essential for patient cure, but at the onset of the disease, the masses are too small to be efficiently identified by imaging, considering the problems of spatial resolution and definition. At this stage, contrast agents cannot be sufficient to overcome this drawback and it is necessary to amplify the signals, by means of multifunctional nanosystems. Biosensors can be a good means for early detection but pose some problems, related to the blood concentration of significant pathological biomarkers and to the sensitivity of detection systems. For example, prostate-specific antigen (PSA) concentration can give an indication of the health state with respect to prostate cancer, but it has widely different baseline expression in the population and is non-specific. Although nanotechnology cannot be really helpful to overcome this problem, it can play an important role in increasing the sensitivity of sensors.

Many detection platforms have been studied and are under development, such as the well-known DNA microarrays. In this context, the nanoscale is the key to improve detection sensitivity. Micro- and nano-cantilevers are examples of this technology,[3] where, by using both the surface functionalisation and the cantilever oscillation properties, it is possible to detect the presence of biomarkers using Hook's law (which relates bending, oscillation frequency and weight). These systems are the so-called MEMS or NEMS (micro- or nano-electrical mechanical sensors).

Thanks to the multi-functionalisation, nanosystems can be employed as biosensors by exploiting their peculiarity, such as paramagnetism. QDs have been used in the contemporary detection of four different biomarkers to provide spatial information.[4] Nanowires, nanobarcodes and nanotubes play a role as sensors. Many examples concerning CNTs will be reported in the next paragraphs. Scanning probe microscopy, as atomic force microscopy (AFM) and high-resolution atomic force microscopy, improved the design of new biosensoring systems because it overcomes the restriction of optical imaging, related to wavelength and molecular object dimensions. At the state of the art, the design of sensors based on nanoarrays is possible by grafting molecules in nanopatches with nanolithography.[8,9]

3.2 CARBON NANOTUBES: A BRIEF OVERVIEW

It is widely reported in literature that CNTs present interesting electronic and chemical properties, as well as mechanical strength and thermal stability. Their high aspect ratio (length/diameter) makes CNTs extremely important substrates to develop biosensors with high-packaged immobilised biomolecules, and their wide surface area is highly accessible to electrochemical manipulation.

Their dimensions and the possibility of modifying their external surface by functionalisation with suitable chains and molecules (Fig. 3.2) or of filling their cavities with bioactive agents render CNTs optimal as drug vectors, but it is necessary to clarify their toxicity effects. It is possible to roughly summarise that cytotoxicity is higher for longer CNTs,[10] but it is also related to the presence of catalytic metals used in CNT preparation and to the solubility of the analysed compounds. In this scenario, a preliminary purification is often needed and chemical functionalisation is the keystone to achieving good solubility in biocompatible solvents and to decreasing toxicity.

Covalent and supramolecular functionalisations are currently used. The first affects the inherent properties of CNTs but, at the same time, permits to link a high number of biomolecules on the nanotube surface, as is necessary in biosensor design. Many different reactions on CNTs are reported in literature.[11,12]

The attachment of biomolecules, anticancer drugs or radionuclides to CNTs, through covalent or non-covalent functionalisation, transforms CNTs into vectors or electronic devices for biosensoring, leading to multifunctional nanosystems.[13] Nevertheless, the two approaches can be adopted simultaneously: Pantarotto *et al.* reported the functionalisation of CNTs by 1,3-dipolar cycloaddition combined with the attachment of plasmid DNA, through electrostatic forces, for gene delivery in cancer therapy.[14]

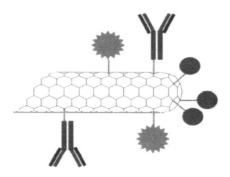

Figure 3.2 An ideal example of multifuncionalised CNTs. See also Colour Insert.

CNTs interact with cells and are able to enter them through different paths, including piercing the cellular membrane as needles. This mechanism is energy-independent and is extremely useful to carry into cells molecules, which have difficulty crossing cellular barriers. In exploiting this penetration mechanism, no specific recognition is needed. Recently, Strano *et al.* reported CNT exocytosis,[15] while systemic distribution and elimination were reported by Lacerda *et al.*, demonstrating renal excretion of functionalised CNTs.[16]

3.3 CARBON NANOTUBES AS DRUG VECTORS IN CANCER TREATMENT

Doxorubicin (DOXO) is a chemotherapic drug that is widely used in clinic for cancer treatment. Although very effective, it induces cardiomyopathy as one of the most severe consequences of lack of selectivity and is subjected to multi-drug resistance (MDR), i.e., the tumour is able to learn how to defend itself against the drug. For these reasons it is very interesting to study the delivery of DOXO through nanocarriers, and up to now, two works have been reported in which CNTs have been used with this purpose. Ali-Boucetta *et al.* described the binding of the drug to copolymer-coated multi-walled carbon nanotubes (MWCNTs), forming a supramolecular complex based on π–π stacking. The cytotoxicity of this derivative has been tested *in vitro* on human breast cancer cells, showing an increased mortality rate in comparison with doxorubicin alone.[17]

Liu *et al.* prepared water-soluble DOXO-CNT complexes using two different kinds of CNTs. In one case, single-walled carbon nanotubes (SWCNTs) were supramolecularly functionalised by the phospholipid- poly(ethylene glycol)

(PEG) surfactant (compound **1** in Fig. 3.3), whereas in another case, oxidised SWCNTs were covalently linked to PEG, through amidation of the carboxylic moieties (compound **2**).[18] The authors suggest π–π stacking, together with hydrophobic interactions, as binding forces between doxorubicin and nanotubes. Moreover, they report a pH-dependent interaction: the loading was performed at pH 9, the condition at which DOXO is deprotonated and has low water solubility, reaching a doxorubicin-to-nanotube weight ratio of ~4:1; a decrease in the pH led to the release of the drug from the SWCNT carrier, because of the consequent higher hydrophilicity of the protonated drug. This mechanism can be very interesting in cancer therapy because of the acidic environment of extracellular tissues in tumours, potentially leading to selective drug release *in vivo*. The *in vitro* cytotoxicity of **1** was tested on human glioblastoma cancer cells. The derivative induced cell death similar to free DOXO at 10 μM concentration, even if the observed IC_{50} value was higher (~8 μM for **1**, ~2 μM for free DOXO).

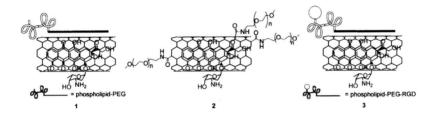

Figure 3.3 Structures of compounds **1–3**. See also Colour Insert.

Furthermore, a targeted doxorubicin delivery was tested using a cyclic arginine–glycine–aspartic acid (RGD) peptide, which acts as recognition moiety for integrin $\alpha_v\beta_3$ receptors, overexpressed in a wide range of solid tumours. The targeting agent was bound to the terminal group on the PEG chain of the non-covalently functionalised SWCNT derivative, and then doxorubicin was loaded (compound **3**). This conjugate, tested on integrin $\alpha_v\beta_3$-positive cells, showed enhanced drug delivery compared with the derivative without RGD as confirmed by the IC_{50} value, which was smaller for **3** (~3 μM) than for **1**.

Methotrexate is another well-known anticancer drug, but it has low bioavailability and toxic side effects. Preliminary studies of CNT-based drug delivery with this compound were performed by Pastorin *et al.*[19] The authors covalently functionalised CNTs with methotrexate and tested the obtained derivatives on human Jurkat T lymphocytes, finding a rapid internalisation of compound **4** (Fig. 3.4) and thereby proving the effectiveness of CNTs as nanocarriers.

Yang *et al.* reported the preparation of a folate-targeted DNA transporter with CNTs that is in principle exploitable to deliver siRNA to cancer cells.[20] They covalently functionalised SWCNTs with an aliphatic chain bearing a positive charge and able to form electrostatic interactions with nucleic acids. Then they bound fluorescently labelled double-stranded DNA (dsDNA) to this derivative and proved by UV a significant enhancement of the loading of dsDNA for the positively charged SWCNTs (compound **5**) compared with the uncharged SWCNTs. They further functionalised the derivative by wrapping folic acid–modified phospholipids (compound **6**). The complexes were tested on mouse ovarian epithelial cells, showing an increased uptake for compound **6**, compared with **5**. Furthermore, the fluorescently labelled dsDNA alone was very poorly internalised, thus demonstrating the role of CNTs as carriers. In a final experiment with HeLa cells overexpressing folate receptors, the internalisation of these derivatives was found to be much higher than in normal HeLa cells, confirming the efficacy of the delivery system.

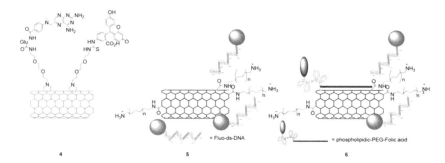

Figure 3.4 Structures of compounds 4–6.

Also taxanes, an important class of antineoplastic drugs, have been conjugated with CNTs to prepare new drug delivery systems. Paclitaxel (PTX) was linked to PEGylated SWCNTs (compound **7** in Fig. 3.5) and then administered by intravenous injection to xenograft tumour mice (PTX-resistant 4T1 murine breast cancer mice model) to test the *in vivo* efficacy of this compound.[21] The results were very promising, showing a tumour growth inhibition of almost 60% for compound **7**, while only 28% and 21% of inhibition was obtained by taxol and PEG-PTX, respectively. A higher apoptosis level and a diminished proliferation of active cells were observed in the presence of the CNT derivative, when compared with the drug itself. It is important to notice that the PEGylated CNT complex did not demonstrate toxicity in mice, as reported in another paper by Liu *et al.*[22] In the pharmacokinetic studies, compound **7** showed an increased plasmatic half-life, coherently with the enhanced hydrophilicity of the conjugate with respect to the drug itself. In fact there was a decrease in protein absorption, which is responsible for an accelerated uptake by macrophages and 2 hours after injection a much higher PTX presence was found in the intestine and the reticuloendothelial system of different organs, such as the liver and spleen. This is predictable behaviour for nanomaterials in general and can raise concerns about toxicity for these organs. Nevertheless, the authors reported a difference between the biodistribution of SWCNTs and and of PTX after treatment with compound **7**, indicating a rapid release of the drug from the conjugate in the various organs and tissues, probably because ester cleavage by esterases. As a consequence, the drug seemed to be rapidly excreted, lowering its toxicity. Importantly, 2 hours after injection, the PTX levels in tumour were 10 times higher for derivative **7** than for taxol, and the tumor-to-normal organ/tissue PTX uptake ratio was also higher, thus indicating a better selectivity for the CNT-delivered drug.

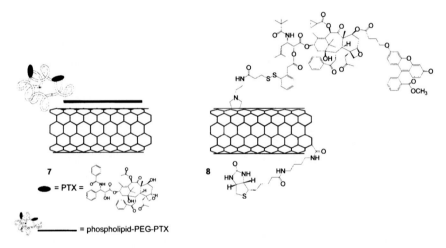

Figure 3.5 Structures of compounds **7** and **8**.

A very interesting work recently published presents another study on taxanes, using compound **8** as an anticancer drug delivery system targeted by biotin towards cancer cells.[23]

This construct is a prodrug, targeted by its biotin unit to cancer cells overexpressing biotin receptors on their surface. Using a control SWCNT derivative without biotin, the authors observed a temperature-dependent but energy-independent internalisation. This is coherent with a non-endocytosis mechanism, in accordance with the hypothesis of needle-like penetration of CNTs through cell membranes. On the other hand, for the biotin conjugate, the mechanism observed was endocytosis, and the degree of internalisation was higher than that obtained using cells not expressing the biotin receptor. Once inside the cell, the prodrug released its cargo upon reduction of the disulphide bond by endogenous thiols such as glutathione (GSH), whose concentrations are typically more than 10^3 times higher in tumour tissues than in blood plasma. In this way the system should act specifically on cancer cells. Furthermore, the authors proved efficient release of the drug by GSH and the binding of the conjugate to the microtubule network, where it exerts its cytotoxic action (inhibiting the mitosis) with an IC_{50} value smaller than that for the drug itself. This could be due to an increase in the drug uptake thanks to the CNT-based system, thus highlighting the effectiveness of this drug delivery method.

Another class of anticancer agents has been investigated as a candidate in drug delivery through CNTs for cancer treatment: platinum analogues. A Pt-based targeted prodrug was prepared using SWCNTs as a longboat to deliver the system.[24] The derivative was prepared by binding folic acid, which is a targeting agent for many cancer cells overexpressing folate receptors, to a Pt(IV) compound and by tethering this conjugate to CNTs. The Pt(IV) derivative is intended as a prodrug because, once inside cells, it can be reduced losing the two axial ligands and leading to the active Pt(II) compound (Scheme 3.1).

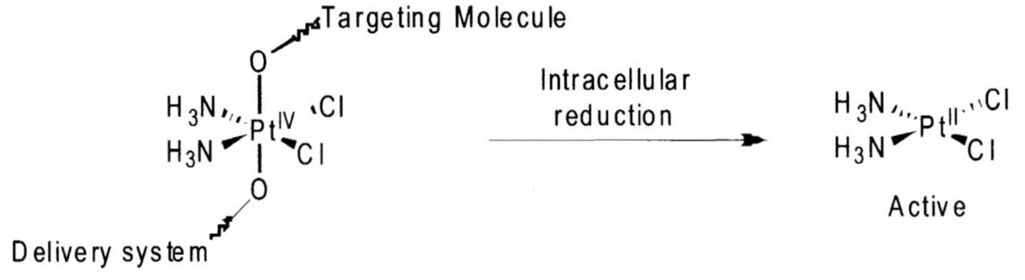

Scheme 3.1 Reaction of Pt reduction.

The authors demonstrated the internalisation via folate–receptor-mediated endocytosis for compound **9** (Fig. 3.6), which killed cancer cells (human choriocarcinoma and human nasopharyngeal carcinoma) with IC_{50} values about 8 times smaller than those for cisplatin alone. Moreover, they detected the formation of the major reaction product of cisplatin with DNA, using a specific monoclonal antibody. This proves the actual ability of the system to act as prodrug. In fact it generates the cytotoxic derivative once internalised and selectively kills cells overexpressing folate receptors.

The same group reported the preparation of another Pt-based SWCNT derivative (**10**), able to deliver the active drug to cells with a concentration 6 times higher than that reached by treatment with the drug itself.[25]

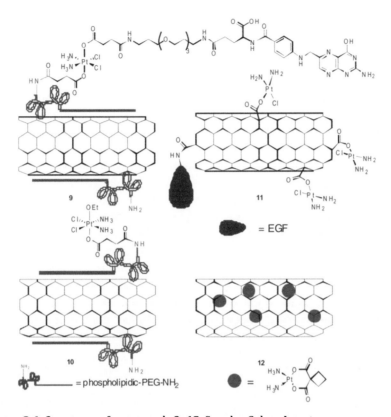

Figure 3.6 Structures of compounds **9–12**. See also Colour Insert.

Recently, Bhirde *et al.* described the use of compound **11**, a cisplatin-delivery system based on a drug-SWCNT conjugate, to target cancer cells *in vivo*.[26] The interaction of the epidermal growth factor (EGF) with its receptor (EGFR), overexpressed on the cell surface of many cancers, has been exploited to target CNTs both *in vitro* and *in vivo*. The authors demonstrated the efficient *in vitro* internalisation of the SWCNT-EGF conjugate by proving that it is actually mediated by the ligand–receptor interaction. Furthermore, they administered QD-functionalised SWCNTs to tumour-bearing athymic mice to study short-term biodistribution. The results showed a much higher accumulation within the tumour mass of the EGF conjugate compared with the control conjugate without EGF, while small amounts of nanotubes were found in the spleen, lungs, liver, kidneys and heart, regardless of the presence of EGF. Finally, the animals were treated with compound **11,** showing a decrease in tumour growth in comparison with the untargeted SWCNT-cisplatin conjugate, demonstrating that SWCNTs can effectively deliver cisplatin *in vivo*, targeted by EGF towards EGFR overexpressing cancer cells.

Another interesting approach for the functionalisation of CNTs, studied also for potential oncology applications, is the filling of the nanotubes. This method was first investigated by Green's group in 1994 and involves the opening of multi-walled nanotube caps by nitric acid treatment and the filling of the inner cavity through a wet chemistry approach.[27] It has been applied to fill the tubes with different materials, including carboplatin.[28] After the cutting procedure, MWCNTs were filled with carboplatin, sonicating them in a solution containing the drug, at different temperatures, and eventually washed in water and ethanol to get rid of the particles deposited on the outer surface. Derivative **12** was characterised through different techniques to confirm the presence of the drug inside the tubes and to identify the best preparation methodology. It was shown that the optimal temperature was 90°C, with a drug loading of 30% (wt). This derivative was then tested on human bladder cancer cells to evaluate its cytotoxicity, showing a decreased cell viability compared with the treatment with the drug alone. Moreover, a synergistic effect in increasing the mortality rate was reported for the simultaneous administration of empty CNTs and carboplatin, while the empty CNTs themselves did not exert any lethal effect. A possible explanation could be an alteration of membrane permeability induced by nanotubes, with a consequent increased uptake of carboplatin.

Not only MWCNTs have been filled with a Pt-based anticancer drug, but also carbon nanohorns (CNHs). They are another class of carbon nanomaterials that appear as spherical aggregates of SWCNTs, with an average diameter of 80–100 nm, and they have the very interesting characteristic of being void of metal particles. Moreover, they are capable of containing molecules.[29,30] Nanometer-size holes can be generated in their walls by oxidation, and drugs can be then incorporated *via* nanoprecipitation. Ajima *et al.* prepared CNHs containing cisplatin, finding a platinum-to-carbon mole ratio of about 1:120, and spectroscopic data suggested that the cisplatin structure was retained during incorporation.[29] Drug release studies performed in phosphate buffer solution (PBS) and in culture medium have shown a decreased dissolution rate with respect to free cisplatin, this being a positive achievement in drug delivery because a slow release reduces the loss of the drug before it reaches the target. Finally, the cisplatin-loaded nanohorns have been proved to reduce the viability of human lung cancer cells, while CNHs alone do not influence the growth of cancer cells.

The same authors performed also a study to compare the effect of the presence of different functional groups at hole edges on drug incorporation and release.[31] CNHs having holes with hydrogen-terminated edges (NHh) and with oxygen-containing groups at the edges (NHox) were prepared, and the two types showed almost no differences in cisplatin incorporation (about

18% in both cases), while the drug release in PBS gave different results. The release rates were different (faster for NHox than for NHh), and the total amount of released cisplatin differed; it was 70% for NHh, but only 15% for NHox. A possible explanation for this is that ions contained in PBS reacted with the –COOH and –OH groups, forming the respective sodium salts (–COO⁻Na⁺ and –O⁻Na⁺), with a decrease in hole diameters, thereby hindering cisplatin release. The difference in the release rate should be due to the hydrophobic characteristic of the NHh holes. In fact aqueous solutions should enter more easily through the hydrophilic oxygen-containing holes of NHox. According to this study, NHh seemed to present better drug release properties than NHox, but the latter gave much better dispersion in PBS than NHh, so that it was a more efficient carrier for drug delivery. Therefore another study was carried out to define the optimal oxidation conditions to get NHox with enlarged hole diameters and an appropriate functionalisation degree of the hole edges.[32] So, the new CNHs should be well dispersible and have a reduced steric hindrance at the same time, and actually the obtained product was able to abundantly (up to 80%) and slowly release cisplatin in PBS. Using a PEG chain linked to a peptide, the same authors further developed this work to coat CNHs previously loaded with cisplatin.[33] This approach was earlier realised using a PEG-DOXO conjugate to improve dispersibility and to avoid aggregate formation in a biological environment.[30] In this case, the complex was used to treat human lung cancer cells, inducing apoptosis, even if at a lower extent than doxorubicin itself.

3.4 DELIVERY OF OLIGONUCLEOTIDES MEDIATED BY CARBON NANOTUBES

The interaction of DNA and CNTs has been studied and demonstrated using different measurements. The condensation of the nucleic acid in the presence of oxidised SWCNTs is dependent on the sequence, being favoured in the case of guanine-cytosine (GC)-rich sequences, with the consequent destabilisation of double and triple helices. From these studies it seems that the nanotubes can induce DNA B-A transition in solution and their interaction with the GC sequences takes place in the major groove.[34] Li *et al.* reported that the presence of SWCNTs (with a diameter of 1.1 nm) induces a stabilisation of quadruplex and of the i-motif adopted by the complementary strand, rich in C (Fig. 3.7).[35] This action can be exploited in an antitumour approach considering that the stabilisation of those structures inhibits the action of telomerases, which elongate and stabilise chromosomes, conferring, in the case of tumour cells, an indefinite proliferative capacity. From some experiments it is possible to suppose that there is a binding of the carbon nanostructures at the 5′ end of the structure with a preferential positioning in the major groove. Moreover, it seems that the presence of carboxylic residues on CNTs compensates for the positive charges of the C–C base pairs. In fact the use of differently funtionalised CNTs (bearing a chain with a free amino unit) did not exert the same effect.

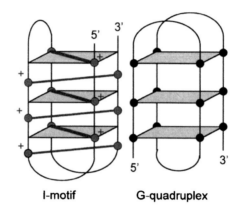

I-motif   G-quadruplex

Figure 3.7 I-motif and G-quadruplex structures. See also Colour Insert.

The same authors reported a stabilisation effect when a poly(rA) chain was used. This sequence has a biological relevance because it is present in all messenger RNAs (mRNAs) at the 3′ end, conferring stability to the structure and influencing the translation and transcription processes.[36] The presence of oxidised SWCNTs (with a diameter of 1.1 nm) induced a self-structuration of A•A⁺ base pairing into duplex at pH 7, where this structure is not naturally formed, as demonstrated by melting and circular dichroism experiments.

Poly(rU) also was used in the presence of CNTs, but in this case to study its translocation into breast cancer MCF-7 cells, successfully obtained with SWCNTs.[37] Myc is a proto-oncogene, fundamental as sensor for mitogenic stimuli in cells; it controls many phenotypic phenomena such as cell growth, proliferation, transformation and apoptosis and is involved in many tumours. Cui *et al.* reported the interaction between antisense-myc conjugated with SWCNTs and HL-60 cells. Part of the complex was found to be attached on the cell surface, but mainly reached the cytoplasm exerting an inhibition effect on cell proliferation.[38]

Oxidised SWCNTs, after activation by thionyl chloride, were allowed to react with the monoprotected hexamethylendiammine, subsequently deprotected. The so-obtained derivatives, bearing positive charges on the carbon surface, were used to complex telomerase reverse transcriptase small interfering RNA (TERT siRNA), able to silence the expression of TERT and to inhibit cell growth. This activity was conserved in both *in vitro* experiments (performed on LLC, TC1 and1H8 cell lines) and *in vivo*.[39] This approach can be really interesting considering the possibility of exploiting the two different actions of CNTs as oligonucleotide vectors and i-motif stabilisers.

Liu *et al.* used non-covalently functionalised SWCNTs bearing positive charges to deliver siRNA into HeLa cells,[40] T cells and primary cells.[41] The oligonucleotide was attached by disulphide bonds to phospholipidic-PEG chains used to wrap CNTs (compound **13**, Fig. 3.8). Depending on the length of the PEG chain, the delivery resulted in being differently effective, corresponding to the differences in efficiency for the internalisation of the nanotubes into cells. The presence of long PEG chain (PEG_{5400}) gave the worst results, while the use of PEG_{2000} was successful, as demonstrated using Raman spectroscopy to localise the presence of the SWCNTs inside cells.

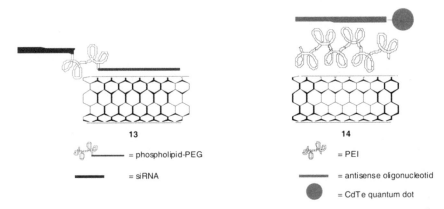

13

$\quad$ = phospholipid-PEG

$\quad$ = siRNA

14

$\quad$ = PEI

$\quad$ = antisense oligonucleotid

$\quad$ = CdTe quantum dot

Figure 3.8 Structures of compounds **13** and **14**.

As previously described (see Section 3.3) Yang *et al.* reported the use of phospholipid-PEG chains derivatised with folic acid to deliver a fluorescently labelled dsDNA, obtaining a selective internalisation in tumour cells overexpressing folate receptors.[20]

Also Jia *et al.* reported the use of CNTs to deliver into cells antisense oligonucleotides able to hybridise mRNA and consequently inhibit protein expression, which is a mechanism of interest in cancer treatment. Oxidised MWCNTs were used to complex polyethyenimine (PEI), which can interact with an antisense sequence by means of electrostatic forces. In this case the oligonucleotide was linked to the cadmium telluride (CdTe) quantum dot. This fluorescent marker was used to follow the cellular trafficking of the complex, demonstrating that, using derivative **14**, the uptake took place by endocytosis and was efficient. Moreover, the system exerted the expected anticancer activity.[42]

A fist approach using a dendrimer to covalently functionalise CNTs for gene delivery was reported in 2007 by Pan *et al.* Although it is not clear which kind of nanotubes they used (it is not specified whether they are single-, double- or multi-walled), the reported results are interesting. The anti-survivin oligonucleotide was anchored onto the polyamidoamine (PAMAM) dendrimer and transfected into MCF-7 cells. This conjugate distributed mainly in the cytoplasm, endosomes and lysosomes of the cells and was able to release the antisense oligonucleotide, which then exerted its apoptotic activity.[43] More recently, Herrero *et al.* reported a study in which a PAMAM dendron was grown on the covalent functionalisation of MWCNTs. Different generations were prepared (G0, G1 and G2, Fig. 3.9), and the free amines were also alkylated by means of glycidyl trimethylammonium chloride, leading to the presence of stable ammonium salts. These compounds have been used to deliver a fluorescent siRNA into HeLa cells, demonstrating the efficiency of dendron-CNT conjugates as nucleic acid carriers, which increased with the number of positive charges on the CNT surface.[44]

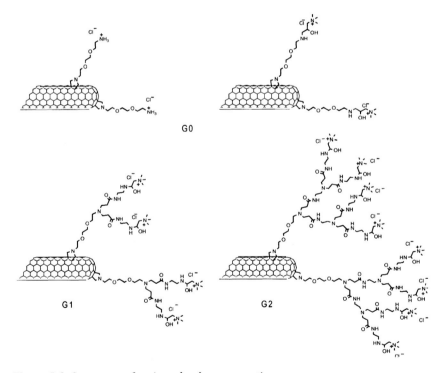

Figure 3.9 Structures of various dendron generations.

The same authors reported the use of MWCNT-NH$_3^+$ to deliver therapeutic siRNA, and its intratumoural administration induced a delay in tumour growth and increased the life span of xenografted animals bearing human lung cancer, demonstrating the efficacy of siRNA delivery by means of CNTs. This system, in fact, showed better results than that obtained by using cationic lyposomes.[45]

All examples reported in literature demonstrate that CNTs, both single- and multi-walled, are good vectors for oligonucleotide delivery and that the transported sequences retain their activity. These results are even more interesting considering that a recent work by Wu *et al.* demonstrated that CNTs exert also an important activity against enzymatic digestion. In fact oligonucleotides, when complexed to SWCNTs, are protected from DNAase I cleavage, increasing their stability in cells.[46]

A unique approach conjugated photodynamic therapy with the use of CNT-oligonucleotide complexes. Zhu *et al.* in fact used SWCNTs as quenchers of singlet oxygen generation (SOG) in the presence of a photosensitiser. An aptamer, i.e., a synthetic DNA/RNA probe able to recognise and bind specific target, was linked to chlorine e6 (Ce6), a well-known photosensitiser. When the aptamer was wrapped on the nanotubes, the excitation of Ce6 did not lead to singlet oxygen ($^1O_2$) generation because the carbon nanostructures, physically close to the chlorine, were able to quench 98% of SOG. When the aptamer was involved in the recognition and binding of its target, i.e., thrombin in the present case, it was released from the nanotubes. In these conditions the SOG was not anymore quenched by CNTs, with the consequent damage of cell structures (Fig. 3.10).[47]

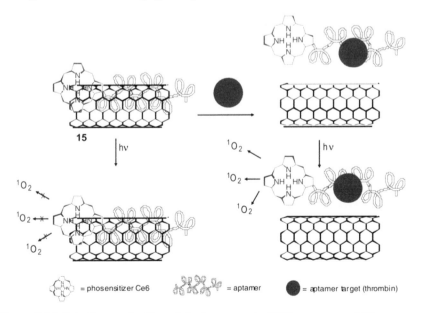

Figure 3.10 Selective production of singlet oxygen in PDT by compound **15**.

It is also necessary to mention some adverse effects on genetic materials reported for CNTs. The genome expression of human skin fibroblast cells, exposed to MWCNTs, was analysed. At cytotoxic doses, the CNTs induced cell cycle arrest and increased apoptosis/necrosis. Expression array analysis indicated that some cellular pathways were altered with the activation of genes involved in cellular transport, metabolism, cell cycle regulation and stress response, inducing the expression of genes indicative of a strong immune and inflammatory response.[48] Moreover, MWCNTs have been proved to induce apoptosis in embryonic stem cells and to cause a twofold increase in the mutation frequency, damaging DNA.[49] One possible explanation, suggested by the authors, is the production of reactive oxygen species in these conditions, but this theory can be considered controversial taking into account that CNTs are considered to act as free radical scavengers.[50,51]

3.5 CARBON NANOTUBES IN RADIOTHERAPY

Up to now the main use of radioisotopes linked to CNTs finds application in imaging studies, to determine the biodistribution of the derivatives following the radioisotope traces. Many different approaches in this sense have been reported using variously functionalised CNTs. In the non-covalent strategy adopted by Liu *et al.*, CNTs were wrapped with PEG chains bearing 1,4,7,10-tetraazacyclododecane-1,4,7,10-tetraacetic acid (DOTA) to chelate ^{64}Cu, a β+-emitting radionuclide used in positron emitting tomography (PET).[52] The covalent approach has been used by McDevitt *et al.* and Singh *et al.*[53,54] In both cases the 1,3-dipolar cycloaddition of azomethine ylides as functiolisation methodology and ^{111}In as γ-emitting radiotracer were employed. McDevitt *et al.* reported the preparation of an SWCNT derivative using DOTA as the chelating agent. Biodistribution studies showed a significative accumulation in the kidney, spleen and liver. In *in vivo* experiments on a murine model of disseminated human lymphoma, there was a selective targeting of the tumour when a specific monoclonal antibody was conjugated (compound **16** in Fig. 3.11), demonstrating the efficacy of this approach.[53] Singh *et al.*, instead, presented CNTs bearing diethylene triamine pentaacetic acid (DTPA) linked to the amine residues, since DTPA is a known chelating agent for ^{111}In (compound **17**). Also in this case the derivative was used to perform biodistribution studies, and it emerged that CNTs were mainly excreted by urine and that after intravenous administration their blood circulation half-life was about 3.5 hours.

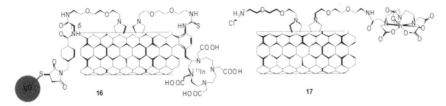

Figure 3.11 Structures of compounds **16** and **17**.

The replacement of indium with some other radionuclides would render these constructs ideal for their application in radiotherapy, although at the moment, to the best of our knowledge, there are no reports on it.

Ultra-short SWCNTs (US-SWCNTs) were adopted for encapsulating [211]At a short-life, α-emitting radionucleus. The dispersion of this complex in buffer released a significant amount of radionuclide, but after treatment with chloramine-T or *N*-chlorosuccinimide, leading to the formation of AtCl, the retention was much higher, with an important overall labelling (At- 19.6% vs AtCl 60.7%). There was no direct evidence of At incapsulation into CNTs, but the authors explored this issue working on I_2 and stated that it is possible to determine the difference in X-ray-induced Auger emission spectrum due to physisorbed or included I_2. From this experiment, the authors revealed the presence of I_2 signals as evidence of the internal complexation and assumed that the same preparation obtained using At should give the same results. These derivatives have not been used for radiotherapy up to now.[55]

Also, the possibility of employing CNTs in boron neutron capture therapy (BNCT) was explored. BNCT is based on the reaction

$$^{10}B + {^1n} \rightarrow {^{11}B} \rightarrow {^7Li} + {^4He}(\alpha)$$

The radiations produced have a very short range in tissues and can seriously damage DNA. So, to be effective in cancer therapy, the boron has to be present into cell nuclei with an opportune concentration. Considering the capacity of CNTs to enter cells and to reach the nuclei, the conjugation of carborane to nanotubes seems to be a successful strategy to explore. The appendage of carborane units on SWCNTs was reported by Yinghuai *et al.*[56] The *nido*-carboranes were derivatised with an azide and allowed to react with the double bonds of the CNT skeleton, leading to the formation of aziridine rings. The derivatives were used in tissue distribution experiments on mice with transplanted EMT6 mammary cancer cells. When a dimethyl sulphoxide

(DMSO) solution of a CNT derivative was used, the maximum boron concentration in tumour tissues, achieved 16 hours after administration, was 27.9 µg boron/g(tissue). In the case of CNT in saline solution, the maximun concentration of boron in tissue was lower (22.8 µg boron/g[tissue]) and it was achieved after 30 hours. In both cases, the tumour-to-blood boron ratio was favourable, with a much higher concentration in tumour cells than in the bloodstream (the ratio was, respectively, 6.13 when DMSO was used and 3.12 when saline solution was used).

These results permit to seriously consider boron-CNT derivatives suitable for BNCT.

3.6 CARBON NANOTUBES IN THERMAL ABLATION

A very interesting application of carbon nanotubes in cancer therapy can arise from their intrinsic optical properties, which can be exploited to kill cancer cells by photothermal destruction. In fact the optical absorbance of this material is very high in the near-infrared region (NIR; 700–1100 nm), while biological systems are transparent to these wavelengths.

It is worth noting that gold nanoshells and nanoparticles represent the only other materials which gave good results in photothermal cancer treatment with NIR radiation.[57-59] In these cases the laser intensity and the radiation time used are often higher than those needed to kill cells in experiments with nanotubes. From this observation it is evident that CNTs are even more promising materials in the field that can pave the way for further explorations to reach the laser energy level of 35–45 mJ/cm^2, established as the safety standard for medical lasers.[60]

Dai *et al.* presented the thermal application of CNTs in 2005 for the first time.[61] Compound **18** (Fig. 3.12), where SWCNTs have been wrapped by phospholipids linked to PEG chains bearing either a fluorescent tag or a folic acid moiety, was administered *in vitro* to normal HeLa cells (as a control) and to HeLa cells overexpressing the folate receptor (FR$^+$ cells). The latter showed a high internalisation of **18**, while little uptake was found in the normal cells. The irradiation by an 808 nm laser (1.4 W/cm^2) for 2 minutes resulted in an extensive death for cells internalising CNTs (FR$^+$ ones) and in normal proliferation behaviour for FR$^-$ cells.

More recently, Shao *et al.* prepared a multi-component targeting system, binding to SWCNTs two different monoclonal antibodies, specific for breast cancer cell antigens (IGF1R and HER2). The double targeting should ensure higher efficacy and selectivity. Derivative **19** has been prepared by

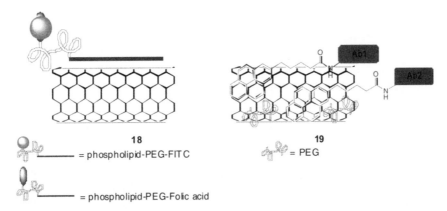

= phospholipid-PEG-FITC

= phospholipid-PEG-Folic acid

= PEG

Figure 3.12 Structures of compounds **18** and **19**. See also Colour Insert.

functionalising CNTs with 1-pyrenebutanoyl succinimmide, which can stack to the aromatic surface of the nanotubes by π–π interaction and which bears a group for the attachment of the antibodies.[62] Cells incubated with **19** were then excited by 808 nm photons at 800 mW/cm² for 3 minutes. This treatment induced nanotube heating, with the consequent death of the cells, which had internalised CNTs. All cells incubated with **19** were destroyed, while 80% of the cells treated with non-specific antibody-SWCNT constructs survived, indicating selective CNT internalisation due to specific receptor recognition on cancer cells.

In another recent work, SWCNTs conjugated with a specific monoclonal Ab were prepared and delivered to human cells bearing a specific antigen for the Ab.[63] Subsequently, they were treated with an NIR laser (808 nm, 5 W/cm² for 7 minutes), resulting in a significant decrease in cell viability.

MWCNTs are expected to be more efficient than SWCNTs in adsorbing NIR radiation and inducing the heating-up process because of more electrons available for absorption per particle and more metallic tubes than SWCNTs. So, they were chosen to treat renal cancer (RENCA) cells, and a 62-fold reduction in cell viability was obtained when the cultures were exposed to NIR laser illumination for 45 seconds. The experiments were also performed *in vivo*, demonstrating that the tumour tissue heated up to 74°C in the presence of MWCNTs (100 μg) and laser illumination (30 seconds, 3W/cm², 1064 nm continuous-wave), with a consistent decrease in tumour volume, while 46°C was reached after the tissue was exposed to simple laser illumination.[64] The same authors studied this effect using nitrogen-doped MWCNTs, which present pyridine-like rings within their structure. This chemical composition results in better conduction properties due to the subsequent filling of the small band gap in MWCNTs.[65]

Moreover, N-doped CNTs have been found to be less toxic than undoped ones.[66] Experiments have been carried out with nanotubes of different lengths, obtained after treatments in sulphuric and nitric acid, to assess the influence of this parameter on the heating.[65] Cells have been incubated for 24 hours with CNTs 1,100 nm and 700 nm in length, respectively, with a nanotube-to-cell ratio of 1,000, and irradiated with an NIR laser at 1,064 nm (3 W/cm^2, 4 minutes). An increase in temperature (~20°C) was observed, with the consequent death of a big percentage of cells (more than 90%). Differently, for 330 nm long nanotubes, no such effects were detected. This is a confirmation of the antenna theory, according to which nanotubes should be at least as long as one-half of the incidental radiation to absorb light efficiently. Since the laser used had a wavelength of 1,064 nm, 330 nm long nanotubes were not long enough to couple effectively.

Gomez-De Arco *et al.* covalently functionalised oxidised SWCNTs with a fluorescent probe and demonstrated internalisation of these compounds by glioblastoma multiforme (GBM) brain tumour cells.[67] The cells were treated with nanotubes and then imaged by Z-stack fluorescence to analyse all their inner volume, showing an actual intracellular localisation, especially in the region surrounding the nucleus. The same cells were not able to internalise the fluorescent molecule alone, thus confirming the ability of CNTs to transport cargos through cellular membranes. Furthermore, cells which had internalised the nanotubes and, as a control, cells without nanotubes were exposed to NIR radiation (808 nm, 0.8–1.5 mW/cm^2, 1–10 minutes). This treatment resulted in cytotoxicity, due to intracellular hyperthermia caused by the heating of the nanotubes (cell survival of 10–20%), while the irradiation in the control samples did not exert any toxicity (cell survival of 95%) thanks to the transparency of the cells to NIR wavelength light.

Photothermal therapy has been also applied to achieve antimicrobial activity[68] by exploiting the ability of CNTs to bind to bacteria.[69] Two different kinds of CNT preparations, one with well-dispersed CNTs and the other containing clusters, have been incubated with bacteria cultures. In both cases, the absorption of CNTs into bacteria was observed. The samples were irradiated by multi-pulse lasers at 532 and 1,064 nm, with pulses of 12 ns, at different laser fluences. Local thermal phenomena led to complete bacteria disintegration at laser fluence values of 2–3 J/cm^2 in all cases.

The ability of CNTs to convert optical energy into thermal energy upon exposure to NIR light was also exploited by Panchapakesan *et al.*, using a different approach.[70] A so-called nanobomb was built up using SWCNTs as agents to kill breast cancer cells. The thermal response of CNTs to NIR irradiation (800 nm, 50–200 mW/cm^2, less than 60 seconds) caused vaporisation of the water trapped in the CNT bundles, thus creating a big pressure and a consequent explosion, with cancer cell death. Interestingly, considering a possible *in vivo* application, the CNTs were completely destroyed with this treatment, thereby avoiding any problem associated with their intrinsic toxicity.

The modulation of NIR pulses can induce a controlled local heating without killing the cells but enhancing the permeability of tumour vasculature, with the possibility of ameliorating the drug delivery in a combined chemotherapy approach[64] or inducing endosomal disruption with the consequent release of cargos (as oligonucleotides) linked to nanotubes.[61]

Another very interesting possibility in the thermal ablation treatment of cancer is given by the use of a non-invasive radiofrequency (RF) field, instead of NIR irradiation.[71] Gannon *et al.* reported the heating of the SWCNT solution when exposed to a RF field. It is important to point out that the RF wavelength exceeds the CNTs' length too much to give rise to resonance and the energy is too small to excite electronic transitions. So, this thermal effect could be due to the resistive conductivity of SWCNTs and their high aspect ratio. In this study, both *in vitro* and *in vivo* experiments have been carried out. For the former, cytotoxicity was observed for CNT-treated cells exposed to a RF field. This effect resulted to be dependent on the SWCNTs concentration and on the duration of RF exposure. For the *in vivo* studies, solutions of nanotubes were administered via intratumor injection into tumour-bearing rabbits. The animals were then treated with an RF field, leading to necrosis of the tumour tissue. In both sets of experiments, control treatments, i.e., a RF field without SWCNTs and SWCNTs without a RF field, were unable to exert any toxicity on the tumours. This finding opens a new perspective in the application of CNTs for cancer treatment.

Magnetic fluid hyperthermia, induced by an alternating current magnetic field in the presence of magnetic nanoparticles, can be exploited for cancer therapy, and CNTs can be really useful in this field. The preparation of CNTs encapsulating ferromagnetic particles was achieved by pyrolysis of ferrocene, resulting in the formation of forests where the tubes are vertically aligned. Adherent human bladder cancer cells (EJ28) were incubated in the presence of a suspension of CNTs with and without cationic lipids. In the first case no adhesion was found, while the presence of lipids helped the entrance of CNTs into cells with a cytoplasmatic localisation.[72] Borowiak-Palen, instead, reported the preparation of CNTs filled with α-Fe (iron allotrope with a body-centred cubic lattice) using a different procedure. The SWCNTs produced by laser ablation were first purified and subsequently filled with $FeCl_3$ and, finally, washed to remove the externally deposited iron.[73] These compounds, variously modified and bearing anticancer drugs, can be used to target lymphatic tissues.[74]

No real application of this approach has been reported up to now, but Pensabene *et al.* reported that by catalytic chemical vapour deposition,

SWCNTs presented Ni and Fe particles entrapped in their structures to selectively "move" cells.[75] Neuroblastoma cells were seeded with a CNT-modified medium under a permanent magnetic field, obtaining cell displacements, with a higher concentration of cells where the magnetic field was stronger. Although the mechanism is not yet understood, it seems to take place during cell duplication and to be driven by the presence of magnetic CNTs, either uptaken by cells or attached to their membranes.

3.7 BIOSENSORS BASED ON CARBON NANOTUBES

As already mentioned, the detection of specific proteins (the so-called biomarkers, indicators of a pathological state) in blood or other biological fluids is really important for the early diagnosis of cancer, and nanotechnology can help the development in this direction.

A biosensor is a device able to detect the presence of specific molecules and to transform the recognition events into analytical outputs, transducing chemical responses into physical, measurable signals, usually current signals. The ideal biosensor should be simple, sensitive, cheap and real-time, and the recent achievements in proteomics are paving the way in this direction.[76]

A specific branch of biosensors is constituted by immunobiosensors: the traditional techniques for protein detections are enzyme-linked immunosorbent assay (ELISA), radioimmunoassay (RIA) and electrophoretic immunoassay. They are rather time-consuming and need sophisticated instrumentation, and moreover, the labelling exigency could affect the detection efficiency. An immunosensor is an analytical device constituted by an antibody or an antigen or a fragment of them combined with a physicochemical transducer responsible for specific properties, such as optical or electrochemical ones or changes in mass or heat (Fig. 3.13). The best sensitivity for this class of devices is, so far, in the picomolar range.

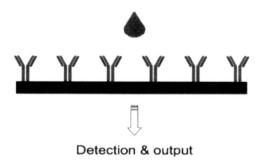

Detection & output

Figure 3.13 Schematic immuno-biosensoring set-up.

CNTs themselves are not able to both perform specific molecular recognition and be a signal transducer, but they seem particularly suitable for these tasks because their electrical conductance is very sensitive and changes with the adsorption of charged molecules on CNTs' surface. As previously mentioned, their high aspect ratio, together with their chemical and electronic properties, makes them critically important in the development of such devices. They can have a high loading of biomolecules, with a consequent high sensitivity as biosensors, but there are still problems about the stability, reusability and repeatability of the systems.[76] The exploitation of electrochemistry is advantageous because it monitors the redox activity of species and is apt in surface science and solid state chemistry. CNTs play a crucial role in the field of sensors because of their ability to facilitate electron transfer when acting as modified electrodes.

CNT-based detection devices can be classified by considering (i) the functionalisation (covalent or non-covalent), (ii) the detection system and (iii) the detected molecule. In fact CNTs can recognise proteins by specific or non-specific interactions. Non-covalent attachment of antibodies to CNTs is strictly related to the size of biomolecules: molecules much larger than SWCNTs (7–8 nm, e.g. immunoglobulin) reveal difficulties in non-specific binding (NSB). It has been reported that tryptophan and histidine are responsible for most interactions in non-specific protein binding.[77] The main reason for this phenomenon is the hydrophobic interaction between CNT sidewalls and proteins, and it is possible to prevent it by using polyethylene oxide and polyethylene glycol, which wrap the tubes and render them more hydrophilic and thus prevent hydrophobic interactions. Dai *et al.* offered a detailed study about non-specific binding of proteins on the SWCNT surface (Table 3.1).[78] Moreover, they suggested that, since the electrical properties are guaranteed by the integrity of an electronic structure, non-covalent functionalisation be preferred, considering that it can prevent the remarkable influence of non-specific binding. They prepared a dense film of SWCNTs by chemical vapour deposition growth on quartz substrates. Subsequent metal evaporation served to obtain the two electrodes of Ti/Au connected by the SWCNT network to use field effect transistor (FET) and quartz microbalance together. The CNTs were non-covalently functionalised through a linker with auto-antigens to generate an immunosensor for autoimmune diseases. This system was used to detect many proteins and the binding was demonstrated by atomic force microscopy, quartz crystal microbalance and conductance measurements in buffer.

By discovering the important effect of non-specific binding, the authors studied the possibility of producing a functionalised derivative for NSB resistance (Table 3.1).

Table 3.1 Protein NSB on differently treated CNTs[a]

Functionalisation	SA	Avidin	BSA	GCD	SpA
CNTs as pristine	√	√	√	√	√
Tween 20-treated	X	X	X	X	X
Pluronic P103-treated	X	X	X	X	X
Triton X-100-treated	X	√	√	√	X
Dextran-treated	√	√	√	√	√

Note: SA = streptavidin, BSA = bovin serum albumin, GCD = α-glucosidase, SPA = staphylococcal protein A; √ = non-specific binding; X = no binding detected.
[a] Adapted from Chen *et al.*[78]

The polyethylene oxide approach seems to be promising in the field of non-specific recognition, and it follows the principle of rendering CNTs' surface hydrophilic, while the linkage of an antibody is strongly recommended to improve the biosensor's specificity and sensitivity.[76] This preparation is based on covalent functionalisation of CNTs with the formation of amide bonds between proteins and CNT surfaces.

There are two main types of immunosensors, the already cited CNT-based field effect transistors (FETs) and the electrochemical immunosensors.

FETs are used for weak-signal amplification, and they are made of a source electrode, a drain electrode and, in between, a channel with a defined physical diameter where a current flow passes along a semiconductor path (Fig. 3.14). A voltage is applied to the gate electrode, and as it varies, the electrical diameter of the channel changes: the conductivity depends on this diameter. FETs amplify signals on the basis of small changes in gate voltage, since they cause a large variation in the current from source to drain. This great sensitivity renders them particularly suitable for immunobiosensor applications.

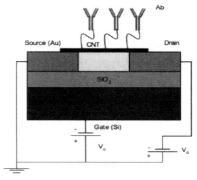

Figure 3.14 Schematic representation of an immunobiosensor.

In this device the conducting channel is realised by a single CNT or a network of CNTs, usually modified by attachment of specific antibodies. Another important characteristic is the property of CNTs to be *p*- or *n*-type semiconductors: SWCNTs themselves are *p*-type semiconductors (conduction in negative gate voltages). In general, both *p*-type and *n*-type semiconductors can be obtained depending on the functionalisation, and the conductance between the transistor and the absorbed molecules (i.e., proteins) can be measured. The sensing efficiency depends on the alignment of CNTs, because the molecule-binding events are controlled by the orientation of the entire system. CNTs can be aligned during the synthesis or after that (i.e., by micropatterning), and they can be aligned as forests or arrays, perpendicular or parallel to the substrates.

In 2005, Kerman *et al.* reported an example of CNT-FETs on a silicon dioxide/silicon (SiO$_2$/Si) substrate based on peptide nucleic acid (PNA) associated with tumour necrosis factor-α gene (TNF-α), for real-time electrical detection of DNA hybridisation with easy discrimination of single-nucleotide polymorphism.[79] Tumour necrosis factor belongs to the cytokine family and is responsible for the inflammatory response and for some anomalies in the regulation process related to cancer (i.e., overexpression), and so it can be considered a biomarker for early cancer detection. PNA is a synthetic oligonucleotide, in which the phosphate and deoxyribose skeleton is replaced by a polypeptide, and it can hybridise with complementary DNA or RNA sequences. There are some advantages in using PNAs: their backbone is neutral and does not suffer from electrostatic repulsions, the base pairing is not influenced by ionic strength and they are not altered by proteases and nuclease degradation.

Source and drain electrodes were separated by a 4 µm gap and the junction was constituted by a single SWCNT, while the solid state gate was replaced by molecules able to modulate CNT conductance. The biomolecules were not adsorbed on the CNT, but on the gold electrode (AuE) with, on the backside, a self-assembled monolayer (SAM) of PNA fully complementary to the wild-type TNF-α sequence. The entire system was fabricated by photolithography and conventional lit-off technique on a *p*+-doped-Si substrate with a layer of natural SiO$_2$ (100 nm thick). The contacts between source and drain were Ti/Au electrodes, and they were built, together with the side gate, onto the patterned SiO$_2$ substrate after the growth of CNTs. The back gate contact in Ti/Au was formed on the Si substrate. All these parts were in contact with a microfluidic chip in polydimethylsiloxane (PDMS), which is often used for these applications. The backside electrode was modified by a SAM of both 3-mercaptopropionic acid, used to link PNA (30 pM) by amidic bond, and a SAM of mercapto propanol, used as a filler to prevent non-specific adsorption on gold (Fig. 3.15).

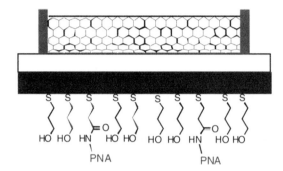

Figure 3.15 After treating the surface with 3-mercaptopropionic acid and subsequent activation via 1-ethyl-3-(3-dimethylaminopropyl)carbodiimide/*N*-hydroxysuccinimide (EDC/NHS), the PNAs were attached to the backside of the electrode.

The PNA-modified Au was placed in contact with the open chamber for the introduction of analysable solutions. The analyser measured the source-to-drain voltage (V_{SD}) versus the source-to-drain current (I_{SD}), with a reference electrode (Ag/AgCl) immersed in phosphate buffer solution.

Different solutions of DNA at different concentrations were used to assess the perfect running and the sensitivity of the device. The increase in conductance for these *p*-type semiconductors was consistent with an increase in negative charge density due to the hybridisation of negatively charged DNA. The experiments performed demonstrated that the device was able to discriminate not only non-complementary DNA but also single-nucleotide polymorphism. The recorded low fluctuations of conductance were ascribed to a non-specific interaction between the DNA and the Au surface. Moreover, the detection was the lowest (6.8 fM) with respect to other existing real-time, label-free DNA detection systems.

As mentioned already, another important biomarker is represented by prostate-specific antigen (PSA), secreted by epithelial cells of prostate and then introduced into the bloodstream. Little quantities of PSA (about 4 ng/mL) are normally produced in healthy people, but its overexpression is strictly related to prostatic cancer, and therefore it has high relevance as a biomarker. In addition, it can be considered a model for the evaluation of the design of new devices because a lot of data are available about it. PSA, once into the bloodstream, partially interacts with plasma proteins, and so the total PSA is given by the combination of the free PSA and the bounded PSA, rendering the quantification more complicated.

Li *et al.* were the first to design a hybrid FET with CNTs and InO_2 nanowires (NWs), combining *p*-type and *n*-type semiconductors. They obtained a complementary electrical response for the detection of PSA with a detection limit of 0.14 nM (5 ng/mL). The device consisted of a source and a drain electrode, placed on a Si gate with a SiO$_2$ layer, where SWCNTs and NWs formed the active channel.[80]

Anti-PSA monoclonal antibody was reacted with activated NWs, while SWCNTs were non-covalently functionalised through π–π interactions with 1-pyrenebutanoic acid and allowed to react with the same antibody after activation of the carboxylic group. The device was incubated with a PSA solution (1 μg/mL) for 15 hours, and after washing the surface, current measurements were performed in air. Because of the interaction with PSA, the conductance increased for NWs and decreased for SWCNTs, which are respectively *negative-* and *positive*-type semiconductors and respond with opposite behaviour. The limit of detection was determined by current response, which was 0.14 nM (5 ng/mL) for NWs and 1.4 nM (50 ng/mL) for SWCNTs. Considering the signal-to-noise ratio for NWs (20), the effective limit of detection was 7 pM (250 pg/mL).

Okuno *et al.* proposed an immunodetector for total PSA by using differential pulse voltammetry (DPV). This device was based on the specific oxidation of tyrosine and tryptophan residues (PSA contains 13 tyrosine residues and 11 tryptophan residues) after the interaction of PSA and anti-PSA monoclonal antibody. The apparatus is *p*-type and Si-based with a layer of SiO$_2$ (150 nm thick), and the working electrode is obtained by patterning the surface with Ti/Pt and, afterwards, with CNTs. The latter were non-covalently functionalised with 1-pyrenebutanoic acid to link the PSA monoclonal antibody.[81]

Yu *et al.* proposed a more complicated system, in which they used an SWCNT forest as a platform for a multi-labelled secondary antibody conjugate, achieving a detection limit of 4 pg/mL (100 amol/mL). In this system, BSA and Tween 20 were used to prevent non-specific binding, and primary anti-PSA antibodies were covalently bound to the SWCNT forest. Shortened MWCNTs, instead, were used to build a system bearing anti-PSA secondary antibodies and horseradish peroxidase (HRP), which act to amplify the signal. This strategy led to a remarkable sensitivity and selectivity in complex samples as human serum and cell lysate, which present many different components, and to a gain in mass sensitivity that was better than all commercial PSA detectors.[82]

A more developed device was prepared by Gruner *et al.* Recognising the demand to engineer devices able to analyse biological fluids at room temperature with fast and specific procedures, the researchers designed a novel capacitor with an SWCNT network and a reference electrode immersed in a liquid electrolyte.[83] The use of CNTs in a capacitor configuration permits obtaining a two-terminal device. CNTs are printed on polyethylene terephthalate (PET) strips and dipped into a low ionic strength buffer, along the reference electrode. The buffer acts as dielectric layer of the capacitor, and the low ionic strength is crucial for successful detection of PSA; in fact this device senses the additional charge induced when PSA interacts with the antibody. For CNTs, the total device capacitance (C_t) is given by the electrochemical double-layer capacitance and by the quantum capacitance (C_q), which, in the case of this 1D material, is extremely significant in the determination of C_t, thus making this device potentially more sensitive than any other. CNTs were non-covalently functionalised with anti-human PSA monoclonal antibody by simple exposure to a 5 μg/mL buffer solution of the antibody. The adsorption was confirmed and quantified (about 15–20 ng on the surface device) by Western blot analysis and ELISA. The device was afterwards exposed to different concentrations of PSA, diluted in calf serum, and a decrease in capacitance change was observed. The proposed device seems to be very promising because it did not show any aspecific recognition of serum proteins except PSA. However, it is necessary to scale the entire system to micro- or nano-dimensions, but it seems feasible with modern micro- and nano-fabrication techniques.

Carcinoembryonic antigen (CEA) is another important biomarker to be detected, its overexpression being indicative of cancer. It is involved in cell adhesion, and generally its expression stops before birth. Even though it is possible to reveal the presence of this marker in the bloodstream of smokers, it can be related to cancer and is often used as a tumour marker relapse after surgical operation. For CEA detection, Park *et al.* proposed an FET produced by a patterned catalyst growth technique to synthesise CNTs directly on the substrate.[84] They utilised a doped silicon substrate with a thick oxide layer (500 nm), and SWCNTs were grown on this surface by chemical vapour deposition. Afterwards, Ti/Au electrodes were obtained and the active channel was constituted by SWCNTs. The construct was immersed in a solution of carbonyldiimidazole (CDI)-Tween 20, which non-covalently wrapped the nanotubes by hydrophobic interactions. The imidazoles reacted with the amines of the anti-carcinoembryonic antigen (10 μM, buffer solution), and all the unreacted terminations were protected by ethanol amine. Functionalisation with CDI-Tween 20 caused a leftward shift in the threshold voltage, and the immobilisation of anti-CEA decreased the conductance for CEA concentrations up to 40 ng/mL (200 nM). The sensitivity was increased using SWCNT-FET, with a higher on–off ratio, reaching values of 54 pg/mL (200 fM).

There are several examples in which CNTs are employed to modify electrodes for electrochemical measurements. This use is strictly related to CNTs' ability to amplify signals, which is a widely reported characteristic.

To detect nucleic acid segments related to the BRCA1 cancer gene, Wang *et al.* introduced the use of a glassy carbon electrode modified with MWCNTs.[85] The casting of a CNT solution on the electrode, as in this approach, offers a more continuous layer than does spin coating. The analysed DNA was obtained by streptavidin-coated microspheres reacted with a single-strand DNA probe and then hybridised with the target and washed with a NaOH solution. After digestion, DNA was finally dissolved in acetate buffer. Cyclic voltammetry and chronopotentiometric adsorptive-stripping measurements were performed to detect the release of purine bases with different electrodes. The first step was a cyclic voltammetry with a solution of free guanine because it seems that the response is mainly due to the redox activity of this nucleobase (Scheme 3.2).

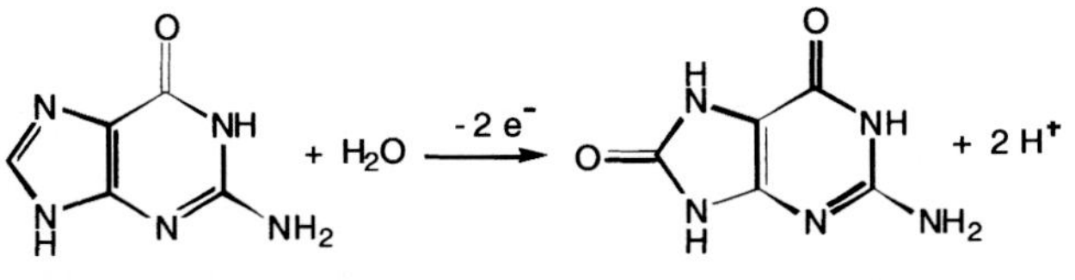

Scheme 3.2 Redox reaction of guanine.

An enhanced signal was recorded, probably due to guanine accumulation at the electrode surface rather than to an accelerated electron transfer. This is related to the high surface-area-to-volume ratio of CNTs and to a stronger affinity of guanine for nanotubes than for the bare glassy carbon electrode. The results showed a larger background and a rising solvent decomposition at the MWCNT-modified electrode; the guanine peak potential (+0.87 V) was shifted with respect to the bare electrode (+0.82 V), while another peak potential (+ 1.18 V), due to adenine, appeared using the modified electrode.

Chronopotentiometric stripping analysis (CPSA) was also presented in the same work. This technique is based on the change of potential over time during chemical oxidation. This effect can be induced either by the accumulation of metals or reductants (as in this case) or by the application of a constant current (i.e., electrochemically). A well-defined hybridisation stripping response for a low level of target DNA (250 μg/L, ppb) was revealed by CPSA performed with such a CNT-modified glassy carbon electrode.[85]

More recently, the same group proposed the detection of hybridised DNA by cyclic voltammetry, amperometry and chronopotentiometry. As previously reported, the electrode was modified by CNTs, cast and dried, but in this case the use of an enzymatic label (alkaline phosphatase) was introduced in the target. When the hybridisation occurred, the phenolic by-products of phosphatase accumulated onto the surface of the CNT-modified electrode and allowed the detection of a very low level of target DNA.[86] This behaviour was confirmed by the detection of DNA segments related to the BRCA1 breast cancer gene, where the signal amplification led to the detection of extremely low target concentrations in solution (100 ppt).

Also, Yao *et al.* proposed CPSA with a CNT-modified electrode as a new methodology to detect human breast cancer (MCF-7) cells and obtain a quantitative response. The authors evaluated the effect of 5-Fluorouracil (5-FU), a well-known anticancer drug, on potentiometric signals. Cell cultures – treated and not treated by 5-FU – were studied, and it emerged that the higher was the anticancer drug dose, the lower was the relative potentiometric signal response.[87] The same group adopted another technique for cytosensing: the quartz crystal microbalance (QCM). They modified a Au electrode with chitosan (known for its high biocompatibility and low mechanical properties) and MWCNTs, which can act as fillers to enhance mechanical strength.[88] The so-obtained electrode has been proved to present a surface favourable for cell loading and adhesion and to work on the QCM. These measurements were accompanied by electrochemical analyses (voltammetry and potentiometry) and a MTT test, demonstrating the possibility of monitoring the presence of breast cancer cells by their adhesion on the modified electrode.

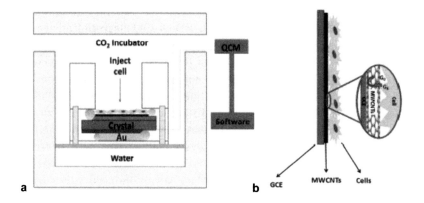

Figure 3.16 (a) A QCM set-up; (b) a modified electrode with attached cells. *Abbreviations*: GCE, glass carbon electrode; G_R, guanidine, reduced form; G_O, guanidine, oxidised form.

Chen *et al.* studied the possibility of selecting the most effective anticancer therapeutic agents by means of the electrochemical behaviour of leukemia cells (K562). The authors modified a glassy carbon electrode with MWCNTs and studied its voltammetric response in the presence of different antitumour agents (5-FU, mitocycin C, adriamycin and vincristine). The cells showed a well-defined anodic peak at the modified electrode, while no peak was observed when a bare electrode was used. It was proved that the anodic peak current increases at a higher concentration of MWCNTs, demonstrating an electron transfer effect for nanotubes, but concentrations higher than 4 mg/mL tend to render the cell membrane unstable.[89] Also in this case the anodic peak is due to purine nucleobase oxidation, especially guanine. The experiments proved that the peak current is strictly related to cell number and to guanine expression in the cytoplasm. In fact, in the presence of the anticancer drugs, which inhibit the growth and viability of cells, this signal is significantly decreased, demonstrating the possibility of using voltammetry to evaluate antitumour drug sensitivity.

The use of electrochemistry implies many advantages with respect to conventional techniques considering that these systems are cheaper and less time-consuming and present wide possibilities in signal amplification with a higher sensitivity.

Other authors, too, screened anticancer drugs by means of sensors. Ovádeková *et al.* modified a screen-printed electrode for the detection of dsDNA by employing MWCNTs, gold nanoparticles (GNPs) and a mixture of the two. Calf thymus dsDNA was introduced with layer-by-layer deposition or by using a mixed coverage. The latter methodology allowed for obtaining more intense responses than layer-by-layer deposition on the GNP- or MWCNT-modified electrode. The authors used this device to detect the activity of different antitumour drugs interacting with DNA, including berberine, a DNA-intercalating alkaloid. The signal decrease was found to be dependent on the concentration of the compound. Moreover, since this drug is quite effective at discriminating between cancer and non-cancer DNA, because of the presence of permanently open loops in tumour cells' DNA, it represents an interesting way to distinguish cancer cells from healthy cells.[90]

3.8 CONCLUSIONS

More efforts are needed to validate the possibility of using carbon nanotubes as drug carriers for cancer treatment in chemotherapy, in photothermal ablation and in radiotherapy. Biocompatibility and toxicity are issues to be addressed in detail, but these first studies are showing very promising results for potential applications in the field of nanomedicine. On the other hand, biosensing has been giving really important results and is not constrained by the need to assess the toxicity of this novel nanomaterial; therefore, it has a tremendous scope for development.

References

1. U.S. National Institute of Health, http://nano.cancer.gov/resource_center/nano_critical.asp.

2. Cuenca, A., Jiang, H., Hochwald, S., Delano, M., Cance, W., and Grobmyer, S. (2006) Emerging implications of nanotechnology on cancer diagnostics and therapeutics, *Cancer*, **107**, 459–466.

3. Ferrari, M. (2005) Cancer nanotechnology: opportunities and challenges, *Nat. Rev. Cancer*, **5**, 161–171.

4. Nie, S., Xing, Y., Kim, G. J., and Simons, J. W. (2007) Nanotechnology applications in cancer, *Annu. Rev. Biomed. Eng.*, **9**, 257–288.

5. Gallego, O., and Puntes, V. (2006) What can nanotechnology do to fight cancer? *Clin. Transl. Oncol.*, **8**, 788–795.

6. Kim, K. (2007) Nanotechnology platforms and physiological challenges for cancer therapeutics, *Nanomedicine*, **3**, 103–110.

7. Sengupta, S., and Sasisekharan, R. (2007) Exploiting nanotechnology to target cancer, *Br. J. Cancer*, **96**, 1315–1319.

8. Liu, G.-Y., Xu, S., and Qian, Y. (2000) Nanofabrication of self-assembled monolayers using scanning probe lithography, *Acc. Chem. Res.*, **33**, 457–466.

9. Staii, C., Wood, D. W., and Scoles, G. (2008) Ligand-induced structural changes in maltose binding proteins measured by atomic force microscopy, *Nano Lett.*, **8**, 2503–2509.

10. Poland, C., Duffin, R., Kinloch, I., Maynard, A., Wallace, W. A., Seaton, A., Stone, V., Brown, S., Macnee, W., and Donaldson, K. (2008) Carbon nanotubes introduced into the abdominal cavity of mice show asbestos-like pathogenicity in a pilot study, *Nature Nanotechnol.*, **3**, 423–8.

11. Bahr, J. L., and Tour, J. M. (2002) Covalent chemistry of single-wall carbon nanotubes, *J. Mater. Chem.*, **12**, 1952–1958.

12. Georgakilas, V., Kordatos, K., Prato, M., Guldi, D. M., Holzinger, M., and Hirsch, A. (2002) Organic functionalization of carbon nanotubes, *J. Am. Chem. Soc.*, **124**, 760–761.

13. Bianco, A., Kostarelos, K., and Prato, M. (2008) Opportunities and challenges of carbon-based nanomaterials for cancer therapy, *Expert Opin. Drug Delivery*, **5**, 331–342.

14. Pantarotto, D., Briand, J., Prato, M., and Bianco, A. (2004) Translocation of bioactive peptides across cell membranes by carbon nanotubes, *Chem. Commun.*, 16–17.

15. Jin, H., Heller, D. A., and Strano, M. S. (2008) Single-particle tracking of endocytosis and exocytosis of single-walled carbon nanotubes in NIH-3T3 cells, *Nano Lett.*, **8**, 1577–1585.

16. Lacerda, L., Soundararajan, A., Singh, R., Pastorin, G., Al-Jamal, K., Turton, J., Frederik, P., Herrero, M., Li, S., Bao, A., Emfietzoglou, D., Mather, S., Phillips, W., Prato, M., Bianco, A., Goins, B., and Kostarelos, K. (2008) Dynamic imaging of

functionalized multi-walled carbon nanotube systemic circulation and urinary excretion, *Adv. Mater.*, **20**, 225–230.

17. Ali-Boucetta, H., Al-Jamal, K., McCarthy, D., Prato, M., Bianco, A., and Kostarelos, K. (2008) Multiwalled carbon nanotube-doxorubicin supramolecular complexes for cancer therapeutics, *Chem. Commun.*, **8**, 459–461.

18. Liu, Z., Sun, X., Nakayama-Ratchford, N., and DaiR, H. (2007) Supramolecular chemistry on water-soluble carbon nanotubes for drug loading and delivery, *ACS Nano*, **1**, 50–56.

19. Pastorin, G., Wu, W., Wieckowski, S., Briand, J., Kostarelos, K., Prato, M., and Bianco, A. (2006) Double functionalisation of carbon nanotubes for multimodal drug delivery, *Chem. Commun.*, 1182–1184.

20. Yang, X., Zhang, Z., Liu, Z., Ma, Y., Yang, R., and Chen, Y. (2008) Multi-functionalized single-walled carbon nanotubes as tumor cell targeting biological transporters, *J. Nanoparticle Res.*, **10**, 815–822.

21. Liu, Z., Chen, K., Davis, C., Sherlock, S., Cao, Q., Chen, X., and Dai, H. (2008) Drug delivery with carbon nanotubes for in vivo cancer treatment, *Cancer Res.*, **68**, 6652–6660.

22. Liu, Z., Davis, C., Cai, W., He, L., Chen, X., and Dai, H. (2008) Circulation and long-term fate of functionalized, biocompatible single-walled carbon nanotubes in mice probed by Raman spectroscopy, *Proc. Natl. Acad. Sci. USA*, **105**, 1410–1415.

23. Chen, J., Chen, S., Zhao, X., Kuznetsova, L. V., Wong, S. S., and Ojima, I. (2008) Functionalized single-walled carbon nanotubes as rationally designed vehicles for tumor-targeted drug delivery, *J. Am. Chem. Soc.*, **130**, 16778–16785.

24. Dhar, S., Liu, Z., Thomale, J., Dai, H., and Lippard, S. J. (2008) Targeted single-wall carbon nanotube-mediated pt(IV) prodrug delivery using folate as a homing device, *J. Am. Chem. Soc.*, **130**, 11467–11476.

25. Feazell, R. P., Nakayama-Ratchford, N., Dai, H., and Lippard, S. J. (2007) Soluble single-walled carbon nanotubes as longboat delivery systems for platinum (IV) anticancer drug design, *J. Am. Chem. Soc.*, **129**, 8438–9.

26. Bhirde, A., Patel, V., Gavard, J., Zhang, G., Sousa, A., Masedunskas, A., Leapman, R., Weigert, R., Gutkind, J., and Rusling, J. (2009) Targeted killing of cancer cells in vivo and in vitro with EGF-directed carbon nanotube-based drug delivery, *ACS Nano*, **3**, 307–316.

27. Tsang, S. C., Chen, Y. K., Harris, P. J. F., and Green, M. L. H. (1994) A simple chemical method of opening and filling carbon nanotubes, *Nature*, **372**, 159–162.

28. Hampel, S., Kunze, D., Haase, D., Krämer, K., Rauschenbach, M., Ritschel, M., Leonhardt, A., Thomas, J., Oswald, S., Hoffmann, V., and Büchner, B. (2008) Carbon nanotubes filled with a chemotherapeutic agent: a nanocarrier mediates inhibition of tumor cell growth, *Nanomedicine*, **3**, 175–182.

29. Ajima, K., Yudasaka, M., Murakami, T., Maigné, A., Shiba, K., and Iijima, S. (2005) Carbon nanohorns as anticancer drug carriers, *Mol. Pharm.*, **2**, 475–480.

30. Murakami, T., Fan, J., Yudasaka, M., Iijima, S., and Shiba, K. (2006) Solubilization of single-wall carbon nanohorns using a PEG-doxorubicin conjugate, *Mol. Pharm.*, **3**, 407–414.

31. Ajima, K., Yudasaka, M., Maigné, A., Miyawaki, J., and Iijima, S. (2006) Effect of functional groups at hole edges on cisplatin release from inside single-wall carbon nanohorns, *J. Phys. Chem. B*, **110**, 5773–5778.

32. Ajima, K., Maigné, A., Yudasaka, M., and Iijima, S. (2008) Optimum hole-opening condition for cisplatin incorporation in single-wall carbon nanohorns and its release, *J. Phys. Chem. B*, **110**, 19097–19099.

33. Matsumura, S., Ajima, K., Yudasaka, M., Iijima, S., and Shiba, K. (2007) Dispersion of cisplatin-loaded carbon nanohorns with a conjugate comprised of an artificial peptide aptamer and polyethylene glycol, *Mol. Pharm.*, **4**, 723–729.

34. Li, X., Peng, Y., and Qu, X. (2006) Carbon nanotubes selective destabilization of duplex and triplex DNA and inducing B-A transition in solution, *Nucleic Acids Res.*, **34**, 3670–3676.

35. Li, X., Peng, Y., Ren, J., and Qu, X. (2006) Carboxyl-modified single-walled carbon nanotubes selectively induce human telomeric i-motif formation, *Proc. Natl. Acad. Sci. USA*, **103**, 19658–19663.

36. Zhao, C., Peng, Y., Song, Y., Ren, J., and Qu, X. (2008) Self-assembly of single-stranded RNA on carbon nanotube: polyadenylic acid to form a duplex structure, *Small*, **4**, 656–661.

37. Lu, Q., Moore, J., Huang, G., Mount, A., Rao, A., Larcom, L., and Ke, P. (2004) RNA polymer translocation with single-walled carbon nanotubes, *Nano Lett.*, **4**, 2473–2477.

38. Cui, D., Tian, F., Coyer, S., Wang, J., Pan, B., Gao, F., He, R., and Zhang, Y. (2007) Effects of antisense-myc-conjugated single-walled carbon nanotubes on HL-60 cells, *J. Nanosci. Nanotechnol.*, **7**, 1639–1646.

39. Zhang, Z., Yang, X., Zhang, Y., Zeng, B., Wang, S., Zhu, T., Roden, R., Chen, Y., and Yang, R. (2006) Delivery of telomerase reverse transcriptase small interfering RNA in complex with positively charged single-walled carbon nanotubes suppresses tumor growth, *Clin. Cancer Res.*, **12**, 4933–4939.

40. Kam, N. W., Liu, Z., and Dai, H. (2005) Functionalization of carbon nanotubes via cleavable disulfide bonds for efficient intracellular delivery of siRNA and potent gene silencing, *J. Am. Chem. Soc.*, **127**, 12492–12493.

41. Liu, Z., Winters, M., Holodniy, M., and Dai, H. (2007) siRNA delivery into human T cells and primary cells with carbon-nanotube transporters, *Angew. Chem. Int. Ed.*, **46**, 2023–2027.

42. Jia, N., Lian, Q., Shen, H., Wang, C., Li, X., and Yang, Z. (2007) Intracellular delivery of quantum dots tagged antisense oligodeoxynucleotides by functionalized multiwalled carbon nanotubes, *Nano Lett.*, **7**, 2976–2980.

43. Pan, B., Cui, D., Xu, P., Chen, H., Liu, F., Li, Q., Huang, T., You, X., Shao, J., Bao, C., Gao, F., He, R., Shu, M., and Ma, Y. (2007) Design of dendrimer modified carbon nanotubes for gene delivery, *Chinese J. Cancer Res.*, **19**, 1–6.

44. Herrero, M. A., Toma, F. M., Al-Jamal, K. T., Kostarelos, K., Bianco, A., Ros, T. D., Bano, F., Casalis, L., Scoles, G., and Prato, M. (2009) Synthesis and characterization of a carbon nanotube-dendron series for efficient siRNA delivery, *J. Am. Chem. Soc.*, **131**, 9843–9848.

45. Podesta, J., Al-Jamal, K., Herrero, M., Tian, B., Ali-Boucetta, H., Hegde, V., Bianco, A., Prato, M., and Kostarelos, K. (2009) Antitumor activity and prolonged survival by carbon-nanotube-mediated therapeutic siRNA silencing in a human lung xenograft model, *Small*, **5**, 1176–1185.

46. Wu, Y., Phillips, J., Liu, H., Yang, R., and Tan, W. (2008) Carbon nanotubes protect DNA strands during cellular delivery, *ACS Nano*, **2**, 2023–2028.

47. Zhu, Z., Tang, Z., Phillips, J. A., Yang, R., Wang, H., and Tan, W. (2008) Regulation of singlet oxygen generation using single-walled carbon nanotubes, *J. Am. Chem. Soc.*, **130**, 10856–10857.

48. Ding, L., Stilwell, J., Zhang, T., Elboudwarej, O., Jiang, H., Selegue, J., Cooke, P., Gray, J., and Chen, F. (2005) Molecular characterization of the cytotoxic mechanism of multiwall carbon nanotubes and nano-onions on human skin fibroblast, *Nano Lett.*, **5**, 2448–2464.

49. Zhu, L., Chang, D., Dai, L., and Hong, Y. (2007) DNA damage induced by multiwalled carbon nanotubes in mouse embryonic stem cells, *Nano Lett.*, **7**, 3592–3597.

50. Lucente-Schultz, R., Moore, V., Leonard, A., Price, B., Kosynkin, D., Lu, M., Partha, R., Conyers, J., and Tour, J. (2009) Antioxidant single-walled carbon nanotubes, *J. Am. Chem. Soc.*, **131**, 3934–3941.

51. Galano, A. (2008) Carbon nanotubes as free-radical scavengers, *J. Phys. Chem. C*, **112**, 8922–8927.

52. Liu, Z., Cai, W., He, L., Nakayama, N., Chen, K., Sun, X., Chen, X., and Dai, H. (2007) In vivo biodistribution and highly efficient tumour targeting of carbon nanotubes in mice, *Nature Nanotechnol.*, **2**, 46–52.

53. McDevitt, M., Chattopadhyay, D., Kappel, B., Jaggi, J., Schiffman, S., Antczak, C., Njardarson, J., Brentjens, R., and Scheinberg, D. (2007) Tumor targeting with antibody-functionalized, radiolabeled carbon nanotubes, *J. Nucl. Med.*, **48**, 1180–1189.

54. Singh, R., Pantarotto, D., Lacerda, L., Pastorin, G., Klumpp, C., Prato, M., Bianco, A., and Kostarelos, K. (2006) Tissue biodistribution and blood clearance rates of intravenously administered carbon nanotube radiotracers, *Proc. Natl. Acad. Sci. USA*, **103**, 3357–3362.

55. Hartman, K., Hamlin, D., Wilbur, D., and Wilson, L. (2007) 211AtCl@US-Tube nanocapsules: a new concept in radiotherapeutic-agent design, *Small*, **3**, 1496–1499.

56. Yinghuai, Z., Peng, A., Carpenter, K., Maguire, J., Hosmane, N., and Takagaki, M. (2005) Substituted carborane-appended water-soluble single-wall carbon nanotubes: new approach to boron neutron capture therapy drug delivery, *J. Am. Chem. Soc.*, **127**, 9875–9880.

57. Hirsch, L. R., Stafford, R. J., Bankson, J. A., Sershen, S. R., Rivera, B., Price, R. E., Hazle, J. D., Halas, N. J., and West, J. L. (2003) Nanoshell-mediated near-infrared thermal therapy of tumors under magnetic resonance guidance, *Proc. Natl. Acad. Sci. USA*, **100**, 13549–13554.

58. Huang, X., El-Sayed, I. H., Qian, W., and El-Sayed, M. A. (2008) Cancer cell imaging and photothermal therapy in the near-infrared region by using gold nanorods, *J. Am. Chem. Soc.*, **128**, 2115–2120.

59. O'Neal, D. P., Hirsch, L. R., Halas, N. J., Payne, J. D., and West, J. L. (2004) Photothermal tumor ablation in mice using near infrared-absorbing nanoparticles, *Cancer Lett.*, **209**.

60. American National Standards Institute, American National Standard for Safe Use of Lasers and American National Standard for Testing and Labeling of Laser Protective Equipment Z136.1.

61. Kam, N., O'Connell, M., Wisdom, J., and Dai, H. (2005) Carbon nanotubes as multifunctional biological transporters and near-infrared agents for selective cancer cell destruction, *Proc. Natl. Acad. Sci. USA*, **102**, 11600–11605.

62. Shao, N., Lu, S., Wickstrom, E., and Panchapakesan, B. (2007) Integrated molecular targeting of IGF1R and HER2 surface receptors and destruction of breast cancer cells using single wall carbon nanotubes, *Nanotechnology*, **18**, 315101.

63. Chakravarty, P., Marches, R., Zimmerman, N. S., Swafford, A. D.-E., Bajaj, P., Musselman, I. H., Pantano, P., Draper, R. K., and Vitetta, E. S. (2008) Thermal ablation of tumor cells with antibody-functionalized single-walled carbon nanotubes, *Proc. Natl. Acad. Sci. USA*, **105**, 8697–8702.

64. Burke, A., Ding, X., Singh, R., Kraft, R., Levi, N., Rylander, M. N., Szot, C., Buchanan, C., Whitney, J., Fisher, J., Hatcher, H., D'Agostino, R., Kock, N., Ajayan, P. M., Carroll, D. L., Akman, S., Torti, F. M., and Torti, S. (2009) Long-term survival following a single treatment of kidney tumors with multiwalled carbon nanotubes and near-infrared radiation, *Proc. Natl. Acad. Sci. USA*, **106**, 12897–12902.

65. Torti, S., Byrne, F., Whelan, O., Levi, N., Ucer, B., Schmid, M., Torti, F. M., Akman, S., Liu, J., Ajayan, P. M., Nalamasu, O., and Carroll, D. L. (2007) Thermal ablation therapeutics based on CNx multi-walled nanotubes, *Int. J. Nanomed.*, **2**, 707–714.

66. Carrero-Sànchez, J. C., Elìas, A. L., Mancilla, R., Arrellìn, G., Terrones, H., Laclette, J. P., and Terrones, M. (2008) Biocompatibility and toxicological studies of carbon nanotubes doped with nitrogen, *Nano Lett.*, **6**, 1609–1616.

67. Gomez-De Arco, L., Meng-Tse Chen, M.-T., Wang, W., Vernier, T., Pagnini, P., Chen, T., Gundersen, M., and Zhou, C. (2008) Optical properties of carbon nanotubes: near-infrared induced hyperthermia as therapy for brain tumors, *Mater. Res. Soc. Symp. Proc.*, **1065E**, 1065-QQ04-07.

68. Kim, J., Shashkov, E., Galanzha, E., Kotagiri, N., and Zharov, V. (2007) Photothermal antimicrobial nanotherapy and nanodiagnostics with self-assembling carbon nanotube clusters, *Lasers Surg. Med.*, **39**, 622–634.

69. Rojas-Chapana, J. A., Correa-Duarte, M. A., Ren, Z. F., Kempa, K., and Giersig, M. (2004) Enhanced introduction of gold nanoparticles into vital acidothiobacillus ferrooxidans bycarbon nanotubebased microwave electroporation, *Nano Lett.*, **4**, 985–988.

70. Panchapakesan, B., Lu, S., Sivakumar, K., Teker, K., Cesarone, G., and Wickstrom, E. (2005) Single-wall carbon nanotube nanobomb agents for killing, *Nanobiotechnology*, **1**, 133–139.

71. Gannon, C., Cherukuri, P., Yakobson, B., Cognet, L., Kanzius, J., Kittrell, C., Weisman, R., Pasquali, M., Schmidt, H., Smalley, R., and Curley, S. (2007) Carbon nanotube-enhanced thermal destruction of cancer cells in a noninvasive radiofrequency field, *Cancer*, **110**, 2654–2665.

72. Mönch, I., Meye, A., Leonhardt, A., Krämer, K., Kozhuharova, R., Gemming, T., Wirth, M., and Büchner, B. (2005) Ferromagnetic filled carbon nanotubes and nanoparticles: synthesis and lipid-mediated delivery into human tumor cells, *J. Magn. Magn. Mater.*, **290–291**(pt 1), 276–278.

73. Borowiak-Palen, E. (2008) Iron filled carbon nanotubes for bio-applications, *Mater. Sci. Poland*, **26**, 413–418.

74. Yang, F., Fu, D., Long, J., and Ni, Q. (2008) Magnetic lymphatic targeting drug delivery system using carbon nanotubes, *Med. Hypotheses*, **70**, 765–767.

75. Pensabene, V., Vittorio, O., Raffa, V., Ziaei, A., Menciassi, A., and Dario, P. (2008) Neuroblastoma cells displacement by magnetic carbon nanotubes, *IEEE Trans. Nanobiosci.*, **7**, 105–110.

76. Veetil, J., and Ye, K. (2007) Development of immunosensors using carbon nanotubes, *Biotechnol. Prog.*, **23**, 517–531.

77. Wang, S., Humphreys, E., Chung, S., Delduco, D., Lustig, S., Wang, H., Parker, K., Rizzo, N., Subramoney, S., Chiang, Y., and Jagota, A. (2003) Peptides with selective affinity for carbon nanotubes, *Nat. Mater.*, **2**, 196–200.

78. Chen, R. J., Bangsaruntip, S., Drouvalakis, K. A., Kam, N. W., Shim, M., Li, Y., Kim, W., Utz, P. J., and Dai, H. (2003) Noncovalent functionalization of carbon nanotubes, *Proc. Natl. Acad. Sci. USA*, **100**, 4984–4989.

79. Kerman, K., Morita, Y., Takamura, Y., Tamiya, E., Maehashi, K., and Matsumoto, K. (2005) Peptide nucleic acid-modified carbon nanotube field-effect transistor for ultra-sensitive real-time detection of DNA hybridization, *Nanobiotechnology*, **1**, 65–70.

80. Li, C., Curreli, M., Lin, H., Lei, B., Ishikawa, F., Datar, R., Cote, R., Thompson, M., and Zhou, C. (2005) Complementary detection of prostate-specific antigen using In_2O_3 nanowires and carbon nanotubes, *J. Am. Chem. Soc.*, **127**, 12484–12485.

81. Okuno, J., Maehashi, K., Kerman, K., Takamura, Y., Matsumoto, K., and Tamiya, E. (2007) Label-free immunosensor for prostate-specific antigen based on single-walled carbon nanotube array-modified microelectrodes, *Biosens. Bioelectron.*, **22**, 2377–2381.

82. Yu, X., Munge, B., Patel, V., Jensen, G., Bhirde, A., Gong, J. D., Kim, S. N., Gillespie, J., Gutkind, J. S., Papadimitrakopoulos, F., and Rusling, J. F. (2006) Carbon nanotube amplification strategies for highly sensitive immunodetection of cancer biomarkers, *J. Am. Chem. Soc.*, **128**, 11199–205.

83. Briman, M., Artukovic, E., Zhang, L., Chia, D., Goodglick, L., and Gruner, G. (2007) Direct electronic detection of prostate-specific antigen in serum, *Small*, **3**, 758–762.

84. Park, D., Kim, Y., Kim, B., So, H., Won, K., Lee, J., Kong, K., and Chang, H. (2006) Detection of tumor markers using single-walled carbon nanotube field effect transistors, *J. Nanosci. Nanotechnol.*, **6**, 3499–3502.

85. Wang, J., Kawde, A., and Musameh, M. (2003) Carbon-nanotube-modified glassy carbon electrodes for amplified label-free electrochemical detection of DNA hybridization, *Analyst*, **128**, 912–916.

86. Wang, J., Kawde, A., and Jan, M. R. (2004) Carbon-nanotube-modified electrodes for amplified enzyme-based electrical detection of DNA hybridization, *Biosens. Bioelectron.*, **20**, 995–1000.

87. Chen, K., Chen, J., Guo, M., Li, Z., and Yao, S. (2006) Electrochemical behavior of MCF-7 cells on carbon nanotube modified electrode and application in evaluating the effect of 5-fluorouracil, *Electroanalysis*, **18**, 1179–1185.

88. Jia, X., Tan, L., Xie, Q., Zhang, Y., and Yao, S. (2008) Quartz crystal microbalance and electrochemical cytosensing on a chitosan/multiwalled carbon nanotubes/Au electrode, *Sens. Actuators B*, **134**, 273–280.

89. Chen, J., Du, D., Yan, F., Ju, H., and Lian, H. (2005) Electrochemical antitumor drug sensitivity test for leukemia K562 cells at a carbon-nanotube-modified electrode, *Chem. Eur. J.*, **11**, 1467–1472.

90. Ovádeková, R., Jantová, S., Letašiová, S., Štepánek, I., and Labuda, J. (2006) Nanostructured electrochemical DNA biosensors for detection of the effect of berberine on DNA from cancer cells, *Anal. Bioanal. Chem.*, **386**, 2055–2062.

Chapter 4

BIOMEDICAL APPLICATIONS III: DELIVERY OF IMMUNOSTIMULANTS AND VACCINES

Li Jian, Gopalakrishnan Venkatesan and Giorgia Pastorin

Department of Pharmacy, Faculty of Science, National University of Singapore, Singapore
g0801096@nus.edu.sg, gopalpharm83@gmail.com and phapg@nus.edu.sg

4.1 INTRODUCTION TO THE IMMUNE SYSTEM

The immune system protects our organism from infections with layered defenses of increasing specificity. B and T cells represent essential components in the immune system, since their stimulation or deficiencies are at the basis of many patho-physiological conditions. More precisely, each B cell presents on its surface a receptor protein (immunoglobulin) called B-cell receptor (BCR), which is responsible for the recognition and immobilisation of antigens circulating in the blood or lymph in their native form (Fig. 4.1).[1]

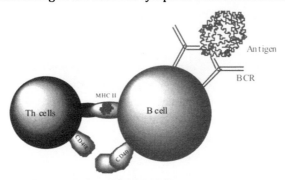

Figure 4.1 Recognition and immobilisation of an antigen at the surface of B cells through a B-cell receptor (BCR), as well as subsequent digestion and presentation of the antigen in the form of processed peptide to T cells (helper), which induce B cells to produce antibodies against the antigen. See also Colour Insert.

Carbon Nanotubes: From Bench Chemistry to Promising Biomedical Applications
Edited by Giorgia Pastorin
Copyright © 2011 Pan Stanford Publishing Pte. Ltd.
www.panstanford.com

Once a B cell encounters a pathogen (antigen), it processes and presents the same antigen (although not in the native shape but in a digested form, i.e., as a peptide fragment) to special subtypes of T cells (called Th cells or helper T cells, i.e., CD4+ T cells) through a class II MHC, thus acting as antigen-presenting cells (APCs). This phenomenon, known as immunological synapse, stimulates CD4+ T cells to secrete cytokines[2] that activate specific B cells towards the production of antibodies against the antigens. Type 1 T helper (Th1) cells produce interferon-gamma (IFN-γ), interleukin (IL)-2 and tumour necrosis factor (TNF)-β, which activate macrophages and are responsible for cell-mediated immunity and phagocyte-dependent protective responses. By contrast, type 2 Th (Th2) cells produce interleukin (IL)-4, IL-5, IL-10 and IL-13, which are responsible for strong antibody production, eosinophil activation and inhibition of several macrophage functions, thus providing phagocyte-independent protective responses. Th1 cells mainly develop after infection by intracellular bacteria and some viruses, whereas Th2 cells predominate in response to infestations by gastrointestinal nematodes.[3] Once stimulated by Th cells, activated B cells produce antibodies (e.g., IgG, IgA or IgE),[4] which assist in inhibiting extraneous pathogens and neutralising viruses and bacterial toxins. Alternatively, cytotoxic CD8+ (CTL) T cells can directly clear virus-infected cells and regulate the immune responses to foreign antigens on the basis of the cytokine profile they secrete via a different mechanism.

In a similar way, specific synthetic peptides, made up of selected amino acid sequences, represent useful tools to elicit an immunogenic response, because they mimic crucial parts of natural proteins while avoiding the redundancy of superfluous residues. These same sequences are extremely advantageous in the development of vaccines, because they can act as antigens (epitopes), provided they induce the production, inside the host, of specific antibodies from B and T cells. In order to succeed, the native structure of the synthetic peptide antigen should be preserved to obtain a protective antibody response. This represents a crucial requirement in the development of vaccines, since it is critical for recognition by antibodies.

4.2 IMMUNOGENIC RESPONSE OF PEPTIDE ANTIGENS CONJUGATED TO FUNCTIONALISED CNTS

The development of effective delivery systems has been rapidly expanding, because of the imperative need to target bioactive molecules and elicit the best pharmacological profiles while avoiding side effects. Carbon nanotubes (CNTs) have emerged as promising vehicles essentially because of their reduced dimensions and the possibility of undergoing chemical or physical modification for an optimal incorporation of therapeutic candidates. For example, soluble functionalised CNTs present high loading capacity of poorly water-dispersible active drugs and antigens on the basis of their high aspect ratio. However, despite the huge amount of synthetic procedures that have been elaborated so far on or with CNTs, very few experimental data are available concerning the use of such nanomaterial for the delivery of vaccines or immunostimulants.

The basic rationale for the employment of CNTs in this context is to link antigens to CNTs. However, once incorporated onto CNTs, several molecules experience a decrease in their activity due to unavoidable changes on their conformation, thus hampering a wide exploitation of this nanomaterial. In fact, in order to induce an immune response, the incorporation of antigens into CNTs should not affect the antigens' natural attributes, so that antibody response can be triggered with high specificity. At the same time, CNTs have to demonstrate that they do not evoke any immunogenic response from the host. In other words, they should guarantee a lack of intrinsic immunogenicity.

Pantarotto *et al.* described different methods of linking bioactive peptides to CNTs through a stable bond, and these methods have been described in the following paragraphs.[5]

4.2.1 Fragment Condensation of Fully Protected Peptides

This procedure, previously reported by Goodman *et al.*,[6] was used to bind the free amino groups present on the sidewalls of single-walled carbon nanotubes (SWCNTs, compound **2**) (after 1,3-dipolar cycloaddition of azomethine ylides) with the strategically protected pentapeptide KGYYG. This peptide was added in a threefold excess upon activation with O-(7-aza-N-hydroxybenzotriazol-1-yl)-1,1,3,3-tetramethyluroniumhexafluorophosphate (HATU) and diisopropylethylamine (DIEA) in dimethylformamide (DMF) for 2 hours. The final complex was isolated by repeated precipitations from a methanol/diethyl ether mixture. The protecting groups were subsequently removed by treating the conjugate with trifluoroacetic acid (TFA) in order to give compound **3** in good yields (Scheme 4.1).

The most challenging part was represented by the confirmation of the successful functionalisation of the tubes with the peptide; in fact the samples obtained could not be injected through a C_{18} chromatographic column of RP-HPLC, since they remained trapped in the pre-column system. As a consequence, the progression of the reaction could be followed by observing the decrease in peptide concentration of the reaction mixture at HPLC (hence via indirect quantification).

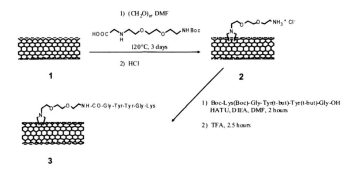

Scheme 4.1 Functionalisation of CNTs via fragment condensation.

Another characterisation became feasible after the incorporation of N^{15}-labelled Gly at the C-terminal part of the sequence, which allowed for a homonuclear and heteronuclear two-dimensional (2D) NMR analysis to be performed. A broad correlation peak in the decoupled ^{15}N-^{1}H spectrum of the fully protected compound **3**, showing a maximum peak height at 119.6/7.40 ppm, was indicative of a uniform distribution of the peptide around the nanotubes' sidewall. A series of bi-dimensional experiments then permitted all the resonances of the peptide moiety to be assigned (Fig. 4.2). A decrease in and a broadening of the signal intensities were observed for the amino acid residues approaching the aromatic tube walls. All the predictable sequential RH_i-NH_{i+1} cross peaks were confirmed in the ROESY spectrum. Moreover, a spatial correlation between the RH of glycine at position 5 and the amide proton of the triethylene glycol chain confirmed the presence of a covalent bond between the peptide and the CNTs (Fig. 4.2b).

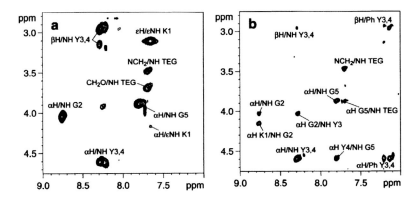

Figure 4.2 Partial (a) TOCSY and (b) ROESY ^{1}H NMR spectra of peptide-CNT **3** in H_2O/t-BuOH-d_9 (9:1) solution. Peptide residues are numbered from Lys1 to Gly5. TEG denotes triethylene glycol. The TOCSY spectrum was recorded while decoupling the ^{15}N heteronucleus. Reproduced from Pantarotto *et al.*[5] with permission.

4.2.2 Selective Chemical Ligation

An interesting example of this type of functionalisation has been provided by Pantarotto *et al.*,[7] who attached a peptide (i.e., the epitope corresponding to a specific sequence of the VPI protein of the foot-and-mouth disease virus [FMDV]) to CNTs and investigated its immunogenic properties. CNTs underwent the 1,3-dipolar cycloaddition of azomethine ylides, while the peptide was coupled to the amino group of functionalised CNTs via a peptide bond (Scheme 4.2).[7] SWCNTs with a free amino group $-NH_2$ and eventually the aminoacid lysine were subsequently modified with a maleimido group, which allowed for linking the FMDV peptide bearing an additional cysteine (at the N terminus) necessary for a selective chemical ligation.

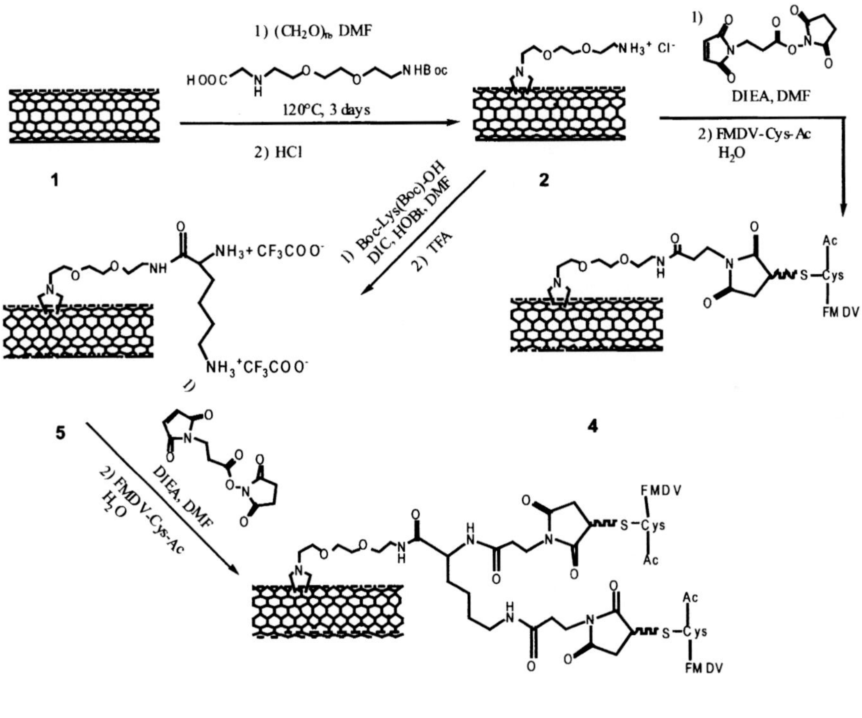

Scheme 4.2 Synthesis of mono- (compound **4**) and bis-(compound **6**)-derivatised CNTs with the epitope of the foot-and-mouth disease virus (FMDV).

The covalent functionalisation of CNTs to conjugate the peptides was a necessary step, as demonstrated by the observation that the simple mixture of the two components did not elicit high antibody titers. When the CNT-peptide complex was incubated with specific monoclonal and polyclonal antibodies, the composite was recognised by the antibodies equally well as the free peptide. Therefore, the experiments revealed that the epitope was appropriately presented after conjugation with mono- (compound **4**) and bis-derivatised (compound **6**) CNTs as measured by BIAcore technology. As concerns this last aspect, surface plasmon resonance (SPR) on a BIAcore instrument enabled the analysis of antigen–antibody interactions in real time and with high sensitivity. Interestingly, CNTs functionalised but not conjugated to the peptide moiety were not recognised. This important finding indicated that no anti-CNT antibodies were detected, suggesting the absence of cross-reactivity with the CNTs.

Moreover, since the peptide under investigation lacked the ability of eliciting an immune response (immunogenicity) in BALB/c mice, additional carrier proteins or T-cell epitope (provided by ovalbumin [OVA] in a Freund's emulsion) were employed. After two co-injections, strong anti-peptide antibody responses were induced, showing higher virus-neutralising capacity than the antibodies elicited after co-immunising the free peptide with OVA (Fig. 4.3).[7] This finding highlights the potential of functionalised CNTs to act as a delivery system capable of presenting critical epitopes at an appropriate conformation to elicit antibodies with the right specificity.

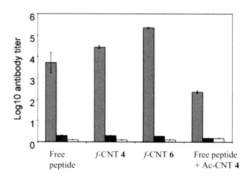

Figure 4.3 Anti-peptide antibody responses following immunisation with peptide and peptide-CNT conjugates. Serum samples were screened by ELISA for the presence of antibodies using FMDV 141–159 peptide conjugated to BSA (cyan bar), control peptide conjugated to BSA (magenta bar) or CNTs 1 functionalised with a maleimido group without peptide (white bar) as solid-phase antigens. Reproduced from Davide Pantarotto *et al.*[7] with permission. See also Colour Insert.

A similar approach was adopted for the incorporation of another peptide to be used against *Plasmodium vivax* in the treatment of malaria.[8] *Plasmodium* is a parasite that contains a very highly conserved trans-membrane protein called AMA-1, which consists of 562 amino acids,[9] with an extracellular N-terminal domain, an ectodomain, a trans-membrane region and a C-terminal cytoplasmatic domain.[10] The ectodomain presents 16 conserved cysteine residues, which are responsible for stabilisation via intramolecular disulphide bridges that seem essential for inducing an antibody response which could inhibit parasite development.[11] Because of its participation in the invasion process, AMA-1 has been suggested as candidate for a vaccine against *Plasmodium* pre-erythrocytic blood stages[12]; in other words, it has shown the ability to act as a target of antibodies that prevent parasites from invading red blood cells *in vitro* and to induce protective immunity against challenge with parasites in several animal models (*in vivo*).[13] To that purpose, a promising malaria peptide from the AMA-1 protein sequence was conjugated to multi-walled CNTs (MWCNTs) and injected in mice. Unfortunately, the CNT-peptide complex did not improve the antibody titers as much as the free peptide did, most probably because of other factors (e.g., surface charges) that might have promoted the co-adsorption of some complement system's down-regulating agents (e.g., factor H).[14] At the same time, the use of acetylated CNTs (which are uncharged nanotubes without peptide sequences attached) did not trigger any significant values for antibodies, thereby demonstrating that the immune response was peptide-specific and that CNTs were not intrinsically immunogenic.[5,7]

As regards the protection of treated mice against a challenging dose of antigen, three mice (out of 30 mice) became protected: mice 7 and 9 from the group treated only with the peptide and mouse 23 from the CNT-peptide conjugate group. Interestingly, the protection conferred to the animals was not correlated to the antibody (Ab) titers produced, thus decreasing the relevance of the higher Ab titers reported above with the treatment of peptide alone. In fact some mice presented high antibody titers but became infected following challenge with *Plasmodium berghei*. On the contrary, mice 9 and 23 became protected against infection and presented lower antibody levels (Figs. 4a and 4b). Mouse 7 presented a good antibody level and became protected (Fig. 4a), indicating that both fine

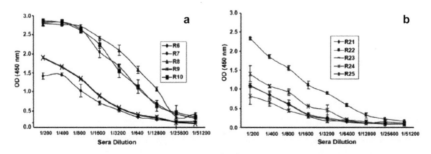

Figure 4.4 Immunised mice sera antibody titers: (a) mice immunised with the 21267 peptide alone; (b) mice immunised with the CNT–peptide 21267 complex. Reproduced from Yandara *et al.*[8] with permission.

specificity and antibody titers are important in ensuring the final protective result. Moreover, the groups of mice immunised with peptide alone or with the peptide-CNT conjugate developed parasitaemia from day 7 onwards, but they survived about 10 days longer than the mice of the control group (which had much lower percentages of parasitaemia). These encouraging results suggested that the groups treated with the peptide developed a late infection and that the presence of CNTs did not hamper this phenomenon.

Finally, results obtained from cytokine profile analysis indicated that the type of immune response induced by the peptide alone or by peptide incorporated onto the CNTs occurred via Th1 cells, because of the representative levels of TNF-α and IFN-γ displayed by these two groups of immunised mice regarding the control, as shown in Figs. 5a and 5b.

Moreover, it is well known that there is an absolute requirement of IFN-γ for protection against malaria in rodents, and it has been suggested that its production may be a good correlation of protective immunity.[15]

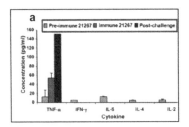 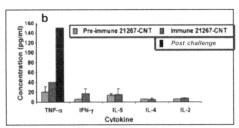

Figure 4.5 Cytokine profiles of malarial challenged mice immunized with the 21267 peptide and its CNT conjugate. (a) Cytokine profile displayed by mice immunised with the 21267 peptide. (b) Cytokine profile displayed by mice immunised with the CNT–peptide 21267 complex. Reproduced from Yandara *et al.*[8] with permission.

4.3 INTERACTION OF FUNCTIONALISED CNTS WITH CpG MOTIFS AND THEIR IMMUNOSTIMULATORY ACTIVITY

The presence of adjuvants is essential to promote and enhance immune responses induced by vaccine antigens. One particular example of these immunostimulants is represented by the synthetic oligodeoxynucleotides (ODN) containing immunostimulatory CpG motifs (ODN- CpG), which have envisaged their extensive use as immunomodulators against tumours or allergies or even as protection against bioterrorist threats.[16-19] CpG motifs are sequences deriving from bacterial DNA. Provided they are internalised inside the cells in sufficient amounts to interact with receptors (called toll-like receptor 9)[20] expressed in endosomal compartments, they can stimulate the immune system efficiently. In fact CpG motifs induce B-cell proliferation and antibody secretion, and they activate APCs to express co-stimulatory molecules and secrete cytokines, including IL-12 and TNF-α.[21,22] In the immunostimulatory process by ODN-CpG, the entrance of such DNA sequences inside the cell and their interaction with endosomal receptors are crucial. This interaction is usually hampered by the low cellular uptake of ODN-CpGs. That is why the use of functionalised CNTs has been recently suggested to increase the ODN-CpG's immunological properties. CNTs bearing positive charges could represent a valuable tool that can interact with negatively charged ODN-CpGs. In another study[23] the same research group demonstrated that it is possible to assess the formation of the supramolecular complexes between CNTs and DNA through transmission electron microscopy (TEM) (Fig. 4.6): CNTs were presented in bundles of different diameters, on which the DNA portion was condensed by forming toroidal clusters, or globular and supercoiled structures.

For immunisation purposes, a constant amount of ODN-CpGs (0.05 μg/culture) was initially conjugated to different concentrations of two types of CNTs (compounds **2** and **5**) and their interaction analysed by SPR (Table 4.1).[24]

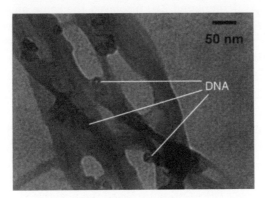

Figure 4.6 TEM image of functionalised CNTs (compound **2**): DNA complexes. Reproduced from Pantarotto *et al.*[23] with permission.

Kinetic analyses revealed that the association rate constant (k_{on}) was higher for compound **4** than for compound **2**. Correspondingly, compound **2** demonstrated a slightly faster dissociation process (k_{off}) compared with compound **4**. This could be due to the avidity factor of the bivalent ligand **4**. However, there were no major differences in the binding affinity (k_d) of ODN-CpG motifs to either functionalised CNTs.

Table 4.1 Kinetics and affinity constants for the interactions of functionalised CNTs with ODN-CpG motifs

Functionalised CNTs	Compound 2a: SWCNTs Compound 2b: MWCNTs	Compound 5: SWCNTs
k_{on} [M^{-1} s^{-1}]	$(5.19 \pm 1.55) \cdot 10^5$	$(1.23 \pm 0.14) \cdot 10^6$
k_{off} [s^{-1}]	$(2.99 \pm 1.46) \cdot 10^{-3}$	$(4.85 \pm 0.35) \cdot 10^{-3}$
K_d [M]	$(5.88 \pm 2.30) \cdot 10^{-9}$	$(3.99 \pm 10.66) \cdot 10^{-9}$

Subsequently, the immunostimulatory properties of the complexes were tested *in vitro*.[24]

In the first series of experiments, both SWCNTs (**2a** and **5**) and MWCNTs (**2b**) were selected. However, the MWCNTs with the free amino group demonstrated much lower affinity for ODN-CpG sequences than did SWCNTs. The difference could be attributable to both dimension and charge distribution of the different CNTs. It was observed that SWCNT-Lys-NH$_3^+$ (compound **5**) and MWCNT-NH$_3^+$ (compound **2b**) displayed the same amount of positive charges. However, a stronger avidity effect could have been involved, resulting from the closer proximity of the two ammonium groups in SWCNT-Lys-NH$_3^+$ than MWCNT-NH$_3^+$ and a consequent double electrostatic interaction for ODN-CpG molecules. Therefore, SWCNTs represented the CNTs of choice for the following experiments, in which various excess ratios of CNTs to a minimal immunostimulatory dose of the ODN-CpG motifs were incubated with naïve mouse splenocytes. In particular, complexes of SWNT-Lys-NH$_3^+$ at 18:1 and 9:1 ratios over ODN-CpG increased its immunostimulatory properties by 58% and 45%, respectively, while lower ratios did not show remarkable results. This is most probably due to the fact that a high excess of positive charges of CNTs over ODN-CpG neutralised the negative charges of the DNA sequences, thus minimising the repulsion effects with the negatively charged cell membrane and promoting ODN-CpG's cellular internalisation.

4.4 IMMUNOGENICITY OF CARBON NANOTUBES

As reported in previous paragraphs, CNTs have been shown to be advantageous as antigen-presenting systems. However, it is still necessary to fully understand the role of functionalised CNTs in the immune response and thoroughly analyse the effect of these structures on the activation of the complement system. As regards this last aspect, the complement system is a group of about 35 soluble and cell-surface proteins in blood which recognise, opsonise and remove invading micro-organisms, altered host cells (such as apoptotic or necrotic cells) and other foreign materials.[25]

Complement activation may occur by any of three mechanisms, which have been classified into "classical", "lectin" and "alternative" pathways. In the classical mechanism, the protein C1q, the recognition subunit of C1, recognises activators essentially on the basis of charge and hydrophobic interactions. In the lectin pathway, the mannan-binding lectin (MBL) binds to targets via neutral sugar residues (e.g., mannose). The alternative pathway starts by the binding of the protein C3b to the pathogen surface, and the subsequent events are analogous to those of the classical pathway (Fig. 4.7).

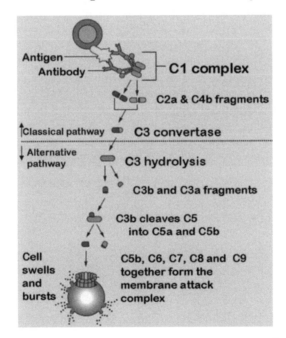

Figure 4.7 Complement activation via classical (up) and alternative (down) pathways. Reproduced from Wikimedia Commons (http://en.wikipedia.org/wiki/Alternative_complement_pathway). See also Colour Insert.

A major premise in the biomedical application of CNTs relies on the possibility that the body's immune system can recognise CNT materials[26] and initiate an immune response. So far, the absence of intrinsic immunogenicity of CNTs as carriers of bioactive meolecules,[23] the translocation of CNTs across the cell membrane and their deposition inside the cytoplasm without being toxic[27] and finally their ability to enhance an immune response when attached to an antigen[7] have reinforced the possibility of using CNTs as therapeutic and vaccine delivery tools.

At the same time, very limited information is available on the interaction between this nanomaterial and the immune system. Green and his research group[14] recently reported an interesting study concerning the ability of different types of CNTs to activate the human serum complement system via the classical pathway. In this investigation, two types of CNTs were used: single-walled (SWCNTs) and double-walled (DWCNTs). The nanotubes were produced through different procedures and reported as arc discharge (arc-SWCNTs), chemical vapour deposition CNTs (CVD-SWCNTs and CVD-DWCNTs) and high-pressure carbon monoxide SWCNTs (HIPco-SWCNTs). The major difference among them consisted of the amount of metal element impurities still present after the purification process: plasma atomic emission spectroscopy analysis revealed that arc-SWCNTs contained 1.4% (w/w) Ni, HIPco-SWCNTs 1% (w/w) Fe, CVD-SWCNTs 0.2% (w/w) Co, and CVD-DWCNTs 1.9% (w/w) Mo. It was observed that all the CNT samples tested activated complement to an extent comparable with zymosan, a well-known complement activator through the classical pathway (Fig. 4.8): more precisely, the mechanism of such activation seemed to involve a direct and selective binding of C1q to CNTs, since the alternative mechanisms (e.g., direct binding of C3b [alternative pathway] to the CNTs, binding of CNTs to fibrinogen, or even binding of C1q or C3b to other plasma/serum proteins,

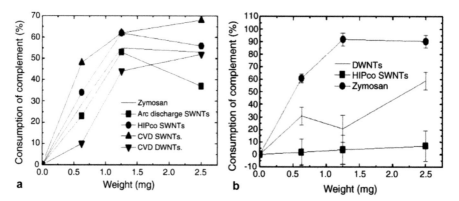

Figure 4.8 (a) Percentage consumption of human serum complement activity via classical pathway due to the presence of different types of carbon nanotubes. (b) Percentage consumption of human serum complement activity via alternative pathway due to the presence of two types of carbon nanotubes. Zymosan samples were used as the positive control. A sample of undiluted human serum incubated at 37°C served as the negative control. Reproduced from Carolina Salvador-Morales *et al.*[14] with permission.

such as IgG and IgM, adsorbed onto CNTs) were not confirmed by this study. The same experiment, but without Ca^{2+} (which blocks the classical pathway activation by inactivating the protein C1qr2s2), was performed to investigate the alternative pathway.

Interestingly, while HIPco-SWCNTs activated the complement via the classical procedure (Fig. 8a) to a much higher extent than DWCNTs, the opposite was observed for the alternative pathway, where DWCNTs were the most activating (Fig. 8b).

On the whole, these results might provide an explanation for the finding of Pantarotto *et al.*,[7] who showed enhanced anti-peptide antibody response by utilising peptides coupled to nanotubes. However, it is important to notice that Pantarotto and collaborators tested functionalised (chemically modified) nanotubes, while in the study on complement activation, pristine materials were tested. Therefore, any comparison between these two investigations might not be accurate. In fact, activation of human complement induced by CNTs might be diminished or eliminated by alteration of surface chemistry. Therefore, further experiments should be performed with modified CNTs.

In any case, the demonstration of complement activation by CNTs might prove to be advantageous for vaccination purposes, as it may enhance antigen-presenting effects. On the other hand, it might also induce an inflammatory response and even lead to the formation of granulomas, as already reported by Lam *et al.*[28]

In another study, Nygaard *et al.*[29] investigated whether SWCNTs and MWCNTs were able to promote allergic immune reactions when given together with the allergen ovalbumin (OVA). They also investigated the acute inflammatory response to CNTs 24 hours after a single intranasal exposure to the particles in the absence of the allergen. The results showed that both SWCNTs and MWCNTs, together with OVA, increased serum levels of allergen-specific IgE antibodies and the secretion of Th2-associated cytokines, which are involved in allergic responses. In general, the increased production of IgE by nanomaterials could be attributed to other factors, including oxidative stress, increased antigen presentation and allergen-carrying effects.[30-32] Therefore their role for the adjuvant effect of CNTs requires further studies. Conversely, treatment with ultra-fine carbon black particles (ufCBPs, used as positive control) increased TNF-α and IgG2a levels, following a behaviour that was observed for MWCNTs but not for SWCNTs. Similarly, ufCBP and MWCNTs, but not SWCNTs, induced an acute influx of neutrophils 24 hours after a single exposure to the particles in the absence of the allergen (Fig. 4.9).

On the whole, these preliminary findings pave the way for future *in vivo* experiments using animal models of disease or vaccination protocols to test the therapeutic or adjuvant potential of such complexes. Moreover, further studies will need to address whether serum proteins can alter the surface potential of CNTs, their size, the stability of complexes and the formation of aggregates.

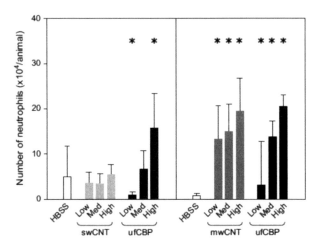

Figure 4.9 Neutrophil cell numbers in BALF collected 24 hours after a single intranasal exposure of mice to buffer (Hank's balanced salt solution [HBSS], open bars) and low, medium and high doses of SWCNTs (light gray bars), MWCNTs (dark gray bars) and ufCBPs (black bars). Mean values and SD for groups of eight mice are shown. The asterisk denotes a statistically significant difference compared with the HBSS group. Reproduced from Nygaard *et al.*[29] with permission.

4.5 CONCLUSIONS

CNTs can serve as vaccine delivery and adjuvant vehicles by virtue of their nanoparticulate nature. The hydrophobic nature of CNTs contributes to vaccine delivery capability by facilitating the interaction of CNTs with antigens or immunostimulatory molecules and uptake of the vaccine particles by immunocompetent cells.

Although the application of CNTs in the field of vaccines is still at its early stage, CNTs have shown the ability to provide a platform for the attachment of adjuvants and antigens. In any case, it is important to evaluate in detail each system and application, in order to identify eventual toxicity, immunogenicity, stability, biocompatibility and costs for a competitive scale-up of these composites. Although extremely challenging, further research should be encouraged to examine the potential of CNTs to deliver immunostimulants and vaccines on the basis of the still-not optimal delivery systems available on the market.

References

1. Li, J., Barreda, D. R., Zhang, Y.-A. , Boshra, H., Gelman, A. E., LaPatra, S., Tort, L., and Sunyer, J. O. (2006) B lymphocytes from early vertebrates have potent phagocytic and microbicidal abilities, *Nature Immunol.*, **7**, 1116–1124.

2. Mosmann, T. R., and Coffman, R. L. (1989) Th1 and Th2 cells: different patterns of lymphokine secretion lead to different functional properties, *Annu. Rev. Immunol.*, **7**, 145–173.

3. Romagnani, S. (1999) Th1/Th2 cells, *Inflamm. Bowel Dis.*, **5**, 285–294.

4. Leist, T. P., Cobbold, S. P., Waldmann, H., Aguet, M., and Zinkernagel, R. M. (1987) Functional analysis of T lymphocyte subsets in antiviral host defense, *J. Immunol.*, **138**, 2278–2281.

5. Pantarotto, D., Partidos C. D., Graff R., Hoebeke J., Briand J.-P., Prato M., and Bianco, A. (2003) Synthesis, structural characterization, and immunological properties of carbon nanotubes functionalized with peptides, *J. Am. Chem. Soc.*, **125**, 6160–6164.

6. Goodman, M., Felix, A., Moroder, L., and Toniolo, C. (2002) *Methods of Organic Chemistry (Houben-Weyl)*, vol. E22a, Thieme, Stuttgart, Germany.

7. Pantarotto, D., Partidos, C. D., Hoebeke, J., Brown, F., Kramer, E., Briand, J.-P., Muller, S., Prato, M., and Bianco, A. (2003) Immunization with peptide-functionalized carbon nanotubes enhances virus-specific neutralizing antibody responses, *Chem. Biol.*, **10**, 961–966.

8. Yandara, N., Pastorin, G., Prato, M., Bianco, A., Patarroyoa, M. E., and Lozano, J. M. (2008) Immunological profile of a *Plasmodium vivax* AMA-1 N-terminus peptide-carbon nanotube conjugate in an infected *Plasmodium berghei* mouse model, *Vaccine*, **26**, 5864–5873.

9. Cheng, Q., and Saul, A. (1994) Sequence analysis of the apical membrane antigen I (AMA-1) of *Plasmodium vivax*, *Mol. Biochem. Parasitol.*, **65**, 183–187.

10. Kocken, C. H., Withers-Martinez, C., Dubbeld, M. A., van der Wel, A., Hackett, F., Blackman, M. J., and Thomas., A. W. (2002) High-level expression of the malaria blood-stage vaccine candidate *Plasmodium falciparum* apical membrane antigen 1 and induction of antibodies that inhibit erythrocyte invasion, *Infect. Immun.*, **70**, 4471–4476.

11. Hodder, A. N., Crewther, P. E., Matthew, M. L., Reid, G. E., Moritz, R. L., Simpson, R. J., Anders, R. F. (1996) The disulfide bond structure of *Plasmodium* apical membrane antigen-1, *J. Biol. Chem.*, **271**, 29446–29452.

12. Chesne-Seck, M. L., Pizarro, J. C., Vulliez-Le Normand, B., Collins, C. R., Blackman, M. J., Faber, B. W., Remarque, E. J., Kocken, C. H. M., Thomas, A. W., and Bentley,

G. A. (2005) Structural comparison of apical membrane antigen 1 orthologues and paralogues in apicomplexan parasites, *Mol. Biochem. Parasitol.*, **144**, 55–67.

13. Saul, A., Lawrence, G., Allworth, A., Elliott, S., Anderson, K., Rzepczyk, C., Martin, L. B., Taylor, D., Eisen, D. P., Irving, D. O., Pye, D., Crewther, P. E., Hodder, A. N., Murphy, V. J., and Anders, R. F. (2005) A human phase 1 vaccine clinical trial of the *Plasmodium falciparum* malaria vaccine candidate apical membrane antigen 1 in Montanide ISA720 adjuvant, *Vaccine*, **23**, 3076–3083.

14. Salvador-Morales, C,. Flahaut, E., Sim, E., Sloan, J., Green, M. L., and Sim, R. B. (2006) Complement activation and protein adsorption by carbon nanotubes, *Mol. Immunol.*, **43**, 193–201.

15. Marsh, K., and Kinyanjui, S. (2006) Immune effector mechanisms in malaria, *Parasite Immunol.*, **28**, 51–60.

16. Miconnet, I., Koenig, S., Speiser, D., Krieg, A., Guillaume, P., Cerottini, J. C., and Romero, P. (2002) CpG are efficient adjuvants for specific CTL induction against tumor antigen-derived peptide, *J. Immunol.*, **168**, 1212–1218.

17. Klinman, D. M., Currie, D., Gursel, I., and Verthelyi, D. (2004) Use of CpG oligodeoxynucleotides as immune adjuvants, *Immunol. Rev.*, **199**, 201–216.

18. Klinman, D. M. (2004) Immunotherapeutic uses of CpG oligodeoxynucleotides, *Nat. Rev. Immunol.* 4, 249–258.

19. Klinman, D. M., Verthelyi, D., Takeshita, F., and Ishii, K. J. (1999) Immune recognition of foreign DNA: a cure for bioterrorism? *Immunity*, **11**, 123–129.

20. Hemmi, H., Takeuchi, O., Kawai, T., Kaisho, T., Sato, S., Sanjo, H., Matsumoto, M., Hoshino, K., Wagner, H., Takeda, K., and Akira, S. A. (2000) A toll-like receptor recognizes bacterial DNA, *Nature*, **408**, 740–745.

21. Krieg, A.M., (2002) CpG motifs in bacterial DNA and their immune effects, *Annu. Rev. Immunol.*, **20**, 709–760.

22. Krieg, A. M. (2000) The role of CpG motifs in innate immunity, *Curr. Opin. Immunol.*, **12**, 35–43.

23. Pantarotto, D., Singh, R., McCarthy, D., Erhardt, M., Briand, J.-P., Prato, M., Kostarelos, K. and Bianco, A. (2004) Functionalised carbon nanotubes for plasmid DNA gene delivery, *Angew. Chem. Int. Ed.*, **43**, 5242–5246.

24. Bianco, A., Hoebeke, J., Godefroy, S., Chaloin, O., Pantarotto, D., Briand, J.-P., Muller, S., Prato, M., and Partidos, C. D. (2005) Cationic carbon nanotubes bind to CpG oligodeoxynucleotides and enhance their immunostimulatory properties, *J. Am. Chem. Soc.*, **127**, 58–59.

25. Sim, R. B., and Tsiftsoglou, S. A. (2004) Proteases of the complement system, *Biochem. Soc. Trans.*, **32**, 21–27.

26. Descouts, P. A. R. (1996) Nanoscale probing of biocompatibility of materials, in *Nanoscale Probes of the Solid/Liquid Interface* (ed. Gewirth, A., and Siegenthaler, H.), Kluwer Academic Publishers, Dordrecht, the Netherlands, pp. 317–343.

27. Pantarotto, D., Briand, J. P., Prato, M., and Bianco, A. (2004) Translocation of bioactive peptide across the cell membranes by carbon nanotubes, *Chem. Commun.*, **1**, 16–17.

28. Lam, C. W., James, J. T., McCluskey, R., and Hunter, R. L. (2004) Pulmonary toxicity of single-wall carbon nanotubes in mice 7 and 90 days after intratracheal instillation, *Toxicol. Sci.*, **77**, 126–134.

29. Nygaard, U. C., Hansen, J. S., Samuelsen, M., Alberg, T., Marioara, C. D., and Løvik, M. (2009) Single-walled and multi-walled carbon nanotubes promote allergic immune responses in mice, *Toxicol. Sci.*, **109**, 113–123.

30. de Haar, C., Kool, M., Hassing, I., Bol, M., Lambrecht, B. N., and Pieters, R. (2008) Lung dendritic cells are stimulated by ultrafine particles and play a key role in particle adjuvant activity, *J. Allergy Clin. Immunol.*, **101**, 1246–1254.

31. Parnia, S., Brown, J. L., and Frew, A. J. (2002) The role of pollutants in allergic sensitization and the development of asthma, *Allergy*, **57**, 1111–1117.

32. Wan, J., and Diaz-Sanchez, D. (2007) Antioxidant enzyme induction: a new protective approach against the adverse effects of diesel exhaust particles, *Inhalation Toxicol.*, **19**(suppl. 1), 177–182.

Chapter 5

BIOMEDICAL APPLICATIONS IV: CARBON NANOTUBE–NUCLEIC ACID COMPLEXES FOR BIOSENSORS, GENE DELIVERY AND SELECTIVE CANCER THERAPY

Venkata Sudheer Makam, Jason Teng Cang-Rong, Sia Lee Yoong and Giorgia Pastorin

Department of Pharmacy, Faculty of Science,
National University of Singapore, Singapore
phapg@nus.edu.sg

5.1 INTRODUCTION

Even though nucleic acids have been typically associated with genetic information and with a fascinating helical structure as proposed by Watson and Crick in 1953, their role has been exploited well beyond such preliminary definition. More precisely, deoxyribonucleic acid (DNA) and ribonucleic acid (RNA, which relays the information stored in DNA to synthesize proteins) not only contain the genetic information that allows all modern living organisms to function, grow and reproduce, but also have demonstrated to be involved in several patho-physiological conditions in both developed and primitive species.

All humans have unique gene sequences, and genetic tests can suggest a predisposition to certain illnesses, such as breast cancer, cystic fibrosis or liver diseases, thus providing better cure or even prevention. On the other hand, the analysis of DNA could ascertain the authenticity of children born from promiscuous couples, on the basis of the fact that the offspring inherits part of the genomic information from both parents. As a consequence, and

Carbon Nanotubes: From Bench Chemistry to Promising Biomedical Applications
Edited by Giorgia Pastorin
Copyright © 2011 Pan Stanford Publishing Pte. Ltd.
www.panstanford.com

especially after the first assembly of the genome was completed in 2000, many researchers have been focusing their efforts on the identification of crucial genomic sequences able to provide new avenues for advances in medicine and biotechnology. More precisely, gene therapy has become extremely popular because of the possibility to replace defective genes with functional ones to treat a disease, especially hereditary illnesses. To that purpose, and because all viruses bind to their hosts and introduce their genetic material as part of their replication cycle, doctors and molecular biologists have tried to replace such genetic material inside the viruses with information that enables the production of more copies of the lacking gene or protein. In other words, they used viruses as vehicles to transfer useful genetic information inside the host. Similarly, RNA catalysts have been used as mediators for a variety of organic reactions and inorganic-particle growth.[1-5] Moreover, small interfering RNAs evoke a natural defense mechanism called RNA interference, shutting off viral infection through tagging viral RNA. However, despite these incredible discoveries, several problems have prevented the spreading of gene therapy using viral vectors, including difficulties in the targeting, undesired effects and disruption of other vital genes already in the genome. Therefore, recent advances in vector technology have suggested the use of vehicles alternative to viral vectors: for example, the combination of positively charged carriers (e.g., liposomes, lipids and nanoparticles) with negatively charged DNA sequences has been extensively studied to achieve an efficient gene transfection without the use of viral genes. Recently, it has been suggested that carbon nanotubes (CNTs) play an important role in the complexation and delivery of gene-encoding nucleic acids; their use as gene delivery vectors seems promising because they are readily produced, are stable for long-term storage and have shown suspendibility in aqueous solutions with low toxicity *in vitro*.[6,7] Additionally, they can be functionalised with several cationic groups to tailor the delivery. Even with these positive aspects, their combination with nucleic acids has been more focused on the dispersion of CNTs (Section 5.2) and on the construction of sensors, nanocircuits and nanocomposites (Section 5.3) rather than on real gene or cancer therapy (Section 5.4).

In the following paragraphs we have reported the most relevant publications concerning the use of CNT–nucleic acid complexes for the tubes' dispersibility, sensing, gene delivery and cancer therapy.

5.2 INTERACTION OF CNTS WITH NUCLEIC ACIDS

5.2.1 CNTs and DNA

5.2.1.1 DNA favours dispersion and separation of CNTs

DNA molecule contains three constituents, namely phosphate acid groups, basic groups and sugar units. The basic groups are adenine, guanine, cytosine and thymine, while in RNA thymine is replaced by the uracil residue. The structure of the DNA consists of two molecular chains, in which one chain is tightly bound to the other to form a double helix and they are held together by many hydrogen bonds. Further, DNA is classified as a natural polymer, and it has shown interesting solubilising properties of numerous materials as well as high sensitivity towards complementary sequences. On the other hand, a wide variety of applications of matrices made of CNTs for the detection of bio-organic and inorganic compounds has also been reported.[8-11]

One of the first articles describing the interaction between CNTs and DNA was published in 2003 by Zheng *et al.*[12]: it demonstrated that nucleic acids could easily adhere to the surface of CNTs on the basis of a π-stacking interaction that dispersed the bundles into individual tubes, while exposing the sugar–phosphate groups to water. This was confirmed by electronic adsorption and near-infrared (NIR) fluorescence. In a few simulation studies, the authors also indicated that there were many permissible ways in which short single-stranded DNA (ssDNA) could bind to the nanotube surface. These included helical wrapping or simply surface adsorption of a linearly extended structure, with poly-thymine (poly(T)) showing the highest dispersion efficiency (Fig. 5.1). It was suggested that both DNA chain flexibility and backbone charge might have contributed to such high dispersion efficiency. Very recently, it was even shown that experimental binding energies of nucleobases with single-walled carbon nanotubes (SWCNTs) in aqueous solution decrease in the order G > T > A > C.[13]

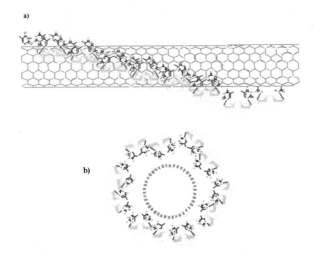

Figure 5.1 Binding model of a carbon nanotube wrapped by a poly(T) sequence. (a) The bases (red) orient to stack with the surface of the nanotube and extend away from the sugar–phosphate backbone (yellow). (b) The DNA wraps to provide a tube within which the carbon nanotube can reside, thus converting it into a water-soluble object. See also Colour Insert.

In another experiment, the same authors demonstrated how to separate CNTs on the basis of the tubes' characteristics and diameter, by simply adsorbing DNA onto their external sidewall.[14] Anion exchange chromatography showed that the best separation of the tubes was obtained with sequences rich in two alternating nucleotides, deoxyguanylate (dG) and deoxythymidylate (dT), expressed as d(GT)$_n$ (with n = 10 to 45). Such sequences were responsible for the formation of self-assembled helical wires around individual nanotubes involving hydrogen-bonding interactions among different strands (Fig. 5.2a). More precisely, two complementary (antiparallel) d(GT)$_n$ strands tended to form a double-stranded strip, which subsequently surrounded CNTs in a densely packed wrapping around their sidewalls. The mechanism that enabled the separation of the tubes was attributed to electrostatic interactions among the CNT-DNA hybrids. In fact CNT-DNA complexes carried a net negative charge on the basis of the deprotonated backbone phosphate groups on the DNA (Fig. 5.2b); the negative charges interacted with the positive charges present in the ion exchange resin, while the eluting salt solution depended on the effective linear charge density.

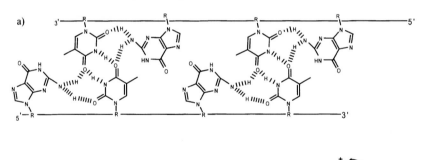

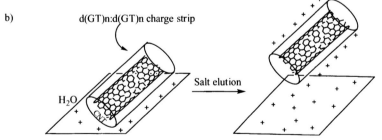

Figure 5.2 (a) Proposed hydrogen-bonding interactions between two d(GT)$_n$ strands that lead to the formation of a "d(GT)$_n$:d(GT)$_n$ charge strip". (b) Schematic for anion exchange separation process. At lower salt concentration, the surface-bound state is favoured in which positive ions on the resin attract the negative surface charge on the DNA-CNT. With increasing salt concentration, the surface and DNA-CNT interactions are screened, favouring elution. Image modified from Zheng *et al.*[14]

The chromatographic elution showed that in case of metallic CNTs, the linear charge density was reduced from that of the DNA alone, and it was responsible for early fractions enriched in smaller-diameter and metallic tubes, as demonstrated by a more pronounced absorption in the metallic M_{11} (400–600 nm) band, and weaker absorption in the semiconductor E_{11} band (900–1,600 nm); instead, for semiconducting tubes, since the polarisability of the nanotubes was lower and more influenced by the diameter of the tubes, the separation resulted in late fractions enriched in larger-diameter and semiconducting tubes. Therefore, ion exchange chromatography (IEC) enabled the separation of CNTs with different electronic properties by coating DNA sequences with specific length.

To confirm the obtained results, an elution model was associated with these experiments[15] in order to elucidate the most significant factors contributing to the separation of CNT-DNA hybrids. The advantage of such investigation is that the elution model semiquantitatively captured the available experimental observations made by Zheng *et al.*[16] and it predicted that the IEC separations are very sensitive to SWCNTs' radius, a_t, in addition to the CNTs' dielectric constant (Fig. 5.3). In fact it confirmed that metallic hybrids are expected to elute before semiconducting hybrids of the same radius size. Since in the simulation studies the binding distance between the DNA and SWCNTs, $a_h - a_t$, was kept constant, the helix angle increased with SWCNT radius. Therefore, higher salt concentrations were generally required to elute larger diameter hybrids when the DNA intercharge segment bond angle was held constant.

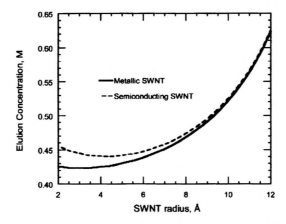

Figure 5.3 Salt concentration at hybrid elution as a function of the SWCNT radius for metallic, ϵ_t = 5000 (solid line), and semiconducting, ϵ_t = 4 (dashed line), SWCNTs. Reproduced from Lustig *et al.*[15] with permission.

Moreover, this analysis suggested that the helix angle is the most influential property in IEC separation. SWCNTs can shield DNA backbone charges, which may contract the helical pitch. The tubes' polarisability may further modulate the helical pitch because of induced dipolar interactions between the DNA backbone and the SWCNTs.

Alternatively to the partitiong approach on the basis of their diameters and chirality, CNTs can also be separated through the size-exclusion chromatography (SEC) process,[17] which helps purify DNA-wrapped carbon nanotubes (DNA-CNTs) from graphitic impurities and sort them into fractions of uniform length (Fig. 5.4). The advantage of this technique is the narrow length variation of 10% in each of the measured fractions, in which length decreases monotonically from 500 nm in the early fractions to 100 nm in the late fractions.

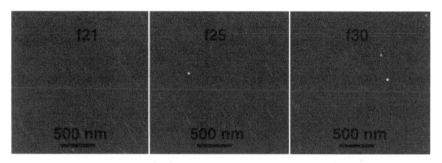

Figure 5.4 AFM images of three representative SEC fractions deposited onto alkyl silane-coated SiO$_2$ substrates. Reproduced from Huang *et al.*[17] with permisssion.

In another study, photoluminescence (PL) spectra from HiPco nanotube dispersed with salmon DNA (SaDNA) showed that the genomic sequence mediates selective stabilisation of (6,5) SWCNTs, providing >86% of (6,5) chirality enrichment without requiring additional separation through ion exchange or density gradient columns.[18] More precisely, the supernatant collected after ultracentrifugation of the SWCNT-SaDNA samples showed that the relative emission intensity of semiconducting (6,5) SWCNTs was significantly enhanced (Fig. 5.5a), while the redispersed precipitate showed peaks corresponding to the remaining ((7,6), (8,4) and (7,5)) chiral tubes (Fig. 5.5b). Therefore, selective solubilisation of (6,5) tubes in aqueous media was observed using SaDNA. Such evidence was further supported by radial breathing mode (RBM) peaks from resonance Raman spectra (RRS) [19], which were obtained from the 633 nm excitation laser line. It was observed that the relative intensity of semiconducting SWCNT-SaDNA peaks was increased, suggesting that SaDNA preferentially stabilised semiconducting

SWCNTs in the aqueous supernatant, while leaving the metallic counterpart in the precipitate.

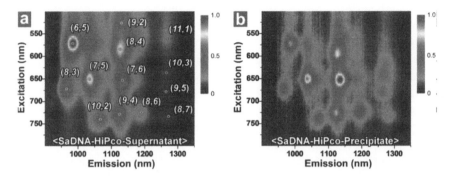

Figure 5.5 Normalised PLE emission contour plot (a, b) from SaDNA dispersed SWCNTs in the ultracentrifuged supernatant (a) and the redispersed-precipitate fractions (b), respectively. Reproduced from Kim *et al.*[18] with permission. See also Colour Insert.

Together with the separating properties, another important factor to consider when dealing with CNT-DNA conjugates is the kinetics of hybridisation once DNA is adsorbed onto the tubes. For that reason, Jeng *et al.* reported a mechanistic model to monitor DNA hybridisation on nanotubes using an energy increase in the NIR fluorescence of the (6,5) nanotubes [20]. Such amplification is determined by the addition and hybridisation of complementary DNA (cDNA) to a suspension of ssDNA previously adsorbed to SWCNTs. It was observed that kinetics were slow when DNA was adsorbed onto the tubes, while they were much faster when DNA was free in solution, as described by a simple second-order rate expression. However, hybridisation on SWCNTs could not be described using the same expression, so an adsorption step (faster) had to be included before the hybridisation reaction (slower). It was found that the activation energy of SWCNT-DNA hybridisation was higher than that of free DNA, providing confirmation that SWCNTs impeded the hybridisation reaction, while dynamic light scattering (DLS) excluded that such phenomenon was simply due to sample aggregation. Moreover, DNA adhered strongly to the nanotubes, which remained stable over time, thus suggesting that the DNA molecules never left the surface completely. As a consequence, it seems that one base at a time detached from the SWCNT surface for hybridisation, after which it was re-adsorbed onto the tubes. Subsequently, there was an energy decrease from adsorption of two DNA sequences (the ssDNA and its complementary cDNA) to a nanotube, between 3.6 and 7.1 kcal/mol, which is very close to the 7.5 kcal/mol in the activation energies of hybridisation for SWCNT-adsorbed DNA.

In another study, molecular dynamics simulations showed that DNA binds to the external surface of an uncharged or positively charged SWCNTs within a few hundred picoseconds.[21] This result seemed in contradiction with the previous study by Jeng, but only apparently, since the rapid interaction of DNA with the tube was limited to a very small portion. More precisely, Zhao *et al.* found that most of the hydrophobic sites of a DNA double helix (dsDNA) were wrapped inside the helices and were not available for interaction. Therefore, the dsDNA did not bind the nanotubes except for the exposed hydrophobic base groups at the ends of the DNA segment (Fig. 5.6). Conversely, hydrophobic groups of ssDNA needed more time to interact, since they were all (not only the extremities) attracted towards the hydrophobic surface of uncharged SWCNTs, while the hydrophilic backbone was exposed to the aqueous environment. This also justified why ssDNA binds SWCNTs more tightly than dsDNA. [15, 22-25]

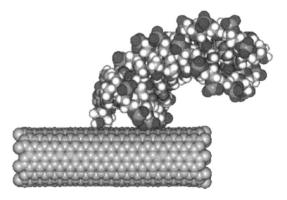

Figure 5.6 Simulation of double-stranded DNA onto uncharged CNTs. Reproduced from Zhao and Johnson[21] by permission. See also Colour Insert.

5.2.1.2 Supramolecular CNT-DNA complexes

The beneficial role of DNA towards carbon nanotubes has resulted also in the possibility of efficiently integrating CNTs into multifunctional supramolecular structures or devices, especially for their use in electrochemical sensors[26-28] by chemical incorporation of ssDNA chains for hybridisation with redox-labelled complementary (cDNA) chains. An interesting structure has been reported by Dai *et al.*,[29] who exploited the DNA hybridisation to realise DNA-directed self-assembling of multiple CNTs and gold nanoparticles (Fig. 5.7). The successful realisation of the complexes formed was confirmed by atomic force microscopy (AFM), although the potential application of such multifunctional material was not clearly stated. In that case DNA was covalently bound to previously oxidised CNTs, and a similar procedure was adopted by another group for the incorporation of an uncharged DNA analogue, called peptide nucleic acid (PNA) (Fig. 5.8).[30] The use of such neutral sequence, besides maintaining the hybridisation properties of ssDNA, offers additional advantages, including resistance of PNA-cDNA complexes towards enzymatic degradation and improved thermal stability in comparison with ssDNA-cDNA complexes, because of absence of electrostatic repulsions. And yet, methods that can characterise the interaction with CNTs, especially in solution, are still lacking. That is why linear dichroism, recently applied on SWCNT-PNA complexes solubilised in sodium dodecyl sulfate,[31] seems particularly useful in demonstrating that the nucleic acid bases lie flat on the nanotube surface with the backbone wrapping around the nanotube at an oblique angle in the region of 45°. Although the analysis is still qualitative, it was able to discriminate between the signals provided by the single-stranded PNA and the larger ones obtained with sonicated double-helix DNA.[32] The authors suggested that the difference might be due to additional small interactions coupled with the intrinsically greater rigidity of the duplex DNA (dsDNA).

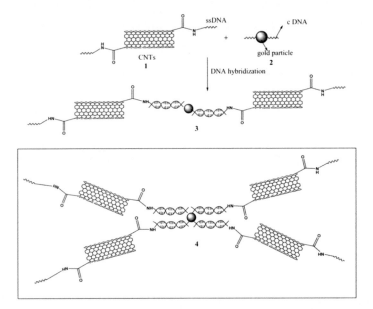

Figure 5.7 DNA-directed self-assembling of multiple carbon nanotubes and nanoparticles.

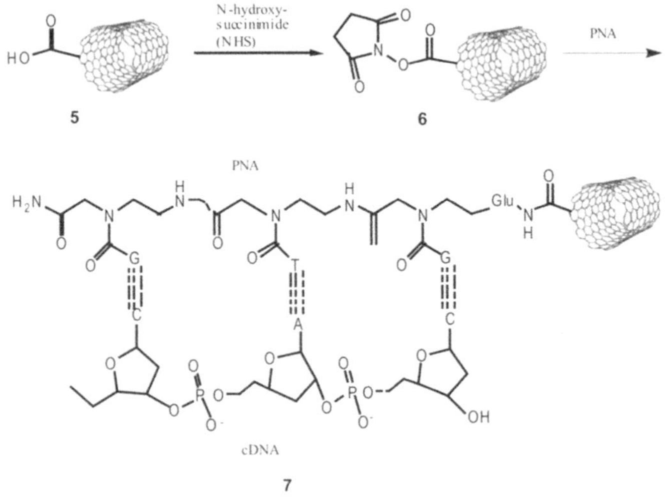

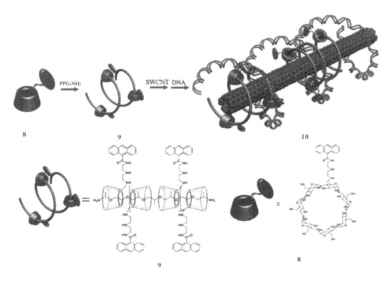

Figure 5.8 A cDNA fragment with a single-stranded, "sticky" end hybridised by Watson–Crick base pairing to the PNA-CNT.

Another supramolecular complex, based on the combination of SWCNTs with anthrylcyclodextrin (ACD) and poly-pseudorotaxane (PPR) (Fig. 5.9), was shown to cleave dsDNA under visible light radiation.[33] The success of the ensemble was guaranteed by the natural tendency of the cyclodextrins (CDs) to adhere to CNTs and the ability of the anthryl group to intercalate within the DNA's groove.

Figure 5.9 Preparation of a SWCNT–ACD-PPR conjugate and its DNA wrapping. See also Colour Insert.

For a qualitative assessment of the interactions between the SWCNT–ACD-PPR conjugate and complementary DNA sequences, temperature-dependent fluorescence titration experiments were performed. The results showed that the association of ACD-PPR with complementary DNA gave negative enthalpic changes ($\Delta H° = -23.6$ kJ/mol) and positive entropic changes (T$\Delta S°$ = 2.7 kJ/mol). The values might be associated with the van der Waals and hydrophobic interactions, which seemed to be the main driving forces and also justified more favourable contributions once the highly hydrophobic SWCNTs were added. It was also proved that the SWCNT–ACD-PPR conjugate not only acted as a DNA carrier but also exhibited a good ability to cleave DNA, similar to the already reported photoinduced DNA-cleavage mechanism for cyclodextrin–fullerene conjugates.[34–36]

A supramolecular architecture was recently proposed by Liu *et al.*,[37] who strategically modified chitosan (CHIT) with CD and multi-walled carbon nanotubes (MWCNTs) to provide a multifunctional gene delivery system (Fig. 5.10). CHIT itself has been considered a promising non-viral gene carrier, on

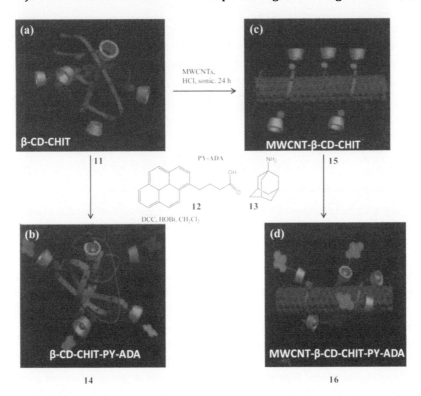

Figure 5.10 Preparation of the supramolecular architecture MWCNT-β-CD-CHIT-PY-ADA. Modified from Liu *et al.*[37] with permission. See also Colour Insert.

the basis of its highly cationic density suitable for DNA condensation. CDs, instead, represent a class of cyclic oligosaccharides with six to eight D-glucose units with intriguing hydrophobic cavities. Upon inclusion of β-CD with CHIT, adamantane-modified pyrene (ADA-PY) (Fig. 5.10a,b) and MWCNTs (Fig. 5.10c), the synergistic effect between aromatic residues (especially in pyrene molecules PY) and cationic groups (mainly in CHIT) greatly improved CHIT's DNA-condensing efficiency.

As shown in Fig. 5.10, β-CDs preferred accommodating the cage of adamantane inside their cavity, while exposing pyrene grafts outside. In the presence of such assemblies, DNA chains were condensed to uniform hollow loops, as shown in Fig. 5.11, prepared using AFM.

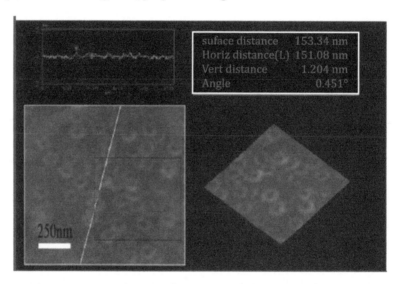

Figure 5.11 AFM images of DNA in the presence of the supramolecular assemblies. Reproduced from Liu *et al.*[37] with permission. See also Colour Insert.

The exploitation of multifunctional structures for their use as electroanalytical devices induced Shie *et al.* to combine MWCNTs with both DNA and cytochrome *c* protein (cyt *c*), which has shown a unique binding property with protein redox partners.[38] The formation of a biocomposite film among MWCNTs, DNA and cyt *c* consisted on the initial attachment (either covalent, between the amino groups of DNA and the activated carboxylated functions on CNTs, or non-covalent, by simple hydrophobic interaction of DNA with the tubes' side walls), and subsequent attachment of cyt *c* on the DNA.[39] This second step was based on the electrostatic interaction between the positively charged groups (e.g., lysine–NH_3^+, or Fe^{2+} and Fe^{3+} in cyt *c*) and the negatively charged phosphates of the DNA backbone[40] (Fig. 5.12).

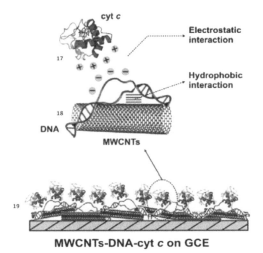

Figure 5.12 Possible interaction between MWCNTs, DNA and cyt *c* for the formation of MWCNT–DNA–cyt *c* biocomposite film modified electrodes. Reproduced from Shie *et al.*[38] with permission. See also Colour Insert.

The results of the experiments performed on several electrodes (e.g., glassy carbon electrode [GCE], gold [Au], indium tin oxide [ITO] and screen-printed carbon electrode [SPCE]) proved that the deposition of cyt *c* on the MWCNT–DNA biocomposite film was more stable and uniform on Au than on other electrodes. The presence of both MWCNTs and DNA in the biocomposite film increased the electron transfer rate constant (K_s) up to 21% and decreased the degradation of cyt *c* during the cycling, thus suggesting an effective approach for the development of voltammetric and amperometric sensors. The biocomposite film also exhibited a promising, enhanced electrocatalytic activity towards the reduction of halogen oxyanions such as IO_3^-, BrO_3^- and ClO_3^- and oxidation of biochemical compounds such as ascorbic acid and L-cysteine.

In a very recent article, Zhou *et al.* exploited the solubilising power of DNA towards CNTs to investigate the complexes formed by the incorporation of quantum dots (QDs) and provide unique fluorescence properties.[41] DNA consisted of guanine (G) and thymine (T) repeating units as linkers between QDs and CNTs, and they provided characteristic bands at Fourier transform infrared–attenuated total reflectance (FTIR-ATR) spectra (3,182 and 1,630 cm^{-1}), correspondent to the stretching and bending modes, respectively, of the primary aromatic amines of the guanine bases. The same sharp bands disappeared after amide bonding between carboxyl-QDs and amine-DNA molecules physisorbed onto SWCNTs (Fig. 5.13).

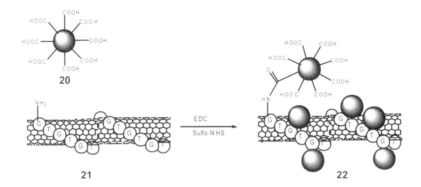

Figure 5.13 Conjugation of DNA-wrapped SWCNTs and COOH-QDs. *Abbreviations*: EDC, 1-ethyl-3-(3-dimethylaminopropyl) carbodiimide hydrochloride; sulfo-NHS, sulfo-*N*-hydroxysuccinimide.

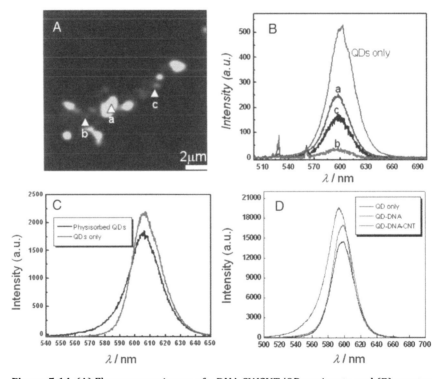

Figure 5.14. (A) Fluorescence image of a DNA-SWCNT/QD conjugate and (B) spectra from different sample locations (a–c). (C) Fluorescence spectra from unconjugated QDs and QDs physisorbed onto DNA-SWCNTs. (D) Fluorescence spectra from unconjugated QDs, DNA-conjugated QDs in the absence of SWCNTs, and DNA-SWCNT/QD conjugates. Reproduced from Zhou *et al.*[41] with permission. See also Colour Insert.

Interestingly, the attachment of QDs did not necessarily reduce the water solubility of the DNA-wrapped SWCNT conjugates, because of sufficient remaining surface COOH groups on QDs even after conjugation (Fig. 5.13). Another advantage of these well-dispersed conjugates is that such samples can be optically excited to give rise to sharp peaks of dispersed SWCNTs in the UV/vis/NIR spectrum. This allowed the investigation of the influence of DNA molecules on the spectral shift of QDs. The emission spectra of small droplets of sample solutions of QDs alone, DNA-QD conjugates with no SWCNTs, and DNA–SWCNT or DNA–QD conjugates were compared and reported (Fig. 5.14).

The QD-DNA exhibited a negligible blue shift, suggesting that DNA had little effect on QDs' spectral properties, while further conjugation with SWCNTs resulted in a noticeable blue shift of at least 4 nm, which confirmed that carboxyl QDs were conjugated onto DNA-SWCNTs. Taken together, these results promote the fabrication of QD-CNT hybrids for advanced nanoelectronics and nanosensors. At the same time, it is one of the very few cases in which the free amino group in guanine is covalently bonded to the QD, although the DNA chain is only physically wrapped around the tubes. On the contrary, Bianco *et al.* demonstrated that it is possible to synthesise SWCNT-adenine complexes by amidation reactions between acid/amine-functionalised (*f*-)SWCNTs and amine/acid-functionalised adenine derivatives (Fig. 5.15).[42] These procedures offer the possibility of generating controlled horizontal alignment of the nanotubes on highly oriented pyrolytic graphite (HOPG) surfaces and of complexing them with metal ions, thus giving rise to a highly patterned supramolecular assembly of SWCNTs. Nucleobases can assemble into organised structures via hydrogen-bonding interactions,[43-52] the mechanism of which involves adsorption followed by rearrangement of molecules on the surface to generate hierarchial assemblies.[53-55]

The tri-ethylene glycol (TEG) monomethyl ether chains seemed to be responsible for the parallel orientation of the complex *f*-SWCNT 2b (Figs. 5.15 and 5.16), through an interdigitation process that keeps the nanotubes apart, while adenine residues at the tips undergo self-hydrogen bonding interactions.[48,49] This behaviour can explain the elongated shape of the nanotubes under AFM (Fig. 5.16a), while SWCNT 2a, being without TEG chains, forms fibrils of tubes (Fig. 5.16b).

As regards the ability of adenine (attached to the CNT surface) to coordinate the formation of Ag(I) nanoparticle-CNT conjugates, *f*-SWCNT 2a derivatives were used to avoid interferences from oxygen and nitrogen atoms deriving from the TEG chain of *f*-SWCNT 2b. Beadlike Ag(I) nanoparticles present over the nanotubes were observed, probably because of the interaction of adenine with Ag(I) ions (Fig. 5.17). Interestingly, the same procedures with AgNO$_3$ solution and SWCNT–COOH 1a showed only physical adsorption of heavy, non-aligned clusters of Ag(I) on the surface of nanotubes.

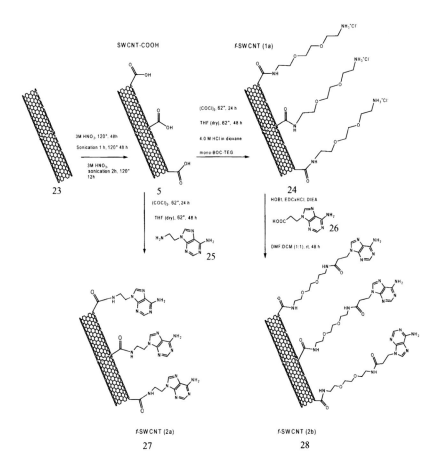

Figure 5.15 Scheme of synthesis of functionalised SWCNTs complexed with adenine nucleobases. *Abbreviations*: TEG, tri-ethylene glycol; EDCxHCl, 1-ethyl-3-(3-dimethy laminopropyl)carbodiimide hydrochloride (1.5 eq.); HOBt, *N*-hydroxybenzotriazole (1.5 eq.); DIEA, diisopropylethylamine (3 eq.); DMF, dimethylformamide; DCM, dichloromethane.

Another covalent conjugation between DNA and CNTs was proposed by Hamers,[56] who treated oxidised SWCNTs with thionyl chloride first and then ethylenediamine to produce free amino groups that were subsequently reacted with the heterobifunctional cross-linker succinimidyl 4-(*N*-maleim idomethyl)cyclohexane-1-carboxylate (SMCC) and finally linked to thiol-terminated DNA (Fig. 5.18). This procedure helped improve the quality of the samples in several ways: First of all, the covalent binding reduced the Ni contamination from 19% to 4 % by weight, and the yttrium contamination from 4.8% to 0.75%. Moreover, the use of SMCC as a covalent linker led to a high affinity for water, dispersing the tubes uniformly, while SWCNTs that were

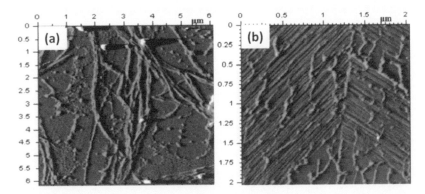

Figure 5.16 AFM images of SWCNT–adenine hybrids without (*f*-SWCNT 2a, left) and with (*f*-SWCNT 2b, right) the tri-ethylene glycol chain. Reproduced from Singh *et al.*[42] with permission. See also Colour Insert.

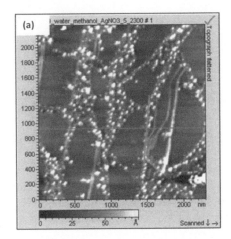

Figure 5.17 AFM image of *f*-SWCNT **2a** on HOPG showing the attachment of Ag(I) nanoparticles all over the nanotube network. Reproduced from Singh *et al.*[42] by permission. See also Colour Insert.

amine-terminated and treated with DNA but without a linker aggregated and precipitated in the buffer solution. Finally, the covalent method did not affect the hybridisation properties of the conjugated DNA, as demonstrated by the specific interaction with a complementary sequence. Interestingly, the same results were obtained after denaturation with 8.3 M urea solution and re-hybridisation experiments. The results obtained in this study demonstrated that the covalent conjugation of DNA allowed an excellent accessibility of the sequences, which most probably are not wrapped around or intercalated within the nanotubes. Therefore, they provide a good starting point for the fabrication of supramolecular structures and reversible biosensors.

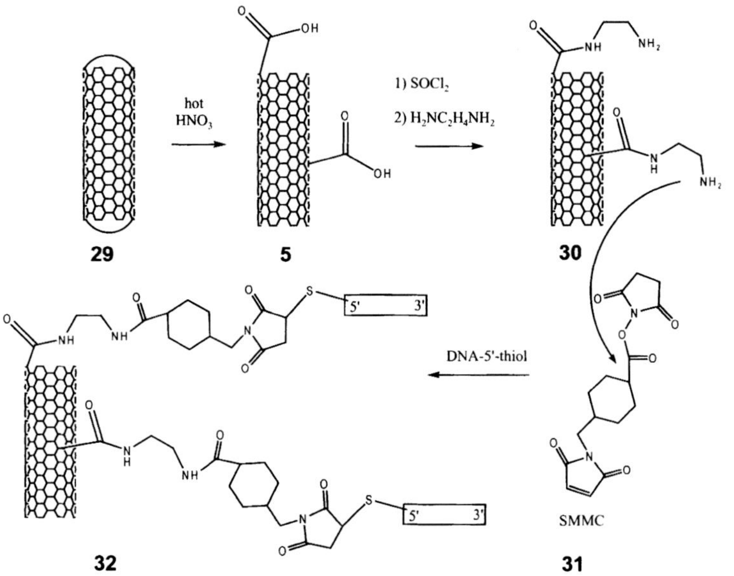

Figure 5.18 Scheme of SWCNTs covalently bonded to DNA through the SMMC linker.

5.2.1.3 Mechanisms of DNA wrapping and internalisation of CNTs

A recent manuscript has attempted to disclose the mechanism involved in the wrapping of DNA around CNTs, showing that the guanine/cytosine (GC) content was directly correlated to the extent of condensation around CNTs. In other words, SWCNTs condensed the GC-DNA major groove more easily, while the adenine/thymine (AT)-DNA major groove appeared too wide for SWCNTs binding and finally the minor groove did not bind SWCNTs. More precisely, the interaction between oxidised SWCNTs and GC-DNA was so strong that it affected not only the DNA's hydration properties but also the whole DNA structure. Indeed, circular dichroism spectra showed that the condensation onto CNTs induced a transition of DNA from its native, right-handed "B" form to an "A" form. Notably, the transition from the B-DNA double helix to the A form is essential for biological function[57-59] as shown by the existence of the A form in many protein–DNA complexes, thus suggesting the use of CNTs as sensors in living cells based on the transition of DNA secondary structure.

In another article, not only the mechanism of DNA wrapping around the tubes but the whole CNT-DNA complex taken up by cells were explored.[60]

CNTs consisted of pristine, un-functionalised SWCNTs that they were non-covalently conjugated with DNA. The authors suggested an endocytotic pathway for cell internalisation on the basis of absent uptake at 4°C or in presence of NaN_3 (two known conditions that inhibit endocytosis). Moreover, after pre-treating the cells with either sucrose (hypertonic treatment) or a K^+-depleted medium prior to exposure to SWCNT conjugates, cell cytometry showed a remarkable reduction of internalised cells, thus indicating the particular clathrin pathway for an endocytotic cellular uptake of SWCNTs. It is important to note that although these results were confirmed by additional assays and controls, they were specific for the samples used in the experiments: since only pristine and oxidised SWCNTs were employed in this investigation, the results might not correspond to the cellular uptake of MWCNTs or other functionalised nanotubes, especially in view of the contradictory results obtained by Pantarotto *et al.*[61] In the latter case, the researchers used CNTs bearing a TEG chain and a peptide, and their findings did not confirm those obtained by the group of Dai, since endocytosis was not involved. Therefore, further investigations are needed to disclose the real internalisation mechanism of CNTs, which do differ remarkably on the basis of several unique characteristics and preparative protocols. To support this statement, Becker *et al.*[62] demonstrated that there is a length dependence in the uptake of DNA-wrapped SWCNTs, so the eventual internalisation process is influenced by this aspect more than other factors, including nanotubes' functionalisation or chirality. In fact the results showed a selective rejection of certain SWCNTs by the incubated cells on the basis of their length. Interestingly, the longer fractions of (335 ± 27) nm and (253 ± 26) nm, respectively, collected through SEC, were internalised without affecting the viability of the cells. Conversely, shorter SWCNT fractions (below 253 nm) exhibited decreased metabolic activity and cell survival, suggesting that shorter tubes could be more toxic to cells than longer SWCNTs. This was established for several cell lines, including A549 (human alveolar basal epithelial cells), MC3T3-E1 (clonal murine calvarial) and A10 (embryonic rat thoracic aorta medial layer myoblasts) cells. However, despite this apparently general phenomenon of length-dependent uptake in these *in vitro* experiments, the exact threshold process is not understood yet and it could vary with cell type and nanotube samples.

5.2.2 Carbon nanotubes and RNA

The use of ribonucleic acid (RNA) has shown several advantages for potential therapeutic applications: contrary to DNA, RNA cannot integrate directly into the host chromosome, and hence it is less prone to become mutagenic.[63]

In addition, strategic manipulations of RNA have shown the ability to avoid recognition by the mammalian immune system.[64] Moreover, unlike DNA (which needs to be transcribed), RNA is directly functional in cells and, once released from a carrier (e.g., CNTs), it may be translated to yield a protein or act as antisense RNA to suppress protein synthesis, or act as interfering RNA (RNAi) to silence a target gene.

With the purpose of using RNA for advanced biomedical applications, Rao *et al.* proposed single-molecule fluorescence microscopy as a useful method to investigate nucleic acids because the method is non-invasive and rapid in image acquisition.[65] The group also exploited resonance Raman analysis to study CNTs, since the nanotubes, especially SWCNTs, show characteristic peaks. Therefore, they suggested the combination of these two techniques as an ideal tool for evaluating the non-specific binding of RNA polymer (poly(rU)), consisting of 500–2,600 nucleotides labelled with OliGreen fluorescent dye corresponding to Uracil (U) residues and isolated CNTs on a silicon substrate. Fluorescence images indicated poly(rU) molecules as individual small blobs, most probably because of the tertiary structure of poly(rU) into loops. Moreover, since poly(rU) molecules appeared to be distributed along the contour of SWCNTs, the results suggested that the π-stacking and hydrophobic interactions between the poly(rU) molecules and the tubes was likely to be stronger than the van der Waals and hydrophobic interactions between the bases and the silicon substrate. Micro-Raman spectra showed the characteristic radial breathing mode (RBM) within 190–290 cm^{-1} and the *G* band centered at 1,593 cm^{-1}. No remarkable shifts in the peak positions were seen in the poly(rU)-bound SWCNT spectra, because of the non-specific binding of RNA to the tubes, which should not provide any significant charge transfer. Conversely, the SWCNT-poly(rU) hybrids exhibited a marked decrease in the intensity of the RBM band and a small enhancement in the *D* band intensity compared with SWCNTs alone, thus confirming an effective binding of poly(rU) onto SWCNTs.

In comparison with the previously reported covalent binding scheme by Kam *et al.*[60] and Pantarotto *et al.*,[61] the non-specific binding between SWCNTs and poly(rU) offers more flexibility for the tracking and release of the load carried by SWCNTs upon delivery. In order to study the translocation of SWCNTs inside the cells, Lu *et al.* performed a radioisotope labelling assay on the tubes with thymidine (methyl-[^{3}H]), while concentrations of SWCNTs were kept lower than 0.4 mg/mL.[66] The uptake of SWCNT-poly(rU) complexes was suggested to be a result of the amphipathic character of both cell membrane and SWCNT-poly(rU) hybrids. Once internalised inside MCF7 cells, endosomes in the cytoplasm could have stored SWCNTs after poly(rU) translocation inside the nucleus (as indicated by their presence inside the cells upon 4 hours of incubation without effect on cell viability), which presumably occurred through a passive ratchet diffusion.[67] As regards this last aspect, the authors suggested cell mitosis as a contributing factor for the nuclear internalisation: in fact in the last mitotic stage (telophase), the nuclear membrane reformed, and hence it could have incorporated the SWCNT-poly(rU) complex.

Finally, in another study, Silva *et al.* exploited the combination of CNTs and RNA for a completely different purpose[68]: they demonstrated that the treatment of SWCNTs with RNA helped stably suspend and purify unfunctionalised CNTs without the use of co-factors or surfactants. Even more interesting, the authors also proved that the subsequent treatment of the RNA-wrapped CNTs with the enzyme ribonuclease (RNase) very effectively removed the "temporarily added" RNA, thus allowing for a complete recovery of pure CNTs for further manipulations.

Taken together, all these findings concerning the interaction of CNTs with nucleic acids seem very promising, as they combine crucial molecules in biochemistry with nanotechnology, thus encouraging interesting applications in biotechnology.

5.3 SENSORS AND NANOCOMPOSITES

Derivatised SWCNTs integrated into field effect transistor (FET) circuits (SWCNT-FETs) are attractive as electronic-readout molecular sensors because of their fast response, high sensitivity and compatibility with dense array fabrication.[69] Non-covalent functionalisation should be applied in order to preserve the electronic properties of the device, especially in case of liquid or gas-phase sensors. Regarding this last aspect, the molecular mechanism that detects a particular odour is not known, but the response seems to be specific for the base sequence of the ssDNA), which has a high affinity for SWCNTs. As a proof of that, Staii *et al.*[70] used several odours to characterise the sensor response, including methanol, propionic acid, trimethylamine (TMA), dinitrotoluene (DNT, as liquid solution prepared by dissolving 50 mg/mL of the material in dipropylene glycol) and dimethyl methylphosphonate (DMMP). A small tank of saturated vapour of each odour was connected to a peristaltic pump so that the flow of air directed over the device could be electrically diverted into one of the odour reservoirs for a set time (typically 50 s), after which the flow reverted to plain air. In Fig. 5.19 the authors reported the signals before (blue points) and after (red points) coating with ssDNA. In the presence of ssDNA adsorbed onto SWCNTs, DNT and DMMP (which are agents simulating explosive vapour and nerve gas, respectively) gave an odour response while bare devices did not, thus suggesting that ssDNA layer increased the binding affinity for these molecules to the device, with a concomitant increase in sensor response. Interestingly, diverse odours elicited different current responses from ssDNA/SWCNT-FET sensors, but the responses were highly reproducible across different devices and specific to the base sequence of the ssDNA used. Moreover, the signal-to-noise levels of the measurements indicated that detection of concentrations less than 1 ppm should be possible and, even more encouraging, it was observed that the ssDNA chemical recognition layer was reusable through at least 50 cycles without refreshing or regeneration. Therefore, such findings represent a significant progress towards the realisation of an effective sensor array for electronic olfaction.

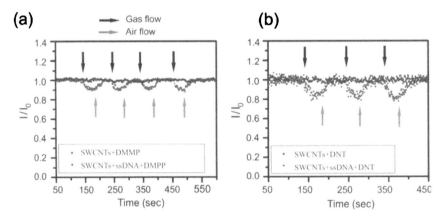

Figure 5.19 (a) Change in the device current when sarin simulant DMMP is applied to SWCNT-FETs before and after ssDNA functionalisation. (b) Sensor response to DNT. Reproduced from Staii *et al.*[70] with permission. See also Colour Insert.

DNA oligonucleotides can also represent a valuable tool able to identify the interaction of small-molecule ligands targeting specific proteins or protein receptors. Wu *et al.*[71] showed that the oligonucleotides not only act as coding sequences for the linked organic molecules but offer immediate signal amplification via polymerase chain reactions (PCRs). A unique advantage is that ssDNA, once covalently conjugated to a small molecule, is protected from degradation by exonuclease I (Exo I), provided that the small molecules are bound to their protein targets, as shown in Fig. 5.20a. Most probably, such terminal protection is due to steric hindrance of the bound protein molecule, which prevents Exo I from approaching and cleaving the phosphodiester bond adjacent to the 3′ terminus. At the same time, it also enables the exploitation of a sensitive electrochemical biosensor strategy for detecting the binding of small molecules to proteins (Fig. 5.20b). The readout system is based on COOH-terminated, long-chain alkanethiols, such as 16-mercaptohexadecanoic acid (MHA), which form a hydrophobic self-assembled monolayer (SAM) in aqueous solutions. SWCNT-DNA complexes can be digested by the Exo I enzyme, which forms precipitates of "naked" SWCNTs that adsorb onto the MHA-SAM and generate a signal through their electron transfer properties. However, when small-molecule-binding proteins are present, the negatively charged SWNT-wrapping ssDNA is bound to the protein target, thus preventing the degradation of the ssDNA by Exo I. In this case, no signal is generated because the SWCNT-DNA complexes create a repulsion with anionic SAM and SWCNTs do not deposit on the electrode. These results provide very useful insights into the development of sensitive, specific and efficient platforms for quantitatively screening the small-molecule–protein interactions, and they can be exploited, for example, to detect the interaction of folate with its protein target, folate receptor (FR), a biomarker that is often over-expressed in various tumours.[72–74]

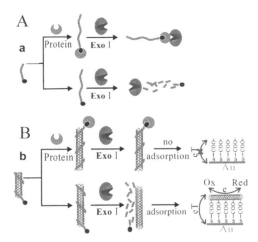

Figure 5.20 Terminal protection assay of small-molecule-linked ssDNA. Small-molecule-linked ssDNA is hydrolysed successively into mononucleotides from the 3′ end by Exo I, while protected from the hydrolysis when the small molecule moiety is bound to its protein target (A). SWNT-wrapping ssDNA terminally tethered to the small molecule is degraded by Exo I, rendering SWNTs assembled on MHA-SAM, which mediates electron transfer between electroactive species and the electrode. Protein binding of small-molecule-linked ssDNA prevents digestion of ssDNA, precluding adsorption of DNA-wrapped SWNTs on MHA-SAM with no redox current generated (B). Reproduced from Wu *et al.*[71] with permission. See also Colour Insert.

Another example of an effective fluorescent sensing platform has been proposed on the basis of the non-covalent conjugation of SWCNTs and dye-labelled ssDNA.[75] CNTs were able to either quench or, in the presence of a target, restore the fluorescence signal. To that purpose, a 23-base oligonucleotide

(P1) or a human R-thrombin (Tmb) binding aptamer (P2) was labelled with a fluorescein derivative. The mechanism at the basis of such a quenched/ enhanced signal is that, in the SWCNT-ssDNA complex, the dye attached to ssDNA is in close proximity to the nanotubes, which are thus able to quench the dye's fluorescence remarkably. In addition, the interaction of P2-SWCNTs with Tmb was effectively transduced by fluorescence enhancement, even in the presence of a low concentration of the target. The limit of sensitivity for Tmb detection was calculated to be ~1.8 nM (i.e., 2×10^{-9} M), which is around 10 times lower than that of the common dye-quencher pair-labelled aptamers,[76,77] thus indicating that the SWCNT–aptamer approach could be exploited for selective target protein detection.

An even better detection limit of 1×10^{-12} M was proposed by Lee *et al.*,[78] who prepared CNTs carrying several horseradish peroxidase (HRP)-labelled sequences for signal amplification of oncogene nucleic acids sequences deriving from human acute lymphocytic leukemia (ALL). To that purpose, oxidised SWCNTs were covalently bound to the 5′ end of pre-labelled detection probes (DPs) in the presence of *N*-(3-dimethylaminopropyl)-*N*′-ethyl carbodiimide hydrochloride (Fig. 5.21). Additional HRP molecules could also be conjugated to SWCNTs through their free amine groups. Hence, a multifunctional vesicle was obtained (Fig. 5.21b), thus increasing the amount of signal per unit of target and providing a better detection than conventional probes (Fig. 5.21a).

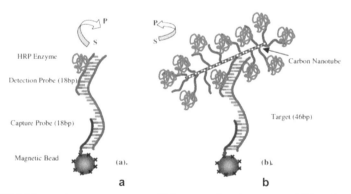

Figure 5.21 Schematic representation of (a) conventional probes and (b) multiple HRP and DP-conjugated CNT-based labels. Reproduced from Lee *et al.*[78] with permission. See also Colour Insert.

The entire "sandwich" hybridisation assay was performed on magnetic beads, on the surface of which capture probes (CPs) were immobilised via streptavidin–biotin or diimide-activated covalent interaction.

Even more important, the labels proposed by this group present the unique characteristic of detecting the target sequences by simple visual inspection:

distinct, dark-coloured aggregates were determined by the sandwich hybridisation of multi-DP conjugated CNT labels (DPs–CNT) and the multi-CP bound beads (CPs–beads) in the presence of a complementary target (TG). This provided a simple and low-cost nucleic acids detection technique in view of early-stage disease and point-of-care diagnostic applications.

Another sandwich structure was proposed by Huang's research group[79] on the basis of the combination between the intrinsic nature of magnetic iron oxide (Fe_3O_4) particles (MPs) and the dispersing properties of MWCNT-DNA, with the final purpose of monitoring the presence of complementary target DNA. In this study, MPs were initially coupled with a 5′-NH_2-modified DNA probe (P1), while MWCNTs were conjugated with a 3′-NH_2-modified DNA probe (P2), thus generating a suspension of MPP1 and MWNTP2 probes. In the absence of target DNA, the light-scattering (LS) signal was intense.[80] Conversely, if complementary DNA (T1) was added, sandwich hybridisation was formed, with the hybrid of P1-P2 (in the form of a double strand)-T1 as filling material (similar to Fig. 5.21b). As a consequence, the LS signal decreased remarkably. Several advantages are associated with this approach: First of all, MP-P1 hybrids could be reused at least 17 times, indicating the high stability and reproducibility of the system. Moreover, since CNT-DNA complexes could disperse in an aqueous medium and have strong LS signals in the UV-vis region, the proposed method did not involve a visual recognition element such as fluorescent/chemiluminescent labels. Therefore, it represents an interesting option for the detection of target DNA sequences.

Besides the identification of target molecules, Kaxiras *et al.* recently proposed an ultrafast and effective DNA sequencing method. This was applied to either a periodic dsDNA (with a nanotube fitting the major groove of the DNA)[81] or an ssDNA, with a resolution of 2 Å.[82] In the first case, the DNA sequence was held fixed, while the CNT, which was rather stiff and did not deform significantly, was allowed to dock to the DNA at the major groove. Conversely, in their second investigation, the authors applied a force (F) on a bead attached to one end of ssDNA in close proximity to CNT (Fig. 5.22a). Therefore, by pulling the ssDNA fragment, the bases along it successively interacted with the CNT, providing a signal that could be measured by a probe sensitive to local electronic states, such as a scanning tunneling spectroscopy (STS) tip. The interactions between the nucleosides and the nanotube are mainly van der Waals forces and mutual polarisation when each base approaches the tube. To maximise the sensitivity of the measurements, a semiconducting (10,0) CNT was used as the substrate, while only the most stable configurations of the bases (shown in Fig. 5.22b and corresponding to 65% of the total) were taken into consideration. The nucleoside bound on the CNT through its base unit was located 3.3 Å away from the CNT's wall, while the sugar residue was more flexible. The results depicted with the quantum approach are illustrated in Fig. 5.23: the interaction mainly involved the π orbitals of the base atoms, especially the NH_2 group at its end, and the π orbitals of the C atoms in the CNT. Upon adsorption, the base plane was positively charged, with electron accumulation (near the base) and depletion near the CNT, determining a net charge transfer of 0.017 e from the adenine base to the CNT. Similar trends were experienced by the other bases, so the STS tips could easily detect their characteristic signals.

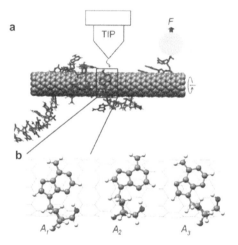

Figure 5.22. (a) Proposed experimental setup for single base measurement: an ssDNA fragment is in partial contact with the CNT and is being pulled at one end. (b) Representative optimal structures of adenine on the (10,0) CNT. The gray, blue, red and white balls represent C, N, O and H atoms, respectively. Reproduced from Meng *et al.*[82] with permission. See also Colour Insert.

Figure 5.23 Isodensity surface of the charge density difference for adenine–CNT. The charge density difference is calculated by subtracting the charge density of the individual adenine (A) and CNT systems, each fixed at its respective position when it is part of the A-CNT complex. Electron accumulation–depletion regions are shown in blue (+) and red (−). Reproduced from Meng *et al.*[82] with permission. See also Colour Insert.

As regards nanoelectrode platform for biosensor development, a vertically aligned CNT array was fabricated by Nguyen's group at the NASA research centre.[83] In this study, the initial treatment in strong acidic conditions, aimed to remove the metal catalyst and introduce carboxylated functions at the opened ends of CNTs, was found to be detrimental for both stability and alignment of the tubes. In order to overcome such limitation, a spin-on glass (SOG) film was deposited between individual CNTs (Fig. 5.24). Pre-treatment with SOG provided structural support to CNTs, retaining their vertical configuration during the subsequent oxidising treatment. The –COOH groups at the tips of CNTs were then activated through (3-dimethylaminopropyl) carbodiimide hydrochloride (EDC), forming a highly reactive *o*-acylisourea active intermediate, which, in the presence of excess of *N*-hydroxysulfo-succinimide (sulfo-NHS), gave a more water-stable sulfosuccinimidyl derivative. Finally, the intermediate underwent a nucleophilic attack by the free amino group of a fluorescence-labelled oligonucleotide, originating the final peptide bond.[84] It is worth mentioning that conjugation of nucleic acid did not occur on the SOG surface in the absence of CNTs, thus confirming the specificity of the interaction and the improved chemical coupling of nucleic acids. Two main advantages are associated with this study: First, the array was made up of CNTs with uniform length, and thus it overcame the inhomogeneous distribution usually encountered with suspended tubes. Second, the addition of SOG rendered the CNTs' interface hydrophilic and thus more suitable for coupling chemistry in aqueous media.

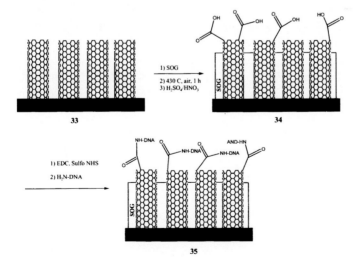

Figure 5.24 Fabrication and pre-treatment of carbon nanoelectrode arrays with SOG for functionalisation with DNA.

As an alternative to the immobilisation of CNTs and their subsequent chemical ligation to DNA, Woolley suggested a different approach, consisting of alignment of dsDNA on a Si substrate and a subsequent treatment with a bifunctional bridging compound (1-pyrenemethylamine hydrochloride [PMA]) and a final incubation with acid-purified SWCNTs.[85] The amine group in PMA was expected to interact electrostatically with the negatively charged phosphate backbone of DNA, while the aromatic pyrenyl group was reported to interact strongly with the surfaces of SWCNTs through π-stacking forces.[86] Although this work presented the potential to facilitate the construction of ordered arrays, it also showed the limitation of ~5% of the total DNA length covered with specifically aligned SWCNTs and about 60% of all SWCNTs deposited on DNA fragments, thus suggesting that the methodology still needs to be optimised.

5.4 CNT–NUCLEIC ACID COMPLEXES FOR GENE DELIVERY AND SELECTIVE CANCER TREATMENT

5.4.1 CNT-DNA complexes for gene therapy

CNT-DNA complexes might become useful tool for gene delivery, provided (i) DNA is condensed by the nanotube, (ii) it is transported inside the cell where it should be delivered and finally (iii) it enters the nucleus prior to transgene expression. Therefore, it is imperative to evaluate the physicochemical properties of the complexes in order to achieve the best results. The formation of such complexes between positive and negative charges was the basis of the experiments performed by Pantarotto *et al.*, who investigated the effect of three cationic nanotubes (Fig. 5.25) on gene expression[87]: in particular, SWCNTs as mono- (SWCNT-NH$_3^+$, compound **36**) or bis-adducts (SWCNT-Lys-NH$_3^+$, compound **37**) and cationic MWCNTs (MWCNT- NH$_3^+$, compound **38**)

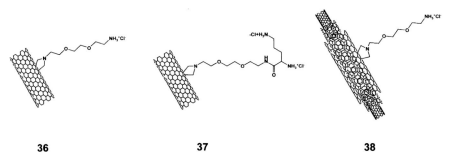

36   **37**   **38**

Figure 5.25. Structure of cationic CNTs: SWCNT-NH$_3^+$ (**36**); SWCNT-Lys-NH$_3^+$ (**37**); MWCNT-NH$_3^+$ (**38**).

were combined with plasmid DNA electrostatically. Their diversified nature in terms of surface functionalisation and charges allowed for discriminating the influencing factors in gene transfection: in fact SWCNT-NH$_3$$^+$ and SWCNT-Lys-NH$_3$$^+$ shared the same dimensions and exposed surface, while SWCNT-Lys-NH$_3$$^+$ had a similar surface charge load as MWCNT- NH$_3$$^+$.

Scanning electron microscopy (SEM) proved very useful in studying the supramolecular lattice formed when functionalised CNTs were exposed to different ratios of plasmid. Images at the microscope showed that parallel bundles of nanotubes formed a framework to which condensed packets of DNA adhered. Interestingly, the DNA conjugated to MWCNTs (Fig. 5.26a–c) appeared more stretched than in SWCNTs, where smaller aggregate particles were formed (Fig. 5.26d–f). However, the risk is that if DNA is too tightly complexed, it may be unable to detach from the nanotube, thereby leading to compromised gene expression. As a proof of that, SWCNT-NH$_3$$^+$ appeared more efficient at gene transfer when complexed to DNA at an 8:1 charge

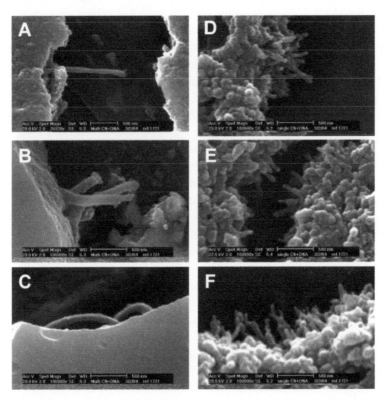

Figure 5.26. SEM images of CNT-DNA complexes formed at a 6:1 charge ratio: (a–c) MWNT-NH$_3$$^+$:DNA; (d–f) SWNT-NH$_3$$^+$:DNA. Reproduced from Singh *et al.*[87] with permission.

ratio, while SWNT-Lys-NH$_3^+$ appeared most efficient at a 1:1 charge ratio. Conversely, in the case of MWCNT-NH$_3^+$, being highly condensed (95%) DNA, the tubes maintained similar levels of transfection efficiency across all charge ratios. Finally, it was noticed that a degree of strong electrostatic interaction was necessary to avoid dissociation on dilution or competition with other molecular species (e.g., blood components) interacting with the CNT-DNA complex.

In addition, Cai *et al.* developed an efficient delivery technique, called nanotube spearing, by which plasmid DNA immobilised onto nickel-embedded nanotubes was speared into targeted cells through the application of an external magnetic field.[88] The authors observed that in order to respond to the applied magnetic field, CNTs needed to be less than 2 µm long. The mechanism behind this technique is nano-penetration of the cell membrane, which determines an efficient molecular delivery of plasmid DNA into splenic B cells, lymphocytes and neurons. Although the investigation was specific for plasmid DNA, the technique showed high versatility, since it could be exploited for the delivery of proteins and RNA segments as well. Moreover, in contrast with previously reported studies, nanotube spearing required very low concentrations of CNTs for an ideal transduction, mainly in the range of 100 fm of tubes. Similarly, it is envisaged that the amount of DNA, which was kept 10^3 times higher than CNTs, could be minimised in future experiments. On the whole, a successful transduction of plasmid DNA into non-dividing (B cells and neurons) and dividing cells was achieved with values comparable to viral vectors, but with the additional advantage of lack of immunogenicity due to this promising nanotube spearing.

The latest paper using plasmid DNA for gene delivery was published by Richard *et al.*,[89] who obtained an enhanced cell transfection *in vitro* when plasmid DNA was conjugated to CNTs previously functionalised with cationic amphiphilic molecules (Fig. 5.27). In fact the amphiphile adsorbed on the CNTs created positive charges on their surface, thus preventing aggregation in biological media.

Although it could be argued that the addition of DNA to cationic amphiphiles onto the CNTs could possibly detach the cationic molecules from the surface, light scattering of the supernatant did not detect any suspended particle, thus indicating that the complexes formed were stable. Moreover, it was observed that the efficiency of transfection of functionalised MWCNTs was higher than that of naked DNA, but much lower than that of functionalised SWCNTs. It is highly possible that the larger-size complexes could be responsible for better internalisation into the cells and thus improved transfection. More precisely, the efficiency of transfection was 100 times higher when using SWCNTs functionalised by the lipid instead of pyrenyl polyamine, because the addition

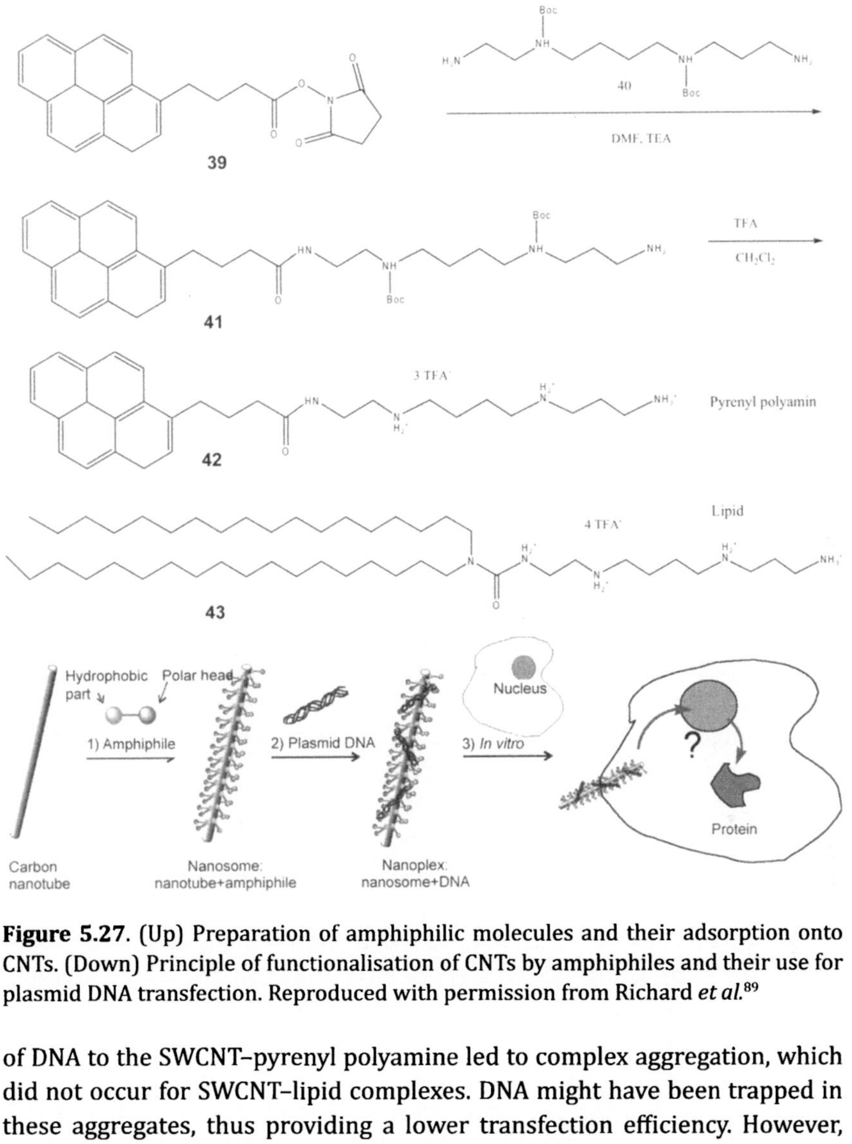

Figure 5.27. (Up) Preparation of amphiphilic molecules and their adsorption onto CNTs. (Down) Principle of functionalisation of CNTs by amphiphiles and their use for plasmid DNA transfection. Reproduced with permission from Richard *et al.*[89]

of DNA to the SWCNT–pyrenyl polyamine led to complex aggregation, which did not occur for SWCNT–lipid complexes. DNA might have been trapped in these aggregates, thus providing a lower transfection efficiency. However, one big limitation is still the possible equilibrium between free and adsorbed molecules. To overcome this problem, the authors suggested the use of other amphiphilic molecules, bearing a diacetylene group in the aliphatic chain. Such molecules, once adsorbed onto the CNTs, could undergo UV irradiation and subsequent polymerisation, stabilising the complex in such a way that the excess of lipid could be removed without removing the lipid from the CNT surface.

5.4.2 CNT-DNA complexes for cancer therapy

As clearly reported throughout this chapter, CNT-DNA complexes present several interesting properties which could be applied not only in biochemical sensors or gene therapy but also in cancer treatment.

A common strategy consists of incorporating folic acid on a system in order to target tumour tissues and trigger selective cancer cell death. As regards this aspect, Ko *et al.* reported an interesting example, by assembling folic acid and DNA into a multifunctional system.[90] The starting material consisted of an ssDNA sequence of 52 bases, with four palindromic segments.[91] Repeated units of ssDNA self-assembled to originate a DNA nanotube (NT) with a diameter of 50–200 nm and a maximum length of 40 µm. To reiterate, no CNTs were complexed with DNA, but only oligonucleotides were assembled into a tubular structure. In order to test such supramolecular architecture for drug delivery, the assemble was non-covalently functionalised with both a fluorescent dye (which represented the drug prototype) and folic acid, so as to target cancer cells on the basis of their high expression of folic acid receptors (FR). More precisely, the free amino groups in DNA were conjugated with folic acid residues in the form of *N*-hydroxysuccinimidic (NHS) esters, thus providing a dual-functionalised DNA-NT. Interestingly, extensive washing and treatment with DNAse I did not reduce the fluorescence of internalised DNA-NT, although its morphology seemed to be affected by the uptake process. In other words, it is possible that the internalisation inside cells required shortening of DNA-NT or changing of the supramolecular structure, but this could not be ascertained because of the limit of the sensitivity (1 µm) of the microscope. However, no sign of cytotoxicity was confirmed even four days after cell incubation with the bulky sample. An important observation was the absence of fluorescence when the dye and the folic acid moiety were not in the same complex and instead formed metastable, single-stranded structures, thus confirming that the self-assembled the DNA-NT composite was a structural requirement for the internalisation process. Finally, it was noticed that the amount of fluorescence was directly correlated with the composition of folic acid, the time of incubation and the concentration of DNA, although there seemed to exist a plateau in terms of both DNA (not specified throughout the article) and folic acid (10%) content.

In a different approach, 700–1,100 nm NIR light was adopted to deliver DNA inside the nucleus of cancer cells, through the strategic use of CNTs as local heaters.[92] Biological systems are known to be completely transparent within this spectral window, while SWCNTs show intense optical absorbance because of the electronic transitions between the first or second van Hove singularities of the nanotubes.[93,94] Such difference can be used to transport bioactive molecules and to selectively kill cancer cells. More precisely, upon exposure of HeLa cells to CNT-DNA solution at 37°C, the SWCNT-DNA conjugates were internalised inside the cytoplasm but not inside the nucleus of cells. On the contrary, when the same experiment was conducted at 4°C, no cellular uptake was observed, thus suggesting endocytosis as the main mechanism involved. However, after NIR radiation (six 10 s on-and-off pulses of 1.4 W/cm^2 laser radiation) confocal imaging revealed co-localisation of fluorescent DNA in the cell nucleus (Fig. 5.28), indicating successful release of DNA from SWCNTs and nuclear translocation after the laser pulses.

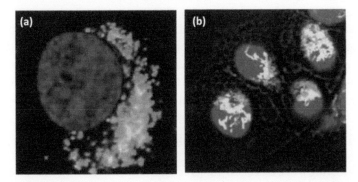

Figure 5.28 Confocal image of *in vitro* HeLa cells after a 12 h incubation in a fluorescent-DNA-SWCNT solution. (a) Dual detection of fluorescent-DNA-SWCNTs (green) internalised into a HeLa cell with the nucleus stained by DRAQ5 (red). (b) Co-localisation (yellow) of fluorescent DNA (green) in cell nucleus (red), after NIR irradiation of 2.5–5 mg/L of SWCNT-DNA, indicating translocation of DNA to the nucleus. Reproduced from Kam *et al.*[92] with]permission. See also Colour Insert.

In order to test whether NIR irradiation was toxic to normal cells, a control experiment was performed on cells under 808 nm laser radiation at 3.5 W/cm^2 power but without exposure to SWCNTs: all cells survived, confirming high transparency of biosystems to NIR light. Conversely, for cells incubated with SWCNTs, extensive cell death was observed after 2 min of radiation under a power of 1.4 W/cm^2. In fact extensive local heating of CNTs inside living cells, caused by continuous NIR absorption, was the most likely origin of cell death. Notably, dead cells "released" SWCNTs to form black aggregates floating in the cell medium solution visible to the naked eye 24 h after irradiation. On the whole, selective NIR radiation triggered cell death without harming normal cells.

The same principle was adopted in the investigation by Gmeiner *et al.*,[95] who used MWCNTs for selective thermal ablation of malignant cells. DNA-encased MWCNTs conferred aqueous solubility to the tubes and produced larger amounts of heat than non-DNA-encased MWCNTs when irradiated under identical conditions. Moreover, for the first time it was demonstrated that such complex was able to eradicate tumour xenografts *in vivo* in a mouse model of human cancer, since complete tumour regression was achieved with a single treatment and without damaging normal tissues. It is important to notice that DNA-encased MWCNTs did not become saturated along the investigation and, on the contrary, could be excited continuously by laser radiation without any reduction in heat generation. In terms of safety, it seems that the presence of DNA incorporated onto the CNTs increased the heating efficiency, thereby suggesting that less material is required to reach the desired effects *in vivo*. Additional advantages are represented by DNA's biocompatibility, which protects from eventual immunogenicity, and the CNT's enhanced dispersibility, which reduces the risks associated with tube aggregation. For all these reasons, CNT-DNA complexes seem to be promising systems that can be applied in several biomedical fields.

5.4.3 CNT-RNA complexes for biomedical applications

An interesting work has been published on small interfering RNA for targeted cancer therapy.[96] The study targeted telomerase, which is an enzyme involved in the stabilisation of chromosomes by the addition of TTAGGG units to the telomere ends.[97] Activation of telomerase has been detected in the majority of malignant tumours, but not in most normal cells. Therefore, small-molecule inhibitors of telomerase activity or knockdown of telomerase expression represents an attractive approach for targeted cancer therapy. In particular, small interfering RNA (siRNA) seems to be a powerful tool to achieve such goals, provided that it is stable and internalized in a sufficiently high amount inside the cells. To that purpose, SWCNTs conjugated with positively charged $-CONH-(CH_2)_6-NH_3^+Cl^-$ were shown to promote the coupling of specific mouse mTERT siRNA to SWCNTs. Different cancer cells, including cervical carcinoma (TC-1), ovarian carcinoma (1H8) and lung carcinoma (LLC) cells (which usually express high levels of mTERT m[messenger]RNA and mTERT proteins), showed suppressed cell growth after incubation with mTERT siRNA-SWNTs+ complex but not with nanotubes alone or mTERT siRNA alone. The mechanism behind this result was attributed to the ability of the complex to knock down mTERT expression, to inhibit cell proliferation and to promote cell senescence *in vitro*. Analogue experiments were subsequently performed to explore the activity with human hTERT and the effects on *in vivo* tumour growth; as expected, injection of hTERT in HeLa cells and mTERT siRNA-SWNT+ into tumour tissue also induced senescence. Noticeably, the authors deduced that because of the rapid response of the cell lines to TERT knockdown, it was possible that other mechanisms were involved in cell growth arrest besides the telomeric shortening, but further studies need to be conducted on this aspect.

An interesting target in gene therapy is represented by cyclin A_2, a protein often over-expressed in many types of cancers, including leukemia. Since it was demonstrated to play a critical role in DNA replication, transcription and cell cycle regulation, it has been hypothesised that its selective inhibition through siRNA can be beneficial against tumour progression. To that purpose, Wang *et al.* condensed ammonium-functionalised SWCNTs with cyclin A_2 siRNA and incubated the complex in human myelogenous leukaemia (K562) cells.[98] Results showed reduced cellular levels through blockage of cell proliferation in S phase as well as promotion of apoptosis. No analogous effects were measured with the controls. Similarly to Dai, the authors demonstrated that CNT transporters can efficiently deliver siRNA into cells and offer remarkable advantages with respect to conventional transfection vectors, which showed little effect in the internalisation of siRNA. On the basis of these encouraging results, together with the evidence that normal cells are less sensitive than transformed cells to siRNA,[99] it could be envisaged that the delivery of siRNA against cyclin A_2 mediated by functionalised SWCNTs is a useful therapeutic strategy for cancer therapy.

A very recent article adopted a similar approach, demonstrating that pegylated CNTs could be coupled with thiol-modified siRNA via sulfosuccinimidyl 6-[3'-(2-pyridyldithio)-propionamido] hexanoate (SPDP-S) to successfully knock down the transient receptor potential channel 3 (TRPC3) involved in insulin-resistant conditions.[100] In the experiments, isolated muscle fibres were transfected with the fluorescent siRNA bond to CNTs. The results obtained disproved the initial hypothesis that inhibition of TRPC3 affected Ca^{2+} influx through such channels. On the contrary, Ca^{2+} influx was not significantly different between muscle fibres cultured with (55±5 nM, n = 9) or without (49±1 nM, n = 9) TRPC3-siRNA for 48 h. However, the knockdown of TRPC3 expression decreased insulin-mediated glucose uptake (Fig. 5.29). It was also suggested that there exist two pools of TRPC3, one constitutively present in the plasma membrane and the other located in the insulin-sensitive glucose transporter 4 (GLUT4)-containing vesicles, which moves to the plasma membrane in response to insulin stimulation. Thus, it was concluded that TRPC3 is a potential target for the treatment of insulin resistance and type II diabetes and that the combination of CNTs with siRNA can represent a valuable therapeutic approach.

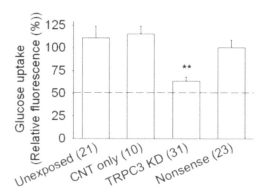

Figure 5.29 Mean ± SE insulin-mediated glucose uptake in isolated muscle fibres. Insulin-mediated glucose uptake (measured with fluorescent glucose) was markedly reduced in TRPC3-knockdown fibres (TRPC3 KD). **P = 0.01. Reproduced from Lanner *et al.*[99] with permission.

Alternatively, Krajcik *et al.* suggested a non-covalent binding of nucleic acid to CNTs to efficiently deliver siRNA inside the cells.[101] In their investigation, aimed to inhibit the extracellular signal-regulated kinases 1 and 2 (ERK1/ERK2) in order to improve contractile function in reconstituted heart tissue, the authors used SWCNTs functionalised with hexamethylenediamine (HMDA) and poly(diallyldimethylammonium) chloride (PDDA) and further complexed them with unmodified siRNA. Once inside the cells, the siRNA sequence was released and induced efficient silencing of target genes (ERK1/ERK2), showing lower cytotoxic side effects than liposomal transfection. It is important to notice that siRNA alone was not able to enter the cells, thus confirming the importance of CNTs as siRNA carriers for a successful gene therapy of heart failure.[102] In the experiments, ERK1 and ERK2 were suppressed by nearly 80%, while maximum silencing results were achieved with a 1:1 PDDA-SWCNT mixture.

It has been also shown that the delivery of siRNA to human T cells to silence the expression of the human immunodeficiency virus (HIV)-specific cell-surface receptors CD4 and/or co-receptors CXCR4/CCR5 can block HIV entry and thus reduce infection.[103-105] However, certain cells are still difficult to transfect by non-viral agents, as in the case of liposomes that have been shown to be incapable of siRNA delivery into T cells.[103] On the contrary, Dai *et al.* demonstrated that SWCNTs were capable of siRNA delivery to afford efficient RNA interference of CXCR4 and CD4 receptors on human T cells through the use of functionalised CNTs.[106] CNTs were initially solubilised through non-covalent adsorption of phospholipids (PLs) and coupled with amine-terminated polyethylene glycol (PEG; PL-PEG$_{2000}$-NH2) (Fig. 5.30).[107]

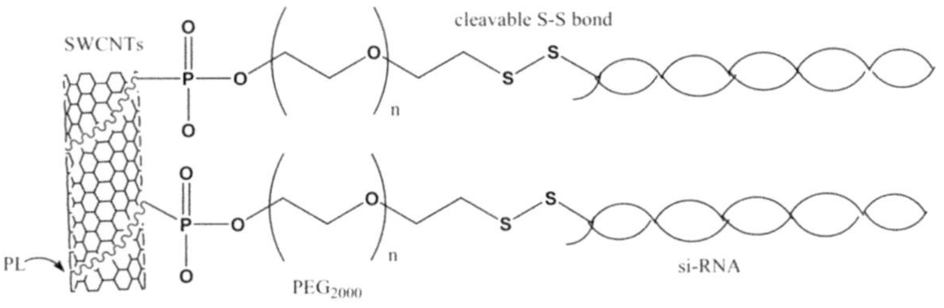

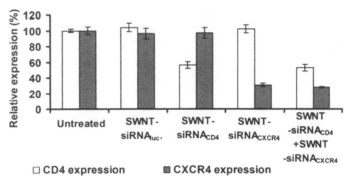

44

Figure 5.30. Functionalisation of SWCNTs with phospholipids (PLs), polyethylene glycol (PEG$_{2000}$) and siRNA through a cleavable S–S bond.

Then RNA was modified with a thiol group to produce cleavable disulfide bonds, and results showed about 90% of silencing efficiency upon incubation for 3 days of T cells with the CNT complexes carrying siRNA$_{CXCR4}$. In addition, since both CD4 receptors and CXCR4 co-receptors are required for HIV infection of human T cells, the delivery of CD4 siRNA (siRNA$_{CD4}$) into T cells by SWNTs was investigated as well. Interestingly, the down-regulation of CD4 receptors on T cells by treatment with SWNT-siRNA$_{CD4}$ had no non-specific effect on the CXCR4 receptors, and vice versa (Fig. 5.31).

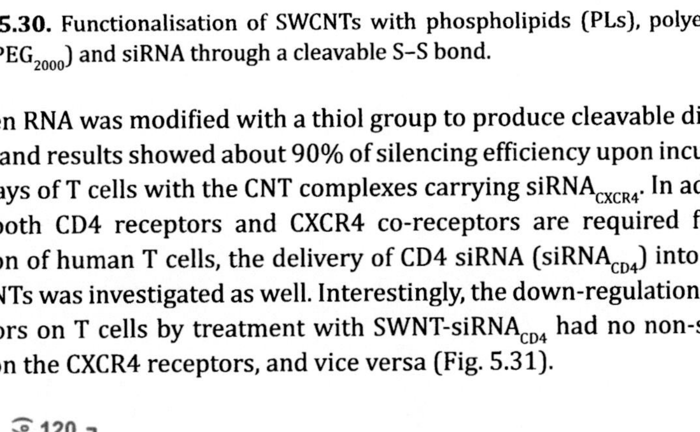

Figure 5.31 CD4 and CXCR4 expression levels on cells treated with an SWNT-siRNA$_{CD4}$ complex. Reproduced from Kam *et al.*[107] with permission.

Another covalent functionalisation of CNTs with siRNA afforded branched structures made up of polyamidoamine (PAMAM) dendrons linked to the surface of MWCNTs.[108] The presence of PAMAM dendrons guaranteed the solubility of the tubes in aqueous media, while the progressive synthetic pathway provided increasing peripheral primary amino groups, with the highest number associated with the second generation of dendrons.

Initially, pristine MWCNTs underwent the 1,3-dipolar cycloaddition of azomethine ylides, forming the zeroth-generation (G0) dendron-MWCNTs. This was followed by incorporation of ethylenediamine and methyl acrylate in a divergent approach (Fig. 5.32), thus deriving generations G1 and G2.

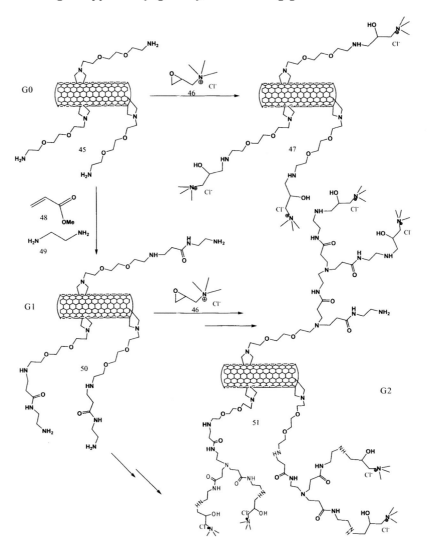

Figure 5.32 Schematic representation of MWCNT-PAMAM dendrons.

Accordingly, a progressive increase in the tubes' diameter was observed as the dendritic structure grew. The fluorescently labelled siRNA was complexed with each conjugate of the synthesised dendron-MWCNT series at a 1:16 mass ratio on the basis of previous optimisation.[109] Interestingly, the cellular internalisation of the siRNA was always nanotube-dependent, while almost no uptake of siRNA alone was observed: in other words, the higher the extent of the branching, the more efficient the delivery of siRNA, thus suggesting promising applications in the field of nanomedicine.

References

1. Illangasekare, M., Sanchez, G., Nickles, T., and Yarus, M. (1995) Aminoacyl RNA synthesis catalyzed by an RNA, *Science*, **267**, 643–647.

2. Lohse, P. A. and Szostak, J. W., (1996) Ribozyme-catalysed amino-acid transfer reactions, *Nature*, **381**, 442–444.

3. Tarasow, T. M., Tarasow, S. L. and Eaton, B. E., (1997) RNA-catalysed carbon-carbon bond formation, *Nature*, **389**, 54–57.

4. Zhang, B. and Cech, T. R., (1997) Peptide bond formation by in vitro selected ribozymes, *Nature*, **390**, 96–100.

5. Gugliotti, L. A., Feldheim, D. L., and Eaton, B. E. (2004) RNA-mediated metal-metal bond formation in the synthesis of hexagonal palladium nanoparticles, *Science*, **304**, 850–852.

6. Alpatova, A. L., Shan, W., Babica, P., Upham, B. L., Rogensues, A. R., Masten, S. J., Drown, E., Mohanty, A. K., *et al.* (2010) Single-walled carbon nanotubes dispersed in aqueous media via non-covalent functionalization: Effect of dispersant on the stability, cytotoxicity, and epigenetic toxicity of nanotube suspensions, *Water Res.*, **44**, 505–520.

7. Prato, M., Kostarelos, K., and Bianco, A. (2008) Functionalized carbon nanotubes in drug design and discovery, *Acc. Chem. Res.*, **41**, 60–68.

8. Yogeswaran, U., and Chen, S. M. (2007) Separation and concentration effect of f-MWCNTs on electrocatalytic responses of ascorbic acid, dopamine and uric acid at f-MWCNTs incorporated with poly (neutral red) composite films, *Electrochim. Acta*, **52**, 5985–5996.

9. Wu, G., Chen, Y. S., and Xu, B. Q. (2005) Remarkable support effect of SWNTs in Pt catalyst for methanol electrooxidation, *Electrochem. Commun.*, **7**, 1237–1243.

10. Wang, J., and Musameh, M. (2004) Electrochemical detection of trace insulin at carbon-nanotube-modified electrodes, *Anal. Chim. Acta*, **511**, 33–36.

11. Wang, J., Li, M., Shi, Z., Li, N., and Gu, Z. (2001) Electrocatalytic oxidation of 3,4-dihydroxyphenylacetic acid at a glassy carbon electrode modified with single-wall carbon nanotubes, *Electrochim. Acta*, **47**, 651–657.

12. Zheng, M., Jagota, A., Semke, E. D., Diner, B. A., McLean, R. S., Lustig, S. R., Richardson, R. E., Tassi, N. G., *et al.* (2003) DNA-assisted dispersion and separation of carbon nanotubes, *Nat. Mater.*, **2**, 338–342.

13. Varghese, N., Mogera, U., Govindaraj, A., Das, A., Maiti, P. K., Sood, A. K., and Rao, C. N. R. (2008) Binding of DNA nucleobases and nucleosides with graphene, *ChemPhysChem*, **10**, 206–210.

14. Zheng, M., Jagota, A., Strano, M. S., Santos, A. P., Barone, P., Chou, S. G., Diner, B. A., Dresselhaus, M. S., *et al.* (2003) Structure-based carbon nanotube sorting by sequence-dependent DNA assembly, *Science*, **302**, 1545–1548.

15. Lustig, S. R., Jagota, A., Khripin, C., and Zheng, M. (2005) Theory of structure-based carbon nanotube separations by ion-exchange chromatography of DNA/CNT hybrids, *J. Phys. Chem. B*, **109**, 2559–2566.

16. Tu, X., Manohar, S., Jagota, A., and Zheng, M. (2009) DNA sequence motifs for structure-specific recognition and separation of carbon nanotubes, *Nature*, **460**, 250–253.

17. Huang, X., Mclean, R. S., and Zheng, M. (2005) High-resolution length sorting and purification of DNA-wrapped carbon nanotubes by size-exclusion chromatography, *Anal. Chem.*, 77, 6225–6228.

18. Kim, S. N., Kuang, Z., Grote, J. G., Farmer, B. L., and Naik, R. R. (2008) Enrichment of (6,5) single wall carbon nanotubes using genomic DNA, *Nano. Lett.*, **8**, 4415–4420.

19. Kuzmany, H., Plank, W., Hulman, M., Kramberger, C., Grüneis, A., Pichler, T., Peterlik, H., Kataura, H., *et al.* (2001) Determination of SWCNT diameters from the Raman response of the radial breathing mode, *Eur. Phys. J. B*, **22**, 307–320.

20. Jeng, E. S., Barone, P. W., Nelson, J. D., and Strano, M. S. (2007) Hybridization kinetics and thermodynamics of DNA adsorbed to individually dispersed single-walled carbon nanotubes, *Small*, **3**, 1602–1609.

21. Zhao, X., and Johnson, J. K. (2007) Simulation of adsorption of DNA on carbon nanotubes, *J. Am. Chem. Soc.*, **129**, 10438–10445.

22. Zheng, M., Jagota, A., Semke, E. D., Diner, B. A., Mclean, R. S., Lustig, S. R., and Tassi, R. E. R. &. N. G. (2003) DNA-assisted dispersion and separation of carbon nanotubes, *Nat. Mater.*, **2**, 338–342.

23. Zheng, M., Jagota, A., Strano, M. S., Santos, A. P., Barone, P., Chou, S. G., Diner, B. A., Dresselhaus, M. S., *et al.* (2003) Structure-Based Carbon Nanotube Sorting by Sequence-Dependent DNA Assembly, *Science*, **302**, 1545–1548.

24. Zheng, M., and Diner, B. (2004) Solution redox chemistry of carbon nanotubes, *J. Am. Chem. Soc.*, **126**, 15490–15494.

25. Strano, Zheng, M., Jagota, A., Onoa, G., Heller, D., Barone, P., and Usrey, M. (2004) Understanding the nature of the DNA-assisted separation of single-walled carbon nanotubes using fluorescence and raman spectroscopy, *Nano Lett.*, **4**, 543–550.

26. Li, J., Ng, H., Cassell, A., Fan, W., Chen, H., Ye, Q., Koehne, J., Han, J., *et al.* (2003) Carbon nanotube nanoelectrode array for ultrasensitive DNA detection, *Nano Lett.*, **3**, 597–602.

27. He, P., and Dai, L. (2004) Aligned carbon nanotube-DNA electrochemical sensors, *Chem. Commun.*, **10**, 348–349.

28. Moghaddam, M., Taylor, S., Gao, M., Huang, S., Dai, L., and McCall, M. (2004) Highly efficient binding of DNA on the sidewalls and tips of carbon nanotubes using photochemistry, *Nano Lett.*, **4**, 89–93.

29. Li, S., He, P., Dong, J., Guo, Z., and Dai, L. (2005) DNA-directed self-assembling of carbon nanotubes, *J. Am. Chem. Soc.*, **127**, 14–15.

30. Williams, K. A., Veenhuizen, P. T. M., Torre, B. G. D. L., and Dekker, R. E. &. C. (2002) Nanotechnology:Carbon nanotubes with DNA recognition, *Nature*, **420**, 761.

31. Rajendra, J., and Rodger, A. (2005) The binding of single-stranded DNA and PNA to single-walled carbon nanotubes probed by flow linear dichroism, *Chemistry*, **11**, 4841–4847.

32. Rajendra, J., Baxendale, M., Rap, L., and Rodger, A. (2004) Flow linear dichroism to probe binding of aromatic molecules and DNA to single-walled carbon nanotubes, *J. Am. Chem. Soc.*, **126**, 11182–11188.

33. Chen, Y., Yu, L., Feng, X. Z., Hou, S., and Liu, Y. (2009) Construction, DNA wrapping and cleavage of a carbon nanotube-polypseudorotaxane conjugate, *Chem. Commun (Camb.)*, 4106–4108.

34. Liu, Y., Liang, P., Chen, Y., Zhao, Y. L., Ding, F., and Yu, A. (2005) Spectrophotometric study of fluorescence sensing and selective binding of biochemical substrates by 2,2'-bridged bis(beta-cyclodextrin) and its water-soluble fullerene conjugate, *J. Phys. Chem. B*, **109**, 23739–23744.

35. Liu, Y., Wang, H., Chen, Y., Ke, C. F., and Liu, M. (2005) Supramolecular aggregates constructed from gold nanoparticles and l-try-CD polypseudorotaxanes as captors for fullerenes, *J. Am. Chem. Soc.*, **127**, 657–666.

36. Liu, Y., Wang, H., Liang, P., and Zhang, H. Y. (2004) Water-soluble supramolecular fullerene assembly mediated by metallobridged-cyclodextrins, *Angew. Chem. Int. Ed.*, **43**, 2690–2694.

37. Liu, Y., Yu, Z. L., Zhang, Y. M., Guo, D. S., and Liu, Y. P. (2008) Supramolecular architectures of beta-cyclodextrin-modified chitosan and pyrene derivatives mediated by carbon nanotubes and their DNA condensation, *J. Am. Chem. Soc.*, **130**, 10431–10439.

38. Shie, J. W., Yogeswaran, U., and Chen, S. M. (2008) Electroanalytical properties of cytochrome c by direct electrochemistry on multi-walled carbon nanotubes incorporated with DNA biocomposite film, *Talanta*, **74**, 1659–1669.

39. Daniel, S., Rao, T. P., Rao, K. S., Rani, S. U., Naidu, G., Lee, H. Y., and Kawai, T. (2007) A review of DNA functionalized/grafted carbon nanotubes and their characterization, *Sens. Actuators B: Chem.*, **122**, 672–682.

40. Chen, S. M., and Chen, S. V. (2003) The bioelectrocatalytic properties of cytochrome C by direct electrochemistry on DNA film modified electrode, *Electrochim. Acta*, **48**, 513–529.

41. Zhou, Z., Kang, H., Clarke, M. L., Lacerda, S. H. D. P., Zhao, M., Fagan, J. A., Shapiro, A., Nguyen, T., *et al.* (2009) Water-soluble DNA-wrapped single-walled carbon-nanotube/quantum-dot complexes, *Small*, **5**, 2149–2155.

42. Singh, P., Kumar, J., Toma, F. M., Raya, J., Prato, M., Fabre, B., Verma, S., Bianco, A., *et al.* (2009) Synthesis and characterization of nucleobase-carbon nanotube hybrids, *J. Am. Chem. Soc.*, **131**, 13555–13562.

43. Sivakova, S., and Rowan, S. J. (2005) Nucleobases as supramolecular motifs, *Chem. Soc. Rev.*, **34**, 9–21.

44. Jatsch, A., Kopyshev, A., Mena-Osteritz, E., and Bäuerle, P. (2008) Self-organizing oligothiophene-nucleoside conjugates: versatile synthesis via "click"-chemistry, *Org. Lett.*, **10**, 961–964.

45. Piana, S., and Bilic, A. (2006) The nature of the adsorption of nucleobases on the gold [111] surface, *J. Phys. Chem. B*, **110**, 23467–23471.

46. Heckl, W. M., Smith, D. P., Binnig, G., Klagges, H., Hänsch, T. W., and Maddocks, J. (1991) Two-dimensional ordering of the DNA base guanine observed by scanning tunneling microscopy, *Proc. Natl. Acad. Sci. U S A*, **88**, 8003–8005.

47. Xu, S., Dong, M., Rauls, E., Otero, R., Linderoth, T. R., and Besenbacher, F. (2006) Coadsorption of guanine and cytosine on graphite: ordered structure based on GC pairing, *Nano Lett.*, **6**, 1434–1438.

48. Fathalla, M., Lawrence, C. M., Zhang, N., Sessler, J. L., and Jayawickramarajah, J. (2009) Base-pairing mediated non-covalent polymers, *Chem. Soc. Rev.*, **38**, 1608–1620.

49. Xu, W., Kelly, R. E. A., Otero, R., Schöck, M., Lægsgaard, E., Stensgaard, I., Kantorovich, L. N., Besenbacher, F., *et al.* (2007) Probing the hierarchy of thymine-thymine interactions in self-assembled structures by manipulation with scanning tunneling microscopy, *Small*, **3**, 2011–2014.

50. Mamdouh, W., Kelly, R. E. A., Dong, M., Kantorovich, L. N., and Besenbacher, F. (2008) Two-dimensional supramolecular nanopatterns formed by the coadsorption of guanine and uracil at the liquid/solid interface, *J. Am. Chem. Soc.*, **130**, 695–702.

51. Gottarelli, G., Masiero, S., Mezzina, E., Pieraccini, S., Rabe, J. P., Samori, P., and Spada, G. P. (2000) The self-assembly of lipophilic guanosine derivatives in solution and on solid surfaces, *Chemistry*, **6**, 3242–3248.

52. Kelly, R., and Kantorovich, L. (2006) Planar nucleic acid base super-structures, *J. Mater. Chem.*, **16**, 1894–1905.

53. Kumar, A. M. S., Fox, J. D., Buerkle, L. E., Marchant, R. E., and Rowan, S. J. (2009) Effect of monomer structure and solvent on the growth of supramolecular nanoassemblies on a graphite surface, *Langmuir*, **25**, 653–656.

54. Kumar, A. M. S., Sivakova, S., Fox, J. D., Green, J. E., Marchant, R. E., and Rowan, S. J. (2008) Molecular engineering of supramolecular scaffold coatings that can reduce static platelet adhesion, *J. Am. Chem. Soc.*, **130**, 1466–1476.

55. Bestel, I., Campins, N., Marchenko, A., Fichou, D., Grinstaff, M. W., and Barthélémy, P. (2008) Two-dimensional self-assembly and complementary base-pairing between amphiphile nucleotides on graphite, *J. Colloid Interface Sci.*, **323**, 435–440.

56. Sarah E. Baker, Wei Cai, Tami L. Lasseter, Kevin P. Weidkamp, and Robert J. Hamers (2002) Covalently bonded adducts of deoxyribonucleic acid (DNA) oligonucleotides with single-wall carbon nanotubes: synthesis and hybridization, *Nano Lett.*, **2**, 1413–1417.

57. Robinson, H., and Wang, A. H. (1996) Neomycin, spermine and hexaamminecobalt (III) share common structural motifs in converting B- to A-DNA, *Nucleic Acids Res.*, **24**, 676–682.

58. Tolstorukov, M. Y., Ivanov, V. I., Malenkov, G. G., Jernigan, R. L., and Zhurkin, V. B. (2001) Sequence-dependent B↔A transition in DNA evaluated with dimeric and trimeric scales, *Biophys. J.*, **81**, 3409–3421.

59. Jose, D., and Porschke, D. (2004) Dynamics of the B-A transition of DNA double helices, *Nucleic Acids Res.*, **32**, 2251–2258.

60. Kam, N. W. S., Liu, Z., and Dai, H. (2006) Carbon nanotubes as intracellular transporters for proteins and DNA: an investigation of the uptake mechanism and pathway, *Angew. Chem. Int. Ed.*, **45**, 577–581.

61. Pantarotto, D., Briand, J. P., Prato, M., and Bianco, A. (2004) Translocation of bioactive peptides across cell membranes by carbon nanotubes, *Chem. Commun. (Camb.)*, 16–17.

62. Becker, M., Fagan, J., Gallant, N., Bauer, B., Bajpai, V., Hobbie, E., Lacerda, S., Migler, K., *et al.* (2007) Length-dependent uptake of DNA-wrapped single-walled carbon nanotubes, *Adv. Mater.*, **19**, 939–945.

63. Donnelly, J., Berry, K., and Ulmer, J. B. (2003) Technical and regulatory hurdles for DNA vaccines, *Int. J. Parasitol.*, **33**, 457–467.

64. Karikó, K., Buckstein, M., Ni, H., and Weissman, D. (2005) Suppression of RNA Recognition by Toll-like Receptors: The Impact of Nucleoside Modification and the Evolutionary Origin of RNA, *Immunity*, **23**, 165–175.

65. Rao, R., Lee, J., Lu, Q., Keskar, G., Freedman, K., Floyd, W., Rao, A., Ke, P., *et al.* (2004) Single-molecule fluorescence microscopy and Raman spectroscopy studies of RNA bound carbon nanotubes, *Appl. Phys. Lett.*, **85**, 4228–4230.

66. Lu, Q., Moore, J., Huang, G., Mount, A., Rao, A., Larcom, L., and Ke, P. (2004) RNA polymer translocation with single-walled carbon nanotubes, *Nano Lett.*, **4**, 2473–2477.

67. Salman, H., Zbaida, D., Rabin, Y., Chatenay, D., and Elbaum, M. (2001) Kinetics and mechanism of DNA uptake into the cell nucleus, *Proc. Natl. Acad. Sci. U S A*, **98**, 7247–7252.

68. Jeynes, J., Mendoza, E., Chow, D., Watts, P., McFadden, J., and Silva, S. (2006) Generation of chemically unmodified pure single-walled carbon nanotubes by solubilizing with RNA and treatment with ribonuclease A, *Adv. Mater.*, **18**, 1598–1602.

69. Qi, P., Vermesh, O., Grecu, M., Javey, A., Wang, Q., Dai, H., Peng, S., Cho, K., *et al.* (2003) Toward large arrays of multiplex functionalized carbon nanotube sensors for highly sensitive and selective molecular detection, *Nano Lett.*, **3**, 347–351.

70. Staii, C., Johnson, A. T., Chen, M., and Gelperin, A. (2005) DNA-decorated carbon nanotubes for chemical sensing, *Nano Lett.*, **5**, 1774–1778.

71. Wu, Z., Zhen, Z., Jiang, J. H., Shen, G. L., and Yu, R. Q. (2009) Terminal protection of small-molecule-linked DNA for sensitive electrochemical detection of protein binding via selective carbon nanotube assembly, *J. Am. Chem. Soc.*, **131**, 12325–12332.

72. Kam, N., O'Connell, M., Wisdom, J., and Dai, H. (2005) Carbon nanotubes as multifunctional biological transporters and near-infrared agents for selective cancer cell destruction, *Proc. Natl. Acad. Sci. USA*, **102**, 11600–11605.

73. Henne, W., Doorneweerd, D., Lee, J., Low, P., and Savran, C. (2006) Detection of folate binding protein with enhanced sensitivity using a functionalized quartz crystal microbalance sensor, *Anal. Chem.*, **78**, 4880–4884.

74. Acharya, G., Chang, C. L., Doorneweerd, D., Vlashi, E., Henne, W., Hartmann, L., Low, P., Savran, C., *et al.* (2007) Immunomagnetic diffractometry for detection of diagnostic serum markers, *J. Am. Chem. Soc.*, **129**, 15824–15829.

75. Yang, R., Tang, Z., Yan, J., Kang, H., Kim, Y., Zhu, Z., and Tan, W. (2008) Noncovalent assembly of carbon nanotubes and single-stranded DNA: an effective sensing platform for probing biomolecular interactions, *Anal. Chem.*, **80**, 7408–7413.

76. Jhaveri, S., Kirby, R., Conrad, R., Maglott, E., Bowser, M., Kennedy, R., Glick, G., Ellington, A., *et al.* (2000) Designed signaling aptamers that transduce molecular recognition to changes in fluorescence intensity, *J. Am. Chem. Soc.*, **122**, 2469–2473.

77. Nutiu, R., and Li, Y. (2005) Aptamers with fluorescence-signaling properties, *Methods*, **37**, 16–25.

78. Lee, A., Ye, J. S., Tan, S., Poenar, D., Sheu, F. S., Heng, C. and Lim, T. (2007) Carbon nanotube-based labels for highly sensitive colorimetric and aggregation-based visual detection of nucleic acids, *Nanotechnology*, **18**, 455102–455120.

79. Hu, P., Huang, C. Z., Li, Y. F., Ling, J., Liu, Y. L., Fei, L. R., and Xie, J. P. (2008) Magnetic particle-based sandwich sensor with DNA-modified carbon nanotubes as recognition elements for detection of DNA hybridization, *Anal. Chem.*, **80**, 1819–1823.

80. Cathcart, H., Nicolosi, V., Hughes, J. M., Blau, W. J., Kelly, J. M., Quinn, S. J., and Coleman, J. N. (2008) Ordered DNA wrapping switches on luminescence in single-walled nanotube dispersions, *J. Am. Chem. Soc.*, **130**, 12734–12744.

81. Lu, G., Maragakis, P., and Kaxiras, E. (2005) Carbon nanotube interaction with DNA, *Nano Lett.*, **5**, 897–900.

82. Meng, S., Maragakis, P., Papaloukas, C., and Kaxiras, E. (2007) DNA nucleoside interaction and identification with carbon nanotubes, *Nano Lett.*, **7**, 45–50.

83. Nguyen, C., Delzeit, L., Cassell, A., Li, J., Han, J., and Meyyappan, M. (2002) Preparation of Nucleic Acid Functionalized Carbon Nanotube Arrays, *Nano Lett.*, **2**, 1079–1081.

84. Hermanson, G. T. (ed.). (2008) *Bioconjugate Techniques*, Academic Press, San Diego, CA.

85. Xin, H., and Woolley, A. T. (2003) DNA-templated nanotube localization, *J. Am. Chem. Soc.*, **125**, 8710–8711.

86. Chen, R., Zhang, Y., Wang, D., and Dai, H. (2001) Noncovalent sidewall functionalization of single-walled carbon nanotubes for protein immobilization, *J. Am. Chem. Soc.*, **123**, 3838–3839.

87. Singh, R., Pantarotto, D., McCarthy, D., Chaloin, O., Hoebeke, J., Partidos, C. D., Briand, J. P., Prato, M., *et al.* (2005) Binding and condensation of plasmid DNA onto functionalized carbon nanotubes: toward the construction of nanotube-based gene delivery vectors, *J. Am. Chem. Soc.*, **127**, 4388–4396.

88. Cai, D., Mataraza, J., Qin, Z. H., Huang, Z., Huang, J., Chiles, T., Carnahan, D., Kempa, K., *et al.* (2005) Highly efficient molecular delivery into mammalian cells using carbon nanotube spearing, *Nat. Methods*, **2**, 449–454.

89. Richard, C., Mignet, N., Largeau, C., Escriou, V., Bessodes, M., and Scherman, D. (2009) Functionalization of single- and multi-walled carbon nanotubes with cationic amphiphiles for plasmid DNA complexation and transfection, *Nano Res.*, **2**, 638–647.

90. Ko, S., Liu, H., Chen, Y., and Mao, C. (2008) DNA nanotubes as combinatorial vehicles for cellular delivery, *Biomacromolecules*, **9**, 3039–3043.

91. Liu, H., Chen, Y., He, Y., Ribbe, A. E., and Mao, C. (2006) Approaching the limit: can one DNA oligonucleotide assemble into large nanostructures?, *Angew. Chem. Int. Ed. Engl.*, **45**, 1942–1945.

92. Kam, N. W. S., O'Connell, M., Wisdom, J. A., and Dai, H. (2005) Carbon nanotubes as multifunctional biological transporters and near-infrared agents for selective cancer cell destruction, *Proc. Natl. Acad. Sci. USA*, **102**, 11600–11605.

93. O'Connell, M. J., Bachilo, S. M., Huffman, C. B., Moore, V. C., Strano, M. S., Haroz, E. H., Rialon, K. L., Boul, P. J., *et al.* (2002) Band gap fluorescence from individual single-walled carbon nanotubes, *Science*, **297**, 593–596.

94. Bachilo, S. M., Strano, M. S., Kittrell, C., Hauge, R. H., Smalley, R. E., and Weisman, R. B. (2002) Structure-assigned optical spectra of single-walled carbon nanotubes, *Science*, **298**, 2361–2366.

95. Ghosh, S., Dutta, S., Gomes, E., Carroll, D., D'Agostino, R., Olson, J., Guthold, M., Gmeiner, W. H., *et al.* (2009) Increased heating efficiency and selective thermal ablation of malignant tissue with DNA-encased multiwalled carbon nanotubes, *ACS Nano*, **3**, 2667–2673.

96. Zhang, Z., Yang, X., Zhang, Y., Zeng, B., Wang, S., Zhu, T., Roden, R. B. S., Chen, Y., *et al.* (2006) Delivery of telomerase reverse transcriptase small interfering RNA in complex with positively charged single-walled carbon nanotubes suppresses tumor growth, *Clin. Cancer Res.*, **12**, 4933–4939.

97. Wang, X., Ren, J., and Qu, X. (2008) Targeted RNA interference of cyclin A2 mediated by functionalized single-walled carbon nanotubes induces proliferation arrest and apoptosis in chronic myelogenous leukemia K562 cells, *Chem. Med. Chem.*, **3**, 940–945.

98. Spänkuch-Schmitt, B., Bereiter-Hahn, J., Kaufmann, M., and Strebhardt, K. (2002) Effect of RNA silencing of polo-like kinase-1 (PLK1) on apoptosis and spindle formation in human cancer cells, *J. Natl. Cancer Inst.*, **94**, 1863–1877.

99. Lanner, J., Bruton, J., Assefaw-Redda, Y., Andronache, Z., Zhang, S. J., Severa, D., Zhang, Z. B., Melzer, W., *et al.* (2009) Knockdown of TRPC3 with siRNA coupled to carbon nanotubes results in decreased insulin-mediated glucose uptake in adult skeletal muscle cells, *FASEB J.*, **23**, 1728–1738.

100. Morin, G. B. (1989) The human telomere terminal transferase enzyme is a ribonucleoprotein that synthesizes TTAGGG repeats, *Cell*, **59**, 521–529.

101. Krajcik, R., Jung, A., Hirsch, A., Neuhuber, W., and Zolk, O. (2008) Functionalization of carbon nanotubes enables non-covalent binding and intracellular delivery of small interfering RNA for efficient knock-down of genes, *Biochem. Biophys. Res. Commun.*, **369**, 595–602.

102. Münzel, F., Mühlhäuser, U., Zimmermann, W. H., Didié, M., Schneiderbanger, K., Schubert, P., Engmann, S., Eschenhagen, T., *et al.* (2005) Endothelin-1 and isoprenaline co-stimulation causes contractile failure which is partially reversed by MEK inhibition, *Cardiovasc. Res.*, **68**, 464–474.

103. Novina, C. D., Murray, M. F., Dykxhoorn, D. M., Beresford, P. J., Riess, J., Lee, S. K., Collman, R. G., Lieberman, J., *et al.* (2002) siRNA-directed inhibition of HIV-1 infection, *Nat. Med.*, **8**, 681–686.

104. Qin, X. F., An, D. S., Chen, I. S. Y., and Baltimore, D. (2003) Inhibiting HIV-1 infection in human T cells by lentiviral-mediated delivery of small interfering RNA against CCR5, *Proc. Natl. Acad. Sci. USA*, **100**, 183–188.

105. Martínez, M. A., Gutiérrez, A., Armand-Ugón, M., Blanco, J., Parera, M., Gómez, J., Clotet, B., Esté, J. A., *et al.* (2002) Suppression of chemokine receptor expression by RNA interference allows for inhibition of HIV-1 replication, *AIDS*, **16**, 2385–2390.

106. Liu, Z., Winters, M., Holodniy, M., and Dai, H. (2007) siRNA delivery into human T cells and primary cells with carbon-nanotube transporters, *Angew. Chem. Int. Ed. Engl.*, **46**, 2023–2027.

107. Kam, N., Liu, Z., and Dai, H. (2005) Functionalization of carbon nanotubes via cleavable disulfide bonds for efficient intracellular delivery of siRNA and potent gene silencing, *J. Am. Chem. Soc.*, **127**, 12492–12493.

108. Herrero, M., Toma, F., Al-Jamal, K., Kostarelos, K., Bianco, A., Da Ros, T., Bano, F., Casalis, L., *et al.* (2009) Synthesis and characterization of a carbon nanotube-dendron series for efficient siRNA delivery, *J. Am. Chem. Soc.*, **131**, 9843–9848.

109. Podesta, J., Al-Jamal, K., Herrero, M., Tian, B., Ali-Boucetta, H., Hegde, V., Bianco, A., Prato, M., *et al.* (2009) Antitumor activity and prolonged survival by carbon-nanotube-mediated therapeutic sirna silencing in a human lung xenograft model, *Small*, **5**, 1176–1185.

Chapter 6

BIOMEDICAL APPLICATIONS V: INFLUENCE OF CARBON NANOTUBES IN NEURONAL LIVING NETWORKS

Cécilia Ménard-Moyon

CNRS, Institut de Biologie Moléculaire et Cellulaire,
Laboratoire d'Immunologie et Chimie Thérapeutiques UPR 9021,
67084 Strasbourg Cedex, France
c.menard@ibmc-cnrs.unistra.fr

6.1 INTRODUCTION

Neurons are essential for the processing and transmission of cellular signals. A neuron consists of a central part known as the soma (or cell body) and long processes called neurites, constituted of axons and dendrites, extending over long distances (Fig. 6.1). The soma contains the nucleus of the cell. The growth of neurites and the formation of synapses are controlled by a highly motile structural specialisation at the tips of the neurites called the growth cones. Each neuron has multiple dendrites that carry signals into the soma and a single axon that carries signals away from the soma towards the next neuronal cell. The movement of ions across the ion channels found on the soma and the axon is driven by electrochemical gradients. It generates electrical signals called action potentials, which normally travel along the axon in one direction, away from the soma and towards the next neuron.[1]

Because of the lack of effective self-repair mechanisms in adults, central nervous system damage results in functional deficits that are often irreversible. The difficult challenge is to find means to cure the disabilities arising from injuries and disorders of nervous systems by stimulating inactive neurons

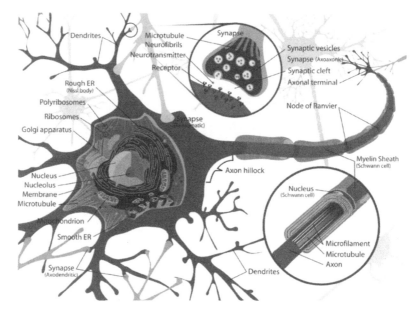

Figure 6.1 Structure of a typical myelinated vertebrate motoneuron. See also Colour Insert.

and regulating their growth in a proper way. In order for neural prostheses to augment or restore damaged or lost functions of the nervous system, they need to be able to perform two main functions: stimulate the nervous system and record its activity. For this purpose, nanotechnology offers new perspectives by providing possibilities of repair.[2] Many crucial steps are necessary for a neuron to rebuild a functional network: (i) survival to the injury, (ii) regrowth of neurites (axons and dendrites) and (iii) reconstruction of active synapses that connect neurons.[3a] Therefore, any therapeutic strategy should promote each of these steps. Nanotechnology and nanomaterials offer exciting promises in neuroscience.[3b,3c] In particular, the unique combination of physical, chemical, mechanical and electronic properties of carbon nanotubes (CNTs) makes them very attractive in basic and applied neuroscience research because of their small size, electrical conductivity, high flexibility, mechanical strength, inertness, non-biodegradability and biocompatibility brought about by surface functionalisation.[4] Nevertheless, methods for manipulating the neuronal growth environment at the nanometer scale are still lacking.

CNTs have been found to be promising substrates for neuroscience applications. Single-walled carbon nanotube (SWNT) bundles and multi-walled carbon nanotubes (MWNTs) possess diameters that can mimic neuronal processes. In particular, the aspect ratio is similar to that of small nerve fibres, growth cone filopodia and synaptic contacts. Moreover, the high electrical conductivity of CNTs should enable the detection of neuronal electrical activity and allow delivering electrical stimulation to neuronal cells in contact with them.

In this chapter, we will present the use of CNTs as substrates for neuronal growth, favouring adhesion of neurons and their survival, growth and differentiation in neurites. We will also detail the use of CNTs to electrically stimulate neuronal cells as well as the influence of CNTs in promoting spontaneous synaptic activity in neuronal networks and in increasing the efficacy of neural transmission. We will then close this chapter by reporting the studies that investigated the mechanisms of the electrical interactions between CNTs and neurons.

6.2 EFFECTS OF CARBON NANOTUBES ON NEURONAL CELLS' ADHESION, GROWTH, MORPHOLOGY AND DIFFERENTIATION

As it will be described in this part, numerous studies have revealed that CNTs can support and control the growth of neuronal cells as well as their morphology. SWNTs and MWNTs have been found to be permissive substrates for neuronal growth characterised by the presence of growth cones, neurite outgrowth and branching.

The first use of CNTs as substrates for nerve cell growth was reported by Mattson and coworkers.[5] In this study, embryonic rat hippocampal neurons were grown on MWNTs dispersed on glass coverslips coated with polyethyleneimine (PEI), which is a common substrate for neuronal growth.[6] Neurons require highly permissive substrates for cell attachment and neurite growth, which include positive-charge modification of glass and plastic culture substrates (e.g., polylysine or polyornithine [PLO])[7] for optimum growth and neurite extension.

Two types of MWNTs were utilised to investigate neuronal growth: pristine MWNTs and MWNTs coated with 4-hydroxynonenal (4-HNE), a molecule that effects neurite growth. 4-HNE is known for its ability to regulate neurite outgrowth in cultured embryonic hippocampal neurons[8] via modulation of intracellular Ca^{2+} levels in cultured hippocampal neurons.[9] Physisorption of 4-HNE on the MWNT surface was achieved by sonicating MWNTs in an acidic solution of 4-HNE. Mattson *et al.* observed that neurons grown on MWNTs survived and continued to grow for at least eight days in culture, indicating that MWNTs support long-term neuronal survival and provide a permissive substrate for neurite outgrowth. It was observed that the direction of growth was not influenced by MWNTs. Scanning electron microscopy (SEM) was used to identify the morphological changes of the neurons grown on the MWNT substrate. Neurons were seen to be attached to the pristine MWNTs while extending one or two neurites. Interestingly, when neurites grew across the MWNTs and then on the glass coverslips coated with PEI, the neurites formed branches on the coverslips but not on the MWNTs. Therefore, the pristine MWNTs did not promote neurite branching, indicating a relatively weak adhesion of growth cones to the surface of non-functionalised nanotubes. This observation suggested that pristine MWNTs were not a suitable support for branch formation. Nevertheless, MWNTs coated with 4-HNE had more and longer neurites (Fig. 6.2). Both neurite outgrowth and branching were enhanced in this case.

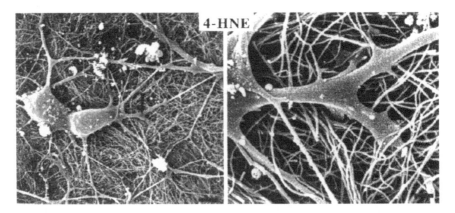

Figure 6.2 SEM images of neurons grown for three days on CNTs coated with 4-HNE. The right image is a high magnification of the neurite designated by the black arrow in the left image. Scale bars: left image, 5 µm; right image, 100 nm. Reproduced from Mattson *et al.*[5] with permission.

The authors pointed out that the enhanced adhesion of growth cones to the MWNTs could be favoured by the presence of 4-HNE that could possibly induce changes in intracellular Ca^{2+} levels. Indeed, Ca^{2+} influx can regulate growth cone motility and neurite elongation.[10]

Further studies were conducted by Haddon *et al.*, and they used MWNTs functionalised with molecules bearing different electrostatic charges to study neuronal growth.[11] The results showed that the neurite outgrowth was controlled by the surface charge of MWNTs.

Three types of functionalised MWNTs were designed to carry negative, neutral or positive charges at physiological pH. The approach relied on

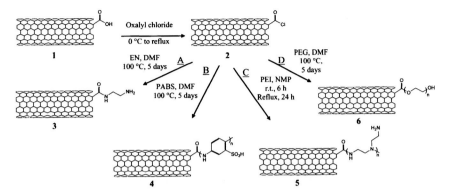

Scheme 6.1 Functionalisation of CNTs with EN, PABS, PEI and PEG.

covalent functionalisation of MWNTs rather than on physisorption (used in the previous study)[5] to allow transient retention of grafted molecules on the nanotube surface. MWNTs were first oxidised using nitric acid, and the resulting COOH functions introduced on the nanotube ends were then activated and coupled with either ethylenediamine (EN) or poly-*m*-aminobenzene sulphonic acid (PABS),[12] as illustrated in Scheme 6.1 (paths A and B).

At physiological pH used to grow neurons, the MWNTs exhibited different surface charges, from negatively charged MWNT-COOH (**1**), neutral/ zwitterionic MWNT-PABS (**4**), to positively charged MWNT-EN (**3**). Finally, MWNT films were deposited on glass coverslips coated with PEI, and hippocampal neuronal cells were cultured on these substrates. In addition to SEM analysis, the morphological features of live (rather than fixed) neurons, which directly reflect their potential capability in synaptic transmission, were characterised by fluorescence spectroscopy. The neurons were labelled with calcein, which is a fluorescent dye that stains only live cells. All neurons that grew on PEI and on MWNT substrates accumulated calcein, which is indicative of their viability. The number of neurites per neuron remained the same for the three substrates (Scheme 6.2a).[11-15] However, the average length of neurites was longer on the positively charged MWNT-EN (**3**), while the number of growth cones was higher on neurons cultured on zwitterionic MWNT-PABS (**4**) or on positive MWNT-EN (**3**). Furthermore, the branching of neurites increased as the substrate became more positive. Thus, by varying the surface charge of CNTs it was possible to control the number of growth cones, neurite outgrowth and branching

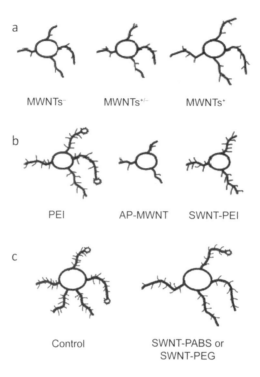

a

MWNTs⁻   MWNTs⁺/⁻   MWNTs⁺

b

PEI   AP-MWNT   SWNT-PEI

c

Control   SWNT-PABS or
SWNT-PEG

Scheme 6.2 The effects of pristine and functionalised CNTs on neurite outgrowth and number of growth cones in comparison with PEI. Reproduced with permission from Hu *et al.*[11] (a), Hu *et al.*[14] (b), and Ni *et al.*[15] (c) with permission.

The use of SWNT composite as scaffold for neuronal growth was also reported by Haddon and coworkers.[13] On the basis of the observations from the previous study showing that neurite outgrowth was enhanced by using positively charged MWNTs,[11] a composite constituted of SWNTs functionalised with PEI polymer was prepared. PEI is commonly used as permissive substrate for neuronal growth, and it is positively charged at physiological pH.[6] The precoating of glass coverslips with PEI in the previous studies was necessary because the adhesion of MWNT films on non-coated glass coverslips was low when the films were exposed to aqueous culture medium.[5,11] Functionalisation of CNTs with PEI should increase their ability to support neurite outgrowth, whereas the precoating of coverslips with PEI could be eliminated.

SWNTs were functionalised using a methodology that was comparable to the previous study.[11] PEI (branched PEI, $n \approx 20$) was covalently grafted on SWNT-COOH by amidation via oxalyl chloride activation of the carboxylic acid functions (Scheme 6.1, path C). Then, SWNT-PEI films were deposited on glass coverslips, and neuronal growth of hippocampal cells was monitored.

The neuronal growth was visualised by SEM, and viability was determined using calcein staining. Neurons were shown to grow on the SWNT-PEI films with neurite outgrowth and branching that are intermediate to those observed for neuronal growth on PEI and as-produced MWNTs (AP-MWNTs). Indeed, the neurite branching on SWNT-PEI was enhanced by comparison with AP-MWNTs, while it was comparable to PEI. The number of neurites and growth cones was similar to that of AP-MWNTs, and the neurite lengths were intermediate to those of neurons grown on pristine MWNTs and PEI (Scheme 6.2b). Hence, the growth parameters were found to be sensitive to the nature of the substrate, in particular the surface charge, since the effects of the SWNT-PEI composite on neuronal growth characteristics are intermediate to those observed when using pristine MWNTs and PEI. This behaviour can be explained by the reduced positive charge of PEI, which is proportional to the percentage of SWNTs in the composite. It should be noted that the results obtained in this study with SWNT-PEI were compared with those obtained with pristine MWNTs, not with unmodified SWNTs. Hence, the authors pointed out that the structural parameters of CNTs, in particular the diameter, could also contribute to the observed effects, in addition to the contribution of PEI.

In summary, neurite outgrowth and branching could be controlled by varying the ratio of SWNTs and PEI in the graft copolymer. This composite could be implemented in building scaffolds for the formation of neuronal circuits to develop neural prostheses.

Haddon *et al.* also investigated the possibility of using water-soluble SWNTs as substrate for neuronal growth.[14] The SWNTs soluble in water were found to induce an increase of neurite length, while a decrease of the number of neurites and growth cones was observed.

SWNTs were functionalised with either PABS[11] or polyethylene glycol (PEG, $n \approx 13$)[15] to form the corresponding graft copolymers (Scheme 6.1, paths B and D). Each polymer imparted water solubility to the SWNTs. The methodology used for the functionalisation was based on amidation of the COOH functions located at the nanotube ends. Hippocampal neuronal cells were cultured on the water-soluble SWNT substrates. The neurons accumulated the vital stain calcein, which is indicative of cell viability and of the biocompatibility of the functionalised SWNTs. The copolymers were found to modulate neurite outgrowth by increasing their length, while reducing the number of neurites and growth cones (Scheme 6.2c). Hence, the neurons treated with the water-soluble SWNTs exhibited sparser, but longer neurites.

The authors suggested that water-soluble SWNTs may modulate intracellular Ca^{2+} homeostasis as Ca^{2+} influx is known to regulate neurite elongation. Indeed, Ca^{2+} channel blockers can cause at low concentrations a simultaneous reduction of growth cone filopodia and an increased elongation of neurite.[8a] The same modifications were induced in this study by the water-soluble SWNTs. To confirm this hypothesis, the intracellular Ca^{2+} levels in cultured individual neurons were monitored. The experiments involved depolarisation of neurons via HiK^{+}-induced intracellular Ca^{2+} accumulation in neurons. PEG-functionalised SWNTs were found to provoke a reduction of the intracellular Ca^{2+} accumulation in a dose-dependent manner. Therefore, the SWNT-PEG copolymer could act as blocker of Ca^{2+} channels by inhibiting the depolarisation-dependent influx of Ca^{2+}. Evidence that CNTs can affect the function of ion channels, such as potassium channels, has already been reported.[16]

The influence of the electrical conductivity of CNT films on the neuronal growth characteristics was recently studied by Haddon and coworkers.[17] The CNT films were prepared by spraying an aqueous dispersion of SWNT-PEG with an airbrush onto glass coverslips heated at 160 °C. The electrical conductivity was controlled by varying the thickness of the nanotube film. Hippocampal rat neurons were then cultured on these substrates. The positive calcein labelling of cells was indicative of cell viability. By comparison with PEI-coated coverslips used as control, the total outgrowth of each neuron (i.e., summed length of all processes and their branches) was superior in neurons grown on the 10 nm thick SWNT-PEG films. However, other parameters of neuronal growth such as the total number of processes and neurites originating from the cell body were unchanged for each neuron. In addition, by increasing the nanotube film thickness to 30 and 60 nm, thus allowing for higher conductivity, these effects disappeared since the neurite outgrowth was unaffected.

In summary, Haddon *et al.* demonstrated that SWNT-based substrates, having a narrow range of electrical conductivity, supported neurite outgrowth, with a decrease in the number of growth cones but an increase in cell body area.

Other research groups also reported the use of CNTs as substrate for neuronal growth. Shimizu *et al.* functionalised MWNTs with neurotrophins (a family of proteins that induce survival, development and functions of neurons), in particular with a nerve growth factor (NGF) or brain-derived neurotrophic factor (BDNF).[18] The aim of this study was to regulate the differentiation and survival of neurons.

Neurotrophins are key proteins for the differential function of neurons. They are endogenous soluble proteins regulating the survival,[19] growth and function[20] of neurons. Neurotrophins belong to a class of growth factors, secreted proteins, which are capable of signalling particular cells to survive, differentiate or grow.[21] They may stimulate the synthesis of proteins in either axonal or dendritic compartments, thus allowing synapses to exert local control over the complement of proteins expressed at individual synaptic sites.[22]

MWNTs were first oxidised using a mixture of sulphuric acid and nitric acid. Diaminoalkyl compounds were then introduced on the resulting MWNT-COOH via amidation. Neurotrophin was covalently bound to the resulting amino-modified MWNTs using a carbodiimide reagent. The influence of neurotrophin-functionalised MWNTs on the neurite outgrowth of embryonic chick dorsal root ganglion neurons was then examined. These MWNTs were found to promote the neurite outgrowth of dissociated neurons, in a similar manner to soluble NGF and BDNF, while the amino-terminated MWNTs did not promote neurite outgrowth.

Romero and coworkers reported the preparation of strong and highly electrically conducting semi-transparent sheets and yarns from pristine MWNTs.[23]

The CNT sheets were fabricated by drawing the CNTs from a sidewall of an MWNT forest produced by chemical vapour deposition (CVD). The thickness of the CNT sheet was approximately 50 nm. The CNT yarn was then prepared by spinning the sheet.[24] The authors demonstrated that the CNT sheets were permissive substrates for primary central and peripheral neuronal culture. Indeed, the results indicated that the CNT substrates were able to promote cell attachment and differentiation of a variety of cell types, including cerebellar and sensory dorsal root ganglion neurons, and to support their long-term growth at a similar level to that achieved on PLO-coated glass substrates used in this study for comparison. The neuronal phenotype was demonstrated by immunofluorescence using β-tubulin,[25] which is a specific neuronal marker. In addition, the extension of neuronal axon growth cones was enhanced. Neurons from murine cerebellum and cerebral cortex, which are usually more sensitive to the substrate and the environment for survival and growth, were used too. Also in this case, the neurons were able to dissociate and extend their neurites onto the CNT sheets, as they did similarly onto PLO-coated glass substrates.[26] Thus, contrary to previous studies reporting that axonal growth is limited on pristine CNTs,[5,11] neurons were shown here to grow on pristine CNT sheets and to extend their neurites with similar characteristics (number and length) than those grown on PLO-treated glass substrates.

Furthermore, the authors demonstrated that the neuronal growth could be directed by CNT yarns. Neonatal dorsal root ganglion neuron explants were prepared by wrapping a single CNT yarn around the ganglia, where the free end suspended in culture media. The sensory neurons adhered and extended along the CNT yarn by intimately following the surface topography of the CNT substrate.

In summary, directionally oriented pristine CNTs configured as sheets or yarns were demonstrated to be viable substrates for neuronal growth.

Lu *et al.* recently reported that thin-film scaffolds constituted of a biocompatible polymer grafted on CNTs can promote neuron differentiation from human embryonic stem cells (hESCs).[27]

The CNT-based composite was prepared by *in situ* polymerisation of acrylic acid onto oxidised CNTs. The neuron differentiation efficiency on poly(acrylic acid)-grafted CNT (PAA-g-CNT) thin films was compared with that of thin films composed of poly(acrylic acid) (PAA) or PLO. The PAA-g-CNT thin films showed enhanced neuron differentiation while maintaining cell viability, as assessed by measuring the metabolic activity of dehydrogenases. SEM images showed that the differentiated neurons on PAA-g-CNT surfaces have more branches, suggesting that they are more mature. In addition, the CNT-based surface exhibited a higher cell adhesion and protein (laminin) adsorption. Laminin is a glycoprotein found in the extracellular matrix, the sheets of protein that form the substrate of all internal organs (i.e., basement membrane). Laminin has been reported to favour neuronal growth.[28] Indeed, laminin is a substrate along which nerve axons can grow both *in vivo* and *in vitro*.

Thus, this work shows for the first time that CNT-based polymer thin films promote the differentiation of hESCs towards neuronal lineage.

The growth of somatosensory neurons on functionalised CNT mats was recently reported by Xie *et al.*[29] MWNTs were functionalised by oxidation using a mixture of sulphuric acid and nitric acid. The CNT mat was then deposited on a track-etch membrane. The CNT substrate was found to be a permissive substrate for the growth of somatosensory neurons from a dorsal root ganglion. Neurite outgrowth and branching were observed by SEM, as well as intertwinement between the neurites and the underlying functionalised CNTs, indicating a strong interaction between both entities at the nanoscale. No obvious neurite growth was observed on neurons plated on a blank polycarbonate membrane. The authors suggested that functional groups on the nanotube surface could act as anchoring seeds which may enhance the adhesion of neurites on the CNT-based substrate. This is consistent with previous studies, such as those reported by Haddon and coworkers, that emphasise the positive impact of the functionalisation of CNTs on permissivity for neuronal attachment and neurite outgrowth.[11,14,15] Neurons adhere to substrates via extracellular proteins such as laminin, whose size is about 70 nm.[30] The authors pointed out that the analogous dimension of CNTs and the roughness of the nanotube surface favour contact with neurites by promoting better adhesion.

Wick *et al.* recently investigated the effects of SWNTs with different degrees of agglomeration on primary cells of the nervous system (mixed neuroglial cultures derived from the chicken embryonic spinal cord [SPC] of the central nervous system or dorsal root ganglia [DRG] of the peripheral nervous system).[31] The cells were exposed to SWNT agglomerates of submicrometre sizes (SWNT-a) and to SWNT bundles (SWNT-b) composed of 10 to 20 tubes. Suspensions of SWNTs-a and SWNTs-b affect glial cells from both tissue types. The level of toxicity was found to be dependent on the agglomeration state of the CNTs. Treatment of mixed neuroglial cultures with up to 30 μg/mL SWNTs significantly decreased the overall DNA content. This effect was more pronounced when cells were exposed to highly agglomerated SWNTs-a compared with better-dispersed SWNTs-b. Indeed, higher concentrations of SWNTs-a were more toxic than the same amount of SWNTs-b. This result is in agreement with previous studies from Wick *et al.* in which the aggregation state of SWNTs might be a crucial factor in terms of toxic effects.[32] Additionally, SWNTs reduced the amount of glial cells in both derived cultures as measured by ELISA. The authors observed that neurons were only affected in DRG-derived cultures with regard to their ionic conductance (diminished inward conductivity) and resting membrane potential (more positive) according to whole-cell patch recordings. On the contrary, the neurite outgrowth and the electrophysiological properties of neurons derived from SPC cultures were not affected.

6.3 ELECTRICAL STIMULATION OF NEURONAL CELLS GROWN ON CARBON NANOTUBE-BASED SUBSTRATES

Electrical stimulation of neuronal cells is widely employed in basic neuroscience research, in neural prostheses[33] and in clinical therapy (e.g., treatment of Parkinson's disease, dystonia and chronic pain).[34] These applications require an implanted microelectrode array (MEA) with the capacity to stimulate neurons. Neuroprosthetic devices currently face various issues, including (i) long-term inflammatory response of the neuronal tissues, resulting in neuron depletion around the electrodes and their replacement with reactive astrocytes that prevent signal transduction; (ii) delamination and degradation of thin metal electrodes; (iii) miniaturisation of the electrodes and (iv) mechanical compliance with neuronal tissues for long-term performance. Currently, semiconductor devices can only partially solve some of these problems. Nevertheless, CNTs are excellent candidates for MEA applications because of their unique set of properties which offer the possibility of constructing small electrodes with high current density.

As will be detailed in this part, microelectrodes coated with CNTs have low impedance and high charge transfer characteristics and provide a rough surface that favours excellent cell–electrode coupling, while remaining chemically inert and biocompatible. Different systems have been utilised to guide neuronal cell growth and have been tested for their function in cultured neuronal networks, such as metal electrodes coated with CNTs, patterned CNT surfaces and CNT-polymer composite thin films.

For instance, Ozkan *et al.* demonstrated the formation of directed neurite growth on patterned, vertically aligned MWNTs.[35] Different substrate geometries and nanotube heights were investigated for their ability to induce guided neuronal growth. Vertical MWNT arrays were functionalised with growth adherents that promote neuron adhesion and viability. In particular, the MWNTs were coated with a growth adherent, poly-L-lysine (PLL), which is known to enhance adhesion of neurons by altering surface charges on the culture substrate.[36] Hippocampal rat cells were cultured on patterns of short (500 nm high) and long (10 µm high) vertical MWNTs. Preferential directed growth of neurons was observed over the long MWNT array, but not over the short MWNT array. Indeed, in the latter case, neurons were seen to grow both on the short MWNT pattern and on the silicon chip, with no selection of one of the two types of substrate. On the contrary, in the former case, neurons showed preferential growth along the MWNT array, although the substrate was covered with PLL. The authors pointed out that the difference observed in terms of scaffolding capability between the short and long MWNT arrays could stem from the flexibility of the CNTs, which depends on the nanotube length. Indeed, to allow proliferation of neurites, the long MWNTs were found to undergo deformation, as observed by SEM. This deformation resulted from the extending neurite that interacted with the edges of the MWNT patterns. The neuronal cell networks were found to be viable after they were stained with Fluo-3 calcium. This fluorescent dye binds to the intracellular free calcium ions, which in principle are in high concentration in the cytoplasm of viable neuronal cells. The results of this study highlight the potential applications of long MWNT array substrates for the development of three-dimensional scaffolds suitable for implants.

Hanein and coworkers published a review on the development of CNT-based MEAs for neuronal interfacing and network engineering applications.[37] They also described the use of CNT patterns as substrate for neuronal growth.[38] The CNT templates on which neurons adhered and assembled were fabricated by photolithography and microcontact printing. The deposition of iron nanoparticles on hydrophilic silicon dioxide or quartz substrates was controlled to allow the growth of regular arrays of CNT islands using CVD. During this process, long CNTs (on the order of 100 µm) can thermally fluctuate and may bind to other nanotubes to form a three-dimensional, entangled network. Neurons and glial cells were then deposited on the patterned quartz substrates. After four days' incubation, neurons were seen to adhere to and grow on the region of the substrate containing CNTs. Interconnections between neighbouring islands were also observed because of the formation of axons and dendrites. The electrical viability of the neuronal networks was then tested and the action potential, generated by electrical stimulations, displayed a normal shape. Thus, this study allowed for the formation of neuronal networks having a normal functionality with pre-defined geometry due to growth on substrates patterned with CNT islands.

An original method to pattern cultured neuronal networks was reported by Hanein *et al.* and was based on the anchoring of neuronal cell clusters constituted of tens of cells.[39] The anchors aimed at stabilising the neuronal cell clusters were CNTs or the strong adhesive substrate poly-D-lysine. The dynamics of the real-time network formation was monitored by placing cultures in an environmental chamber under a microscope. The cell clusters were seen to self-organise by moving away from each other and bridging gaps with concomitant formation of a neurite bundle between neighbouring clusters.

Hanein *et al.* also fabricated CNT-based electrodes by synthesising high-density CNT islands using CVD on a lithographically defined substrate.[40] A mixture of cortical neuron and glial cells from rats were plated and cultured on the surface of the CNT electrode chip. The CNT islands acted as adhesion agents because of the substrate's roughness, thus favouring the migration and adhesion of cells onto the electrode surface. High-fidelity extracellular recordings from cultured neuronal cells were then performed. A typical extracellular signal was obtained with a shape that reflected the first derivative of the intracellular action potential signal. The signal-to-noise ratio was high compared with that of a conventional TiN electrode.

It should be noted that one advantage of the CNT electrodes developed by Hanein *et al.* over standard microelectrodes is the cell-adhesive nature of the nanotubes. Neuronal proliferation can thus be induced without the need for additional adhesion, thereby promoting coating spread between the electrodes.

Electrical stimulation of primary neurons was reported by Wang and coworkers, who used vertically aligned MWNTs as microelectrodes.[41] Lithographic patterning of the catalyst allowed the control of the size and location of the MWNT pillars. The MWNT-based microelectrode was constituted of individual MWNTs having a diameter in the 30–50 nm range and separated by a distance of the order of tens of nanometres. The CNT electrode was 40 μm tall and was made more hydrophilic by coating with an amphiphilic PEG–lipid conjugate. Non-covalent binding of the PEG–lipid conjugate allowed for the preservation of the electronic properties of the CNTs. The MWNT MEA was coated with poly-D-lysine to favour cell adhesion to the nanotube substrates. Embryonic rat hippocampal neurons were deposited on the MWNT microelectrode and were shown to grow and differentiate. The viability and neurite outgrowth were comparable to those of cultures on plastic Petri dish controls. The neurons were then electrically stimulated by the MWNT electrode. The calcium indicator Fluo-4 allowed for the detection of action potentials by observation of intracellular Ca^{2+} level change. Fluo-4 can be loaded into cells and exhibits a large increase in fluorescence intensity on binding free Ca^{2+}. It was also demonstrated that the neurons could be stimulated repeatedly with the MWNT electrode, thus highlighting the long-term endurance of the MWNT electrode.

Recently, Hanein and coworkers investigated the neurite-CNT interactions at the nanoscale and elucidated the nature of the interface between neurons and CNTs.[42] Process entanglement was explained as a neuronal anchorage mechanism to CNT-based surfaces.

In this study, isolated islands of pristine CNTs were fabricated and plated with cells (mammalian cortex neurons from rats and insect ganglion cells from locusts). The arrangement of neurons and glial cells on CNT-based surfaces was characterised by high-resolution scanning electron microscopy (HRSEM), immunostaining and confocal microscopy. The neurons and cells were found to preferentially adhere to CNT islands and extend towards their periphery. This is in agreement with previous reports that have demonstrated preferential adhesion to rough surfaces.[39-41] The processes at the periphery of the CNT islands were curled and entangled, thus leading to enhanced interactions with the CNT surface that facilitated their anchorage. However, the processes that appeared too thick to interact with CNTs had the tendency to intertwine. Therefore, Hanein *et al.* pointed out that the roughness of the surface must match the diameter of the neuronal processes to help them anchor. The authors also suggested that adhesion of neuronal cells was in part achieved through an entanglement process. This mechanical effect may thus contribute to the mechanism by which neurons adhere to rough surfaces.

Recently, Hanein *et al.* also investigated the use of CNT MEAs as an interface material for retinal recording and stimulation applications.[43]

Electrodes were coated via CVD of CNTs, and electrical stimulation of retinal cells was achieved. The signals obtained with the CNT-based electrodes showed a remarkably high signal-to-noise ratio during recordings in comparison with commercial TiN electrodes. This would be particularly interesting for long-term *in vivo* implantation of electrodes. In addition, the authors observed an increase of the signal amplitude over several hours of recording. These results were indicative of an improved electrode–tissue coupling of CNT-based electrodes compared with conventional commercial electrodes. This is consistent with previous studies reported by Wang and coworkers, who validated the effectiveness of CNT electrodes for stimulation applications.[42]

In summary, this work showed that CNT MEAs provide exceptional electrochemical and adhesive properties. This demonstrates the great potential of CNT-based electrodes for retinal implant applications.

The first example of CNT-coated electrodes implanted into different brain areas in rats or monkeys was reported by Keefer *et al.*[44] These electrodes showed the ability to enhance the detection of neuronal signals and to stimulate neurons. Indeed, the results indicated that the CNT-coated electrodes improved the electrochemical and functional properties of cultured neurons. CNT coating enhanced both recording and electrical stimulation of neurons in culture, rats and monkeys by decreasing the electrode impedance and increasing charge transfer. Electrical stimulation experiments were performed with cultured neuronal networks grown on CNT-based electrodes to determine if the CNT coating could alter the capacity to activate neurons. The CNT coating was found to be permissive for neuronal growth and function. Moreover, the performances of the CNT-coated electrodes were much higher than those of gold-coated control electrodes in evoking neuronal responses. This result highlights the potential of CNTs for the development of electrical brain interfaces.

The unique physical, mechanical, chemical and electronic properties of CNTs can be, in part, transferred into CNT composites to combine high electrical conductivity, chemical stability and physical strength with structural flexibility. CNT composite constituted of thin films were prepared by layer-by-layer (LBL) assembly by Kotov and coworkers, as will be detailed later. They demonstrated that CNT composite films were suitable substrates to support growth, proliferation and differentiation, as well as to electrically stimulate neuronal cells.

Applications of CNT substrates in medical implementations, such as implanted scaffolding, would require free-standing structure instead of CNTs attached to supporting glass or plastic. For this purpose, Kotov *et al.* reported the fabrication of free-standing SWNT-polymer thin-film membranes that are biologically compatible with neuronal cell cultures.[45] The membranes constituted of SWNT-polymer composite were prepared by the LBL assembly process. This technique allows for the construction of multilayered composite coatings and free-standing films of SWNTs with versatile architectures, which can be engineered at the nanoscale to attain desirable mechanical, structural, biological and electrical properties. It is based on alternating layers of CNTs and polymers. Coatings made by LBL are generally very mechanically robust.[46] SWNTs were dispersed in an aqueous solution of the amphiphilic poly(*N*-cetyl-4-vinylpyridinium bromide-*co*-*N*-ethyl-4-vinylpyridinium bromide-*co*-4-vinylpyridine). This polymer bears positively charged groups that favour cell adhesion. The biocompatibility of the SWNT composite thin films towards NG108-15 neuroblastoma × glioma hybrid culture cells was assessed using confocal microscopy and calcein. NG108-15 cells are used as neuronal model system since they exhibit, after differentiation, many characteristics of mammalian nerve cells. After long-term incubation (up to 10 days) the cells grew and proliferated on the surface of the SWNT films. The neuronal cells were viable, as illustrated by their continuous intracellular esterase activity, which is determined by the enzymatic conversion of the non-fluorescent cell-permeant calcein dye. The neuronal outgrowth of the NG108-15 cells on the surface of the SWNT films was compared with that of the cells on a control culture dish. Neurons were observed by SEM, and their morphology was assessed by cell labelling with the neuronal lipophilic tracer dialkylcarbocyanine. This dye has the capacity to diffuse through the membrane of cells and to trace along their neurites in order to monitor the surface morphology and growth of neurons. On the substrate comprising SWNTs, many elongated neuronal processes were observed with extension of neurites. Many secondary neurites and elaborated branches were grown from these neurites. Interestingly, many processes were seen to start from the neurites and to terminate at the surface of the SWNT film.

Free-standing SWNT membranes were then prepared in order to be used as scaffold for neuronal growth and differentiation. The preparation involved the etching of an SWNT film from the surface of a glass substrate using HF and the subsequent rolling into a small elongated thread-like sample. The free-standing films showed a tensile strength that was sufficient for soft neural implants. It is noteworthy that the ionic conductivity of these free-standing films is brought about by its polyelectrolyte nature. This is essential for the interaction with ionic fluxes through neuron membrane. Adhesion and differentiation of NG108-15 cells were seen to be similar to those of SWNT films on solid supports. The neurite extension closely followed the morphological shape and curvature of the surface of the free-standing films.

In summary, SWNT composite thin films prepared by LBL and resulting free-standing films can support growth, viability and differentiation of NG108-15 neuroblastoma/glioma hybrid cells. The free-standing SWNT/polymer composite thin film membranes were found to guide the outgrowth of neurites. The combination of different properties, including flexibility, inertness, non-biodegradability and durability, allows us to envisage some applications. For instance, the free-standing films could potentially be integrated into extracellular implants for cell stimulation and pain control and into neuroprosthetic devices for neuronal injuries.

Kotov and coworkers also reported the stimulation of the neurophysiological activity of NG108-15 cells grown on LBL-assembled SWNT-polymer composite films.[47] The LBL SWNT conductive films were prepared from positively charged SWNTs coated with a copolymer, poly(acrylic acid), and were made of 30 multilayers. The NG108-15 cells differentiated and extended long neurites on the surface of the SWNT films. The neurites developed into many secondary processes, branching in many directions on the surface. The SWNT films exhibited a sufficiently high electrical conductivity to electrically stimulate significant ion conductance in excitable neuronal cells that were electrically coupled to the SWNTs. The direction of the electrophysiological response suggested that cells were stimulated by the influx of cations in the cells.

Kotov *et al.* described that mouse cortical neural stem cells (NSCs) can differentiate to neurons, astrocytes and oligodendrocytes with formation of neurites on SWNT-based composite thin films.[48] Note that NSCs are well known for their sensitivity to the environment.

The SWNT-polymer composite thin films were prepared by LBL assembly of six bilayers of SWNT-PEI [(PEI/SWNT)$_6$]. SWNTs were first dispersed in a 1 wt% poly(sodium 4-styrene-sulphonate) solution and then assembled using LBL technique with PEI as polyelectrolyte. Mouse embryonic 14-day neurospheres (i.e., spherical clonal structures of NSCs) from the cortex were seen to attach to the SWNT-polyelectrolyte film substrates and to differentiate, in a similar manner to poly-L-ornithine (PLO)-coated substrates used as control. PLO is a standard substrate widely used for NSC cultures as it is known to alter the surface charge, thus providing a more suitable substrate for the negatively charged neurons.[27] The viability of neurospheres determined using the MTT assay was found to be similar on (PEI/SWNT)$_6$-and PLO-coated substrates. In addition, both types of substrates supported differentiation of neurospheres into the three primary neural cell types: neurons, astrocytes and oligodendrocytes, according to immunostaining experiments. Indeed, to analyse the differentiated phenotypes, neurospheres were immunostained with antinestin-, antimicrotubule-associated protein 2 (anti-MAP2), antiglial fibrillary acidic protein (anti-GFAP) and anti-oligodendrocyte marker O4 (anti-O4). Thus, NSCs grown on LBL-assembled SWNT-polyelectrolyte composite films behaved similarly to NSCs cultured on PLO substrates in terms of biocompatility, neurite outgrowth and expression of neuronal markers.

In summary, this study is the first demonstration of the differentiation of environment-sensitive NSCs on a CNT-composite thin film. The results are promising for further development of CNTs for neural interfaces as the CNT-based substrate is able to promote cell viability and to induce the development of neuronal processes, as well as the appearance and progression of neural markers.

Further studies were conducted by Kotov and coworkers on the electrical stimulation of NSCs by the SWNT composite constituted of laminin.[49]

LBL films (up to 30 bilayers) were prepared from SWNTs wrapped with poly(styrene-4-sulphonate) and laminin, a glycoprotein which is an important protein in the basement membrane. Laminin is commonly used to coat substrates to promote cell adhesion and neurite outgrowth.[50] The SWNT-laminin films were heat-treated to increase their electrical conductivity by enhancing the cross-linking of CNTs among the different layers of the films prepared by LBL. Cell adhesion was clearly seen to depend strongly on the composition of the final layer of the LBL films. Indeed, the film containing SWNTs as top layer was the most suitable for cell adhesion and attachment. Differentiation of NSCs was observed, as indicated by extensive formation of functional neuronal network showing the presence of synaptic connections. By comparison with laminin-coated substrates, longer neuronal outgrowth occurred on the SWNT-laminin substrates that were found to be biocompatible, as indicated by satisfactory cell viability assays. The functionality of the neuronal networks was assessed by immunostaining for the presence of synapsin protein with the above-mentioned neuronal markers. Synapsin was present between the differentiated cells. This indicated that neuronal network was functional. The electrical stimulation of NSCs by the SWNT-laminin films was then investigated. Cell response to chemical and electrical stimuli typically results in a change in the membrane voltage of the cells, referred to as action potential. Detection of action potentials can be achieved via the invasive patch-clamp method.[51] The generation of action potentials upon the application of a lateral current through the SWNT film was deduced from the imaging of the NSCs with a Ca^{2+}-dependent dye and was found to be dependent on the status of the surrounding cells. Once one neuronal cell was excited, the excitation sequentially spread from one cell to another.

In summary, the combination of different properties of the SWNT/laminin thin film makes them a promising candidate for incorporation in neural electrodes. On the one hand, the films were suitable for NSCs' growth and proliferation. On the other, as SWNTs exhibit high electrical conductivity, they were able to electrically stimulate NSCs.

Chen *et al.* have been the first to explore the interface between neurons and CNT probes, both extracellularly and intracellularly (i.e., inside the neural membrane).[52] CNT probes were used to monitor the neuronal activity elicited not only extracellularly but also intracellularly. This work is particularly interesting as it opens the way for intracellular neural probes that minimise damage to the neuron.

In this study, two types of CNT probes were fabricated. MWNT bundles were connected to a silver wire. In one case, CNTs were coated with insulating epoxy, while in the other, CNTs were inserted into a sharp glass pipette, whose tip was used to penetrate the neural membrane. Only the tips of the probes were made of CNTs, contrary to the CNT-coated microelectrodes that were also involved in interfacing neurons.[41,42,53] The CNT probes were examined with the escape neural circuit of crayfish (*Procambarus clarkia*) and were found to have comparable performances to conventional Ag/AgCl electrodes. The CNT probes, with the tip pressed against an axon, were able to detect an action potential extracellularly and some small spikes (i.e., potential waveforms) associated with other neurons in the nerve cord. The results obtained from these extracellular experiments were in agreement with those demonstrated with microelectrodes coated with CNTs.[41,42,54]

The intracellular recording and stimulation with CNT probes has been investigated for the first time. The conduction mechanism was investigated on the basis of impedance measurements and cyclic voltammetry. CNT probes were shown to transmit electrical signals through not only capacitive coupling but also resistive conduction to a comparable extent, contrary to the suggestion that capacitive impedance is the main mechanism to record and stimulate neural activity.[41,42,54] The resistive conduction helps record postsynaptic potentials and equilibrium membrane potentials intracellularly. It also facilitates the delivery of direct-current stimulation. The authors pointed out that the recording capability of the CNTs did not degrade and even improved after delivering direct-current stimuli for a long period of time. This highlights that the CNT probes are suitable for long-term use with a longer endurance compared with conventional Ag/AgCl electrodes, as the latter are inefficient once the silver chloride is reduced to silver. Thus, this work supports the potential use of CNT probes as a promising new neurophysiological tool.

Ballerini and coworkers reported that CNT substrates can boost neuronal electrical signalling under chronic growth conditions.[54] They demonstrated that the growth of primary hippocampal neurons on MWNTs increased the frequency of spontaneous postsynaptic currents, but did not affect the general electrophysiological characteristics of the neurons.

The substrates coated with CNTs were prepared by deposition of a homogeneous dispersion of pure MWNTs obtained by functionalisation using the 1,3-dipolar cycloaddition of azomethine ylides. Defunctionalisation of the resulting pyrrolidine-MWNTs was then induced at 350 °C under nitrogen atmosphere to eliminate the organic functional groups introduced on the nanotube surface. This treatment provided purified non-functionalised MWNTs layered on the glass substrate. Previous attempts to prepare glass coverslips coated with as-produced MWNTs were not satisfactory because of poor reproducibility in terms of neurite growth and elongation. Hippocampal neurons were then seeded on glass coverslips. Attachment and growth of neurons were observed on the substrates covered with purified MWNTs, along with neurite extension. Hence, purified MWNTs layered on glass substrate were found to be permissive substrates to support neuron adhesion, survival and dendrite elongation.

Neural network activity was investigated using single-cell patch-clamp recordings. The frequency of spontaneous postsynaptic currents (PSCs) of neurons deposited on the glass coverslips covered with MWNTs displayed a sixfold increase in comparison with the control substrate constituted of hippocampal neurons seeded directly on glass coverslips. The generation of PSCs was indicative of functional synapse formation. However, the height and half-width of action potentials of the neurons grown on both substrates were not significantly different. In addition, other electrophysiological characteristics of the neurons grown on the MWNT-coated substrates were nearly similar to those grown on the control substrate, in particular the resting membrane potential, input resistance and capacitance values. In summary, purified MWNTs obtained via a functionalisation/defunctionalisation sequence were demonstrated to be suitable growth surfaces for neurons. They boosted neuronal electrical signalling by promoting an increase in the efficacy of neuronal signal transmission. The authors pointed out that this effect was not attributable to differences in neuronal survival, morphology or passive membrane properties, but that it possibly represented a consequence of the properties of the SWNT substrates.

Pappas *et al.* also examined the use of functionalised SWNTs as substrate for neuronal attachment and growth.[55] They demonstrated that neurons were electrically coupled to SWNTs. Using the diazonium salt approach[56] SWNTs were functionalised to introduce 4-benzoic acid or 4-*tert*-butylphenyl functional groups on the nanotube surface. Transparent, conductive SWNT-polyethylene terephthalate (PET) films were prepared via pre-dispersion of the SWNTs in water by using 1% SDS. As already reported,[46] neuroblastoma × glioma NG108 was used as model of neuronal cell. The highest cell growth and neurite extension of NG108 occurred on unmodified SWNT substrates, with decreasing growth on 4-*tert*-butylphenyl SWNT and 4-benzoic acid SWNT substrates. The neuronal cells were viable according to the vital dye calcein test. On the basis of acute attachment assays, SWNT substrates were found to display a more dramatic effect on cell attachment than on cell proliferation. It should be noted that cell adhesion is critical for cell survival. In fact, functional groups on the nanotube surface inhibited cell attachment. These results support those obtained by Haddon and coworkers on the influence of the surface charge of CNTs on neurite outgrowth where negatively charged functional groups at physiological pH or pristine MWNTs induced a reduction of the neurite outgrowth, in comparison with positively charged substrates.[11] The neuronal cells were then electrically stimulated through the SWNT-PET substrates. NG108 and rat primary peripheral neurons showed robust voltage-activated currents. These results suggested that CNTs can be a physiologically compatible substrate to be incorporated into devices for electrical stimulation of neurons.

Gabriel *et al.* reported an alternative and simple method to fabricate CNT-based MEAs.[57] A suspension of purified SWNTs produced by arc discharge was deposited onto platinum electrodes.[58] This method has the advantage to be a post-process procedure that can be used with almost any MEA. The electrical properties of the CNT-based MEAs were found to be superior to those of metal electrodes, as shown by *in vitro* impedance and electrochemical characterisation. To investigate the biocompatibility of the CNT-based electrodes, *in vitro* cytotoxicity tests were carried out with different cell types from the nervous system: neurons (PC12 cells), glial cells (42MG-BA cells), fibroblasts (3T3 cells) and endothelial HEK cells (293T cells). The cell activity, estimated by the MTT assay, was not affected by the presence of CNTs, and no significant decrease in proliferation was detected. In addition, the SWNTs did not display any toxicity and remained attached to the surface of the electrode. Extracellular ganglion cell recordings in isolated superfused rabbit retinas were performed to evaluate the capacity of the CNT-based MEAs. It was the first time that CNTs were used to record electrophysiological activity in cultured retinal explants. The results indicated that the SWNT MEA arrays were more efficient in recording the action potential generated by a population of ganglion cells than Pt electrodes.

The effect of MWNTs functionalised with cell-adhesion peptides on neuronal functionality was recently investigated by Bianco *et al.*[59]

In this study, MWNTs were functionalised via 1,3-dipolar cycloaddition of azomethine ylides, and the resulting pyrrolidine ring was derivatised with two peptides, one including the cell-binding domain RGD sequence (GRGDSP) and the other comprising a domain of the laminin protein (IKVAV), as illustrated in Scheme 6.3. GRGDSP is a fibronectin-derived peptide capable of promoting cell adhesion and neurite outgrowth. RGD-containing peptides are also involved in the mechanism of integrin regulation of neuronal gene expression.[60] Besides, peptides from different domains of laminin were used to stimulate neuronal growth and axon regeneration.[61]

The activity of the two peptide–nanotube conjugates on different types of cells, including tumour cells (Jurkat), primary splenocytes and neurons, was examined. Upon incubation of Jurkat cells with increasing doses of the peptide–nanotube conjugates (up to 100 µg/mL), flow cytometry showed no significant loss of cell viability. Similarly, no cytotoxic effect of the peptide–nanotube conjugates on splenocytes isolated from healthy BALB/c mice was observed, thus highlighting the biocompatibility of the functionalised MWNTs. Furthermore, doubly functionalised peptide–nanotube conjugates containing FITC were prepared and found to be internalised into the neuronal cells.

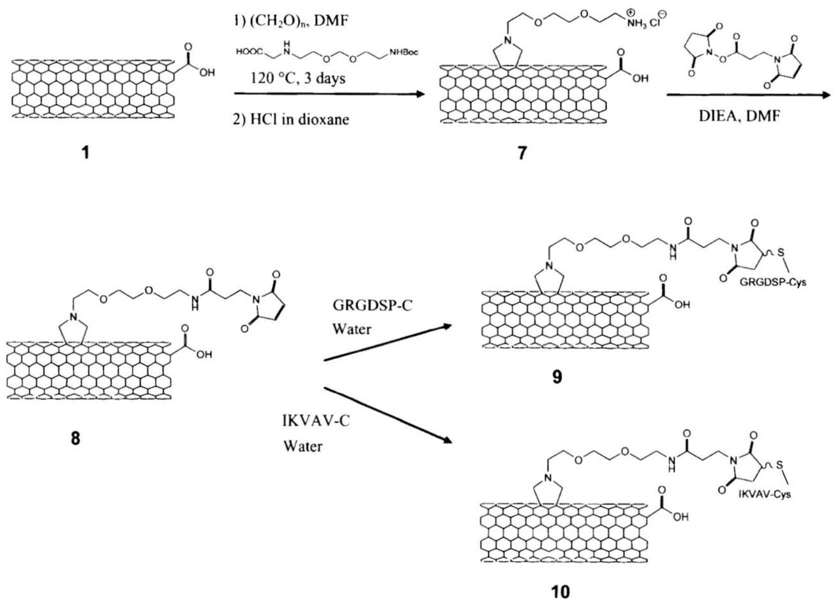

Scheme 6.3 Synthesis of peptide–CNT conjugates.

The electrophysiological responses of rat dissociated hippocampal neurons in culture incubated with the two peptide–nanotube conjugates were then studied. The spontaneous activity of single neurons in voltage clamp configuration was recorded. The neuronal activity was detected as inward currents, corresponding to synaptic events, of variable amplitude. These events represent a mixed population of inhibitory and excitatory spontaneous postsynaptic currents. Some neuronal passive properties were then examined and measured under voltage clamp. They are indicative of neuronal function such as membrane capacitance and input resistance. By quantifying peak amplitude of inward and outward currents to the membrane capacitance value of each recorded cell, inward and outward current densities were obtained. The values were found to be similar in the case of the RGD-containing peptide conjugated to MWNTs when compared with the peptide alone. Thus, the incubation of neurons in the presence of MWNTs functionalised with peptides did not modify neuronal spontaneous activity.

In summary, these data showed that peptide–nanotube conjugates incubated with neuronal cells did not alter the neuronal morphology, viability and basic functions of the neurons. These results are different from those obtained by Haddon and coworkers in which PEG-SWNTs were found to inhibit membrane endocytosis in stimulated neurons, which should lead to a progressive decrease of frequency in spontaneous PSCs, as will be detailed later.[62] Bianco *et al.* pointed out that this difference may be explained by the different types of treatment carried out on the CNTs (i.e., purification and functionalisation with different chemical groups). Thus, these data indicated that differently functionalised CNTs can have different impacts on neuronal activity, thereby highlighting the importance of determining the influence of the functionalisation of CNTs.

6.4 INVESTIGATION OF THE MECHANISMS OF THE ELECTRICAL INTERACTIONS BETWEEN CNTS AND NEURONS

Numerous studies have demonstrated that CNTs can sustain and promote neuronal electrical activity in networks of cultured neuronal cells. But the mechanisms by which they affect cellular function are still poorly understood. Little is known about the details of the interactions between CNTs and neurons. A few studies investigating possible hypotheses have been recently reported.

In vitro models have been developed to investigate the electrophysiological properties of synapses, neurons and networks coupled to CNTs, and to study the electrical interactions between CNTs and membranes. Ballerini and coworkers reviewed various existing experimental *in vitro*, *in vivo* and *in silico* models of neuronal network and compared them by listing their advantages and disadvantages.[3a] The effects of CNT substrates on the electrical behavior of brain networks *in vitro* were reported for the first time.

Studies on the interface between neurons and SWNTs were performed recently by Ballerini and coworkers. They focused on the electrical signal transfer and the synaptic stimulation in cultured brain circuits.[63] Computer simulations were included in the study to model the electrical interactions between neurons connected via SWNTs. The results, strengthened by the modelling of the CNT–neuron junction, showed that SWNTs can directly stimulate brain circuit activity.

In this study, rat hippocampal cells were grown on a film of purified SWNTs, via deposition of functionalised SWNTs and subsequent defunctionalisation. Patch-clamp experiments were then performed to determine the neuronal responses to voltage steps delivered via conductive SWNT glass coverslips. The morphology of the neuronal network was determined by SEM and quantified by immunocytochemistry analysis. SEM confirmed the adhesion and growth of hippocampal neurons on SWNT substrates, accompanied by variable degree of neurite extension. The attachment and proliferation processes were similar to those observed for neuronal growth on control glass surfaces. At high magnification, tight interactions between cell membranes and SWNTs could be visualised. The distributed electrical stimulation of cultured hippocampal neurons was then investigated. Hippocampal neurons grown on SWNTs or on control glass substrates exhibited spontaneous electrical activity. In particular, a significant increase in action potentials and frequency of PSCs was observed for neurons grown on SWNT films when compared with control substrates.

To assess whether SWNTs could perform local network stimulations, short voltage pulses were delivered to SWNTs to induce the appearance of Na^+ fast inward current in the recorded neuron. These stimulations should also induce action potentials in neighbouring neurons that should generate monosynaptic responses to the connected neurons. Indeed, brief SWNT voltage steps effectively delivered PSCs, in 65% of the recorded neurons. To elucidate the electrical interactions occurring in SWNT-neuron networks, the modelling of the neuron-SWNT junction was achieved. It was suggested that the coupling between neurons and SWNTs might be partly resistive. But the authors led to the conclusion that because of the non-idealities of the single electrode voltage clamp, eliciting Na^+ currents through SWNT stimulation did not conclusively prove a resistive coupling between SWNTs and neurons. Moreover, the authors pointed out that whole-cell patch-clamp recordings might yield deceiving results as any resistive coupling between biomembranes and SWNTs is qualitatively undistinguishable from a coupling between SWNTs and the patch pipette.

Ballerini and coworkers further investigated the influence of CNTs on the electrical properties of isolated neurons. In particular, the presence of after-depolarisation events in the cell soma was examined.[64] The aim of this study was to determine if the higher efficiency of the signal transmission of neurons grown on conductive nanotube substrates was linked to the interactions between CNTs and neurons at the nanoscale.

Purified SWNTs were deposited on glass substrates via the previously described process based on functionalisation and subsequent defunctionalisation. Rat hippocampal cells were cultured on the thin film of purified SWNTs. Electrogenesis was studied in single neurons. The authors injected a brief current pulse into the soma to force the neurons to fire a regular train of six action potentials and measured the presence of membrane depolarisation in the soma at the end of the last action potential. Neurons propagate electrical signals, i.e., action potentials, down an axon. Backpropagation of the action potentials to dendrites can occur occasionally. In this case, the propagation of the electrical signals takes place in the direction opposite to that of the flow. It has been shown that rat hippocampal neurons grown on SWNT substrates exhibited a significantly larger after-potential depolarisation (ADP) by comparison with neurons grown on control glass substrates. ADP was found to be dependent on the degree of dendritic branching. Indeed, neurons with minimal dendritic branching grown on the CNT substrate did not display ADP. The backpropagating current induces a voltage change that increases the concentration of Ca^{2+} in the dendrites and can be measured through the presence of a slow membrane depolarisation following repetitive action potentials. Therefore, the interactions between CNTs and membranes of neurons can affect single-cell activity. These experiments were also conducted on two other types of substrates: indium tin oxide substrate, displaying high conductivity, and a non-conductive nanostructured substrate containing peptides that self-assemble into nanofibres. In both cases, no significant enhancement of ADP was observed, indicating that the ADP enhancement effect is specific to CNT substrate.

The authors suggested the electrotonic hypothesis to explain the physical neuron-SWNT interactions and elucidate the mechanisms by which SWNTs might affect the electrical activity of neuronal networks. Electrotonic potential is a non-propagated local potential induced by a local change in ionic conductance. It represents changes to the neuron membrane potential that do not lead to the generation of new current by action potentials. Electrotonic potential is conducted faster than action potential, but it attenuates rapidly and is therefore unsuitable for long-distance signalling.

The authors also examined if an electrotonic shortcut could occur between the soma and the dendrite. The results showed the absence of changes brought about by SWNTs in the dendritic passive time constants. Other hypotheses were drawn to tentatively explain the increased effects of ADP, such as (i) potentiation of Ca^{2+}-mediated currents occurring in neurons grown on SWNT thin films and (ii) channels clustering, induced by mechanical interactions between SWNT bundles and the cell cytoskeleton. The presence of the ADPs, their dependence on trains of action potentials, and the detected sensitivity to calcium channel blockers are indicative of the generation of dendritic Ca^{2+} currents. The discontinuous and tight contacts between CNTs and neuronal membranes were observed by transmission electron microscopy (TEM). The SWNTs were able to modulate the physiology of the neurons. The intimate interactions of the SWNTs with a small area of the neuritic membranes favoured electrical shortcuts between the proximal and distal compartments of the neurons, thus supporting the electrotonic hypothesis. A mathematical model was proposed to simulate how ADP enhancement induced by SWNTs might affect the neuronal activity. More precisely, the aim was to model the membrane voltage input–output relationship between the soma and dendrites. The results indicated that SWNTs may be effectively short-circuiting the dendrites and soma. This shortcut would lead to diverting the electrical activity through the nanotubes, explaining the enhanced ADP effect.

In summary, this study reported that SWNTs might affect neuronal information processing. Indeed, the action potential backpropagation was substantially enhanced in neurons grown on SWNT substrates. However, the exact mechanisms at the origin of this effect are not totally clear. One hypothesis relies on the discontinuous and tight interactions between SWNTs and neuronal membranes.

Finally, following their study on neuronal growth on CNT substrates, Haddon and coworkers investigated the mechanism by which CNTs induce extension of neurite length.[63] They demonstrated that water-soluble SWNTs are able to inhibit stimulated membrane endocytosis in neurons. Following the observation that water-soluble SWNTs modified Ca^{2+} dynamics in neurons, thus leading to a decrease of the depolarisation-dependent influx of Ca^{2+} during cell stimulation, the authors investigated the effect of water-soluble SWNTs on membrane recycling. Indeed, the plasma membrane/vesicular recycling influences the extension of neurites[65] and is regulated by an influx of Ca^{2+} from the extracellular space induced by the depolarisation of neurons.

In this study, hippocampal neuronal cultures were treated with SWNT-PEG. The neurons were exposed to FM1-43 [*N*-(3-triethylammoniumpropyl)-4-(4-(dibutylamino)styryl)pyridinium dibromide],[66] a fluorescent dye that can monitor membrane recycling and that is internalised into cells by endocytosis as it does not passively diffuse across cell membranes. With regard to the constitutive vesicular recycling, taking place in unstimulated cells, the water-soluble graft copolymer SWNT-PEG had no effect on the endocytosis of FM1-43. However, in stimulated neurons (using HIK⁺), the pegylated SWNTs reduced the endocytotic loading of FM1-43 in a dose-dependent way.

In summary, the mechanism responsible for the effects of water-soluble SWNTs on enhancement of selected neurite outgrowth was elucidated. The pegylated SWNTs inhibited regulated/stimulated plasma membrane/vesicular recycling, while the constitutive phenomenon was unaffected.

6.5 CONCLUSIONS AND PERSPECTIVES

CNTs have been demonstrated as a biocompatible substrate that promotes neuronal growth, boosts neural activity and transmits electrical stimulation effectively. Indeed, many studies reported the use of CNTs as substrate to promote neuronal cell adhesion and to induce growth and differentiation of neurons. Functionalisation of CNTs can modulate some processes, such as the outgrowth and branching of neurites. In particular, the surface charge was found to have a strong effect on neuronal growth. Because of their electrical properties, CNTs have also been used as electrodes to stimulate neurons. CNTs were seen to form tight contacts with neuron cell membrane that might favour electrical shortcuts between the proximal and distal compartments of the neurons, thus altering their electrophysiological responses. Therefore, understanding the mechanisms at the origin of the effects of CNTs on neuronal cells is essential for designing functional neuronal circuits.

With regard to other materials or devices used for applications in neuroscience, CNTs show great potential as they possess a unique set of physical, chemical, mechanical and electronic properties.[67] A promising substrate candidate suitable for neuronal growth requires several characteristics, such as light weight, tight binding with neurons, controllable branching of neurites and directed neuron network formation. It should also ensure long-term cell viability. Besides, the combination of high electrical conductivity, corrosion resistance, nanoscale size, strength and flexibility is essential for neuroprosthetic devices for electrical recording and stimulation. CNTs can meet these requirements because of their dimension, high electrical conductivity and biocompatibility. Furthermore, CNTs have the exceptional characteristic to be highly flexible, while being very strong. The properties of CNTs can be, in part, transferred into CNT composites. CNTs are relatively chemically inert, but the possibility of functionalising the nanotube surface offers opportunities to tune their properties. Compared with other devices or materials, CNTs have led to very promising results. Indeed, it has been demonstrated that CNT-based substrates can serve as an extracellular scaffold with a higher capability of directing neurite outgrowth and regulating neurite branching than other types of substrates such as PLL, PEI or PLO, as explained in part 6.2. In terms of recording and stimulating neuronal activity, CNT-based electrodes exhibited better performances because of higher signal-to-noise ratio, longer lifetime and higher tissue–electrode interface in comparison with conventional commercial electrodes.

In terms of future potential applications, CNTs are promising candidates for neural prostheses for restoring the function of damaged neuronal circuits. CNTs could serve as an extracellular scaffold to guide neurite outgrowth governed by their tips and growth cones, and also to regulate neurite branching. These processes would lead to the re-establishment of intricate connections between neurons forming synapses. CNTs could be used to selectively enhance neurite elongation directly at the site of nerve injury to aid in nerve regeneration, thus increasing the chance of connecting the injured sites to sustain functional recovery. Several neurological disorders and injuries, such as Parkinson's disease, epilepsy and stroke, require an implantable device to generate electrical activity in the damaged or diseased tissue.

References

1. (a) De Robertis, E. D. P., and Bennett, H. S. (1955) Some features of the submicroscopic morphology of synapses in frog and earthworm, *J. Biophys. Biochem. Cytol.*, **1**, 47. (b) Bullock, T. H. (1959) Neuron doctrine and electrophysiology: a quiet revolution has been taking place in our concepts of how the nerve cells act alone and in concert, *Science*, **129**, 997. (c) Llinás, R. R. (1988) The intrinsic electrophysiological properties of mammalian neurons: a new insight into CNS function, *Science*, **242**, 1654. (d) Pereda, A. E., Rash, J. E., Nagi, J. I., and Bennett, M. V. L. (2004) Dynamics of electrical transmission at club endings on the Mauthner cells, *Brain Res. Brain Res. Rev.*, **47**, 227. (e) Bullock, T. H., Bennett, M. V. L., Johnston, D., Josephson, R., Marder, E., and Fields, R. D. (2005) The neuron doctrine, redux, *Science*, **310**, 791.

2. Brösamle, C., and Huber, A. B. (2006) Cracking the black box — and putting it back together again: animal models of spinal cord injury, *Drug Discov. Today Dis. Models*, **3**, 341.

3. (a) Giugliano, M., Prato, M., and Ballerini, L. (2008) Nanomaterial/neuronal hybrid system for functional recovery of the CNS, *Drug Discov. Today Dis. Models*, **5**, 37. (b) Silva, G. A. (2006) Neuroscience nanotechnology: progress, opportunities and challenges, *Nature Rev. Neurosci.*, **7**, 65. (c) Zhang, L., and Webster, T. J. (2009) Nanotechnology and nanomaterials: promises for improved tissue regeneration, *Nano Today*, **4**, 66.

4. Prato, M., Kostarelos, K., and Bianco, A. (2008) Functionalized carbon nanotubes in drug design and discovery, *Acc. Chem. Res.*, **41**, 60.

5. Mattson, M. P., Haddon, R. C., and Rao, A. M. (2000) Molecular functionalization of carbon nanotubes and use as substrates for neuronal growth, *J. Mol. Neurosci.*, **14**, 175.

6. Rüegg, U. T., and Hefti, F. (1984) Growth of dissociated neurons in culture dishes coated with synthetic polymeric amines, *Neurosci. Lett.*, **49**, 319.

7. Benson, M. D., Romero, M. I., Lush, M. E., Lu, Q. R., Henkemeyer, M., and Parada, L. F. (2005) Ephrin-B3 is a myelin-based inhibitor of neurite outgrowth, *Proc. Natl. Acad. Sci. USA*, **102**, 10694.

8. (a) Mattson, M. P., and Kater, S. B. (1987) Calcium regulation of neurite elongation and growth cone motility. *J. Neurosci.*, **7**, 4034. (b) Waeg, G., Dimsity, G., and Esterbauer, H. (1996) Monoclonal antibodies for detection of 4-hydroxynonenal modified proteins, *Free Radic. Res.*, **25**, 149.

9. Mark, R. J., Lovell, M. A., Markesbery, W. R., Uchida, K., and Mattson, M. P. (1997) A role for 4-hydroxynonenal, an aldehyde product of lipid peroxidation, in disruption of ion homeostasis and neuronal death induced by amyloid beta-peptide, *J. Neurochem.*, **68**, 255.

10. Kater, S. B., Mattson, M. P., Cohan, C., and Connor, J. (1988) Calcium regulation of the neuronal growth cone, *Trends Neurosci.*, **11**, 315.

11. Hu, H., Ni, Y., Montana, V., Haddon, R. C., and Parpura, V. (2004) Chemically functionalized carbon nanotubes as substrates for neuronal growth, *Nano Lett.*, **4**, 507.

12. (a) Chen, S. A., and Hwang, G. W. (1995) Water-soluble self-acid-doped conducting polyaniline: structure and properties, *J. Am. Chem. Soc.*, **117**, 10055. (b) Roy, B. C., Gupta, M. D., Bhowmik, L., and Ray, J. K. (1999) Studies on water soluble conducting polymer: aniline initiated polymerization of *m*-aminobenzene sulfonic acid, *Synth. Met.*, **100**, 233. (c) Zhao, B., Hu, H., and Haddon, R. C. (2004) Synthesis and properties of a water-soluble single-walled carbon nanotube-poly (*m*-aminobenzene sulfonic acid) graft copolymer, *Adv. Funct. Mater.*, **14**, 71.

13. Hu, H., Ni, Y., Mandal, S. K., Montana, V., Zhao, B., Haddon, R. C., and Parpura, V. (2005) Polyethyleneimine functionalized single-walled carbon nanotubes as a substrate for neuronal growth, *J. Phys. Chem. B*, **109**, 4285.

14. Ni, Y., Hu, H., Malarkey, E. B., Zhao, B., Montana, V., Haddon, R. C., and Parpura, V. (2005) Chemically functionalized water soluble single-walled carbon nanotubes modulate neurite outgrowth, *J. Nanosci. Nanotechnol.*, **5**, 1707.

15. Zhao, B., Hu, H., Yu, A., Perea, D., and Haddon, R. C. (2005) Synthesis and characterization of water soluble single-walled carbon nanotube graft copolymers *J. Am. Chem. Soc.*, **127**, 8197.

16. (a) Park, K. H., Chhowalla, M., Iqbal, Z., and Sesti, F. (2003) Single-walled carbon nanotubes are a new class of ion channel blockers, *J. Biol. Chem.*, **278**, 50212. (b) Chhowalla, M., Unalan, H. E., Wang, Y., Iqbal, Z., Park, K., and Sesti, F. (2005), Irreversible blocking of ion channels using functionalized single-walled carbon nanotubes, *Nanotechnology*, **16**, 2982.

17. Malarkey, E. B., Fisher, K. A., Bekyarova, E., Liu, W., Haddon, R. C., and Parpura, V. (2009) Conductive single-walled carbon nanotube substrates modulate neuronal growth, *Nano Lett.*, **9**, 264.

18. Matsumoto, K., Sato, C., Naka, Y., Kitazawa, A., Whitby, R. L. D., and Shimizu, N. (2007) Neurite outgrowths of neurons with neurotrophin-coated carbon nanotubes, *J. Biosci. Bioeng.*, **103**, 216.

19. Hempstead, B. L. (2006) Dissecting the diverse actions of pro- and mature neurotrophins, *Curr. Alzheimer Res.*, **3**, 19.

20. Reichardt, L. F. (2006) Neurotrophin-regulated signalling pathways, *Philos. Trans. R. Soc. Lond., Ser. B*, **361**, 1545.

21. Allen, S. J., and Dawbarn, D. (2006) Clinical relevance of the neurotrophins and their receptors, *Clin. Sci. (Lond.)*, **110**, 175.

22. Kang, H., and Schuman, E. M. (1996) A requirement for local protein synthesis in neurotrophin-induced hippocampal synaptic plasticity, *Science*, **273**, 1402.

23. Galvan-Garcia, P., Keefer, E. W., Yang, F., Zhang, M., Fang, S., Zakhidov, A. A., Baughman, R. H., and Romero, M. I. (2007) Robust cell migration and neuronal growth on pristine carbon nanotube sheets and yarns, *J. Biomater. Sci. Polym. Ed.*, **18**, 1245.

24. (a) Zhang, M., Fang, S., Zakhidov, A. A., Lee, S. B., Aliev, A. E., Williams, C. D., Atkinson, K. R., and Baughman, R. H. (2005) Strong, transparent, multifunctional, carbon nanotube sheets, *Science*, **309**, 1215. (b) Zhang, M., Atkinson, K. R., and Baughman, R. H. (2004) Multifunctional carbon nanotube yarns by downsizing an ancient technology, *Science*, **306**, 1358.

25. (a) Baas, P. W., and Joshi, H. C. (1992) Gamma-tubulin distribution in the neuron: implications for the origins of neuritic microtubules, *J. Cell. Biol.*, **119**, 171. (b) Moody, S. A., Miller, V., Spanos, A., and Frankfurter, A. (1996) Developmental expression of a neuron-specific beta-tubulin in frog (Xenopus laevis): a marker for growing axons during the embryonic period, *J. Comp. Neurol.*, **364**, 219.

26. (a) Letourneau, P. C. (1975) Cell-to-substratum adhesion and guidance of axonal elongation, *Dev. Biol.*, **44**, 92. (b) Letourneau, P. C. (1975) Possible roles for cell-to-substratum adhesion in neuronal morphogenesis, *Dev. Biol.*, **44**, 77. (c) Corey, J. M., and Feldman, E. L. (2003) Substrate patterning: an emerging technology for the study of neuronal behavior, *Exp. Neurol.*, **184**, S89.

27. Chao, T.-I., Xiang, S., Chen, C.-S., Chin, W.-C., Nelson, A. J., Wang, C., and Lu, J. (2009) Carbon nanotubes promote neuron differentiation from human embryonic stem cells, *Biochem. Biophys. Res. Commun.*, **384**, 426.

28. (a) Freire, E., Gomes, F. C., Linden, R., Neto, V. M., and Coelho-Sampaio, T. (2002) Structure of laminin substrate modulates cellular signaling for neuritogenesis, *J. Cell Sci.*, **115**, 4867. (b) Kleinman, H. K., Cannon, F. B., Laurie, G. W., Hassell, J. R., Aumailley, M., Terranova, V. P., Martin, G. R., and DuBois-Dalcq, M. (1985) Biological activities of laminin, *J. Cell. Biochem.*, **27**, 317. (c) Luckenbill-Edds, L. (1997) Laminin and the mechanism of neuronal outgrowth, *Brain Res. Brain Res. Rev.*, **23**, 1. (d) He, W., and Bellamkonda, R. V. (2005) Nanoscale neuro-integrative coatings for neural implants, *Biomaterials*, **26**, 2983. (e) Sephel, G. C., Burrous, B. A., and Kleinman, H. K. (1989) Laminin neural activity and binding proteins, *Dev. Neurosci.*, **11**, 313.

29. Xie, J., Chen, L., Aatre, K. R., Srivatsan, M., and Varadan, V. K. (2006) Somatosensory neurons grown on functionalized carbon nanotube mats, *Smart Mater. Struct.*, **15**, N85.

30. Ayad, L. (1994) *The Extracellular Matrix Factsbook*, Plenum, New York.

31. Belyanskaya, L., Weigel, S., Hirsch, C., Tobler, U., Krug, H. F., and Wick, P. (2009) Effects of carbon nanotubes on primary neurons and glial cells, *Neurotoxicology*, **30**, 702.

32. Wick, P., Manser, P., Limbach, L. K., Dettlaff-Weglikowska, U., Krumeich, F., Roth, S., Stark, W. J., and Bruinink, A. (2007) The degree and kind of agglomeration affect carbon nanotube cytotoxicity, *Toxicol. Lett.*, **168**, 121.

33. (a) Wilson, B. S., Lawson, D. T., Muller, J. M., Tyler, R. S., and Kiefer, J. (2003) Cochlear implants: some likely next steps, *Annu. Rev. Biomed. Eng.*, **5**, 207. (b) Weiland, J. D., Liu, W., and Humayun, M. S. (2005) Retinal prosthesis, *Annu. Rev. Biomed. Eng.*, **7**, 361. (c) Navarro, X., Krueger, T. B., Lago, N., Micera, S., Stieglitz, T., and Dario, P. (2005) A critical review of interfaces with the peripheral nervous system for the control of neuroprostheses and hybrid bionic systems, *J. Peripher. Nerv. Syst.*, **10**, 229.

34. Benabid, A. L. (2003) Deep brain stimulation for Parkinson's disease, *Curr. Opin. Neurobiol.*, **13**, 696.

35. Zhang, X., Prasad, S., Niyogi, S., Morgan, A., Ozkan, M., and Ozkan, C. S. (2005) Guided neurite growth on patterned carbon nanotubes, *Sens. Actuators B Chem.*, **106**, 843.

36. (a) Qian, L., and Saltzman, W. M. (2004) Improving the expansion and neuronal differentiation of mesenchymal stem cells through culture surface modification, *Biomaterials*, **25**, 1331. (b) Yavin, E., and Yavin, Z. (1974) Attachment and culture of dissociated cells from rat embryo cerebral hemispheres on polylysine-coated surface, *J. Cell Biol.*, **62**, 540.

37. Ben-Jacob, E., and Hanein, Y. (2008) Carbon nanotube micro-electrodes for neuronal interfacing, *J. Mater. Chem.*, **18**, 5181.

38. Gabay, T., Jakobs, E., Ben-Jacob, E., and Hanein, Y. (2005) Engineered self-organization of neural networks using carbon nanotube clusters, *Physica A*, **350**, 611.

39. Sorkin, R., Gabay, T., Blinder, P., Baranes, D., Ben-Jacob, E., and Hanein, Y. (2006) Compact self-wiring in cultured neural networks, *J. Neural Eng.*, **3**, 95.

40. Gabay, T., Ben-David, M., Kalifa, I., Sorkin, R., Abrams, Z. R., Ben-Jacob, E., and Hanein, Y. (2007) Electro-chemical and biological properties of carbon nanotube based multi-electrode arrays, *Nanotechnology*, **18**, 035201.

41. Wang, K., Fishman, H. A., Dai, H., and Harris, J. S. (2006) Neural stimulation with a carbon nanotube microelectrode array, *Nano Lett.*, **6**, 2043.

42. Sorkin, R., Greenbaum, A., David-Pur, M., Anava, S., Ayali, A., Ben-Jacob, E., and Hanein, Y. (2009) Process entanglement as a neuronal anchorage mechanism to rough surfaces, *Nanotechnology*, **20**, 015101.

43. Shoval, A., Adams, C., David-Pur, M., Shein, M., Hanein, Y., and Sernagor, E. (2009) Carbon nanotube electrodes for effective interfacing with retinal tissue, *Front. Neuroengineering*, **2**, 4.

44. Keefer, E. W., Botterman, B. R., Romero, M. I., Rossi, A. F., and Gross, G. W. (2008) Carbon nanotube coating improves neuronal recordings, *Nat. Nanotechnol.*, **3**, 434.

45. Gheith, M. K., Sinani, V. A., Wicksted, J. P., Matts, R. L., and Kotov, N. A. (2005) Single-walled carbon nanotubes polyelectrolyte multilayers and freestanding films as a biocompatible platform for neuroprosthetic implants, *Adv. Mater.*, **17**, 2663.

46. Mamedov, A. A., Kotov, N. A., Prato, M., Guldi, D. M., Wicksted, J. P., and Hirsch, A. (2002) Molecular design of strong SWNT/polyelectrolyte multilayers composites, *Nature Mater.*, **1**, 190.

47. Gheith, M. K., Pappas, T. C., Liopo, A. V., Sinani, V. A., Shim, B. S., Motamedi, M., Wicksted, J. P., and Kotov, N. A. (2006) Stimulation of neural cells by lateral currents in conductive layer-by-layer films of single-walled carbon nanotubes, *Adv. Mater.*, **18**, 2975.

48. Jan, E., and Kotov, N. A. (2007) Successful differentiation of mouse neural stem cells on layer-by-layer assembled single-walled carbon nanotube composite, *Nano Lett.*, **7**, 1123.

49. Kam, N. W. S., Jan, E., and Kotov, N. A. (2009) Electrical stimulation of neural stem cells mediated by humanized carbon nanotube composite made with extracellular matrix protein, *Nano Lett.*, **9**, 273.

50. Freire, E., Gomes, F. C. A., Linden, R., Neto, V. M., and Coelho-Sampaio, T. (2002) Structure of laminin substrate modulates cellular signaling for neuritogenesis, *J. Cell Sci.*, **115**, 4867.

51. Stuart, G. J., and Sakmann, B. (1994) Active propagation of somatic action potentials into neocortical pyramidal cell dendrites, *Nature*, **367**, 69.

52. Yeh, S.-R., Chen, Y.-C., Su, H.-C., Yew, T.-R., Kao, H.-H., Lee, Y.-T., Liu, T.-A., Chen, H., Chang, Y.-C., Chang, P., and Chen, H. (2009) Interfacing neurons both extracellularly and intracellularly using carbon-nanotube probes with long-term endurance, *Langmuir*, **25**, 7718.

53. Nguyen-Vu, T. D. B., Chen, H., Cassell, A. M., Andrews, R., Meyyappan, M., and Li, J. (2006) Vertically aligned carbon nanofiber arrays: an advance toward electrical-neural interfaces, *Small*, **2**, 89.

54. Lovat, V., Pantarotto, D., Lagostena, L., Cacciari, B., Grandolfo, M., Righi, M., Spalluto, G., Prato, M., and Ballerini, L. (2005) Carbon nanotube substrates boost neuronal electrical signaling, *Nano Lett.*, **5**, 1107.

55. Liopo, A. V., Stewart, M. P., Hudson, J., Tour, J. M., and Pappas, T. C. (2006) Biocompatibility of native and functionalized single-walled carbon nanotubes for neuronal interface, *J. Nanosci. Nanotechnol.*, **6**, 1365.

56. Bahr, J. L., and Tour, J. M. (2001) Highly functionalized carbon nanotubes using in situ generated diazonium compounds, *Chem. Mater.*, **13**, 3823.

57. Gabriel, G., Gómez, R., Bongard, M., Benito, N., Fernández, E., and Villa, R. (2009) Easily made single-walled carbon nanotube surface microelectrodes for neuronal applications, *Biosens. Bioelectron.*, **24**, 1942.

58. Gabriel, G., Gómez-Martínez, R., and Villa, R. (2008) Single-walled carbon nanotubes deposited on surface electrodes to improve interface impedance, *Physiol. Meas.*, **29**, S203.

59. Gaillard, C., Cellot, G., Li, S., Toma, F. M., Dumortier, H., Spalluto, G., Cacciari, B., Prato, M., Ballerini, L., and Bianco, A. (2009) Carbon nanotubes carrying cell-adhesion peptides do not interfere with neuronal functionality, *Adv. Mater*, **21**, 2903.

60. Gall, C. M., Pinkstaff, J. K., Lauterborn, J. C., Xie, Y., and Lynch, G. (2003) Integrins regulate neuronal neurotrophin gene expression through effects on voltage-sensitive calcium channels, *Neuroscience*, **118**, 925.

61. Anderson, D. G., Burdick, J. A., and Langer, R. (2004) Smart biomaterials, *Science*, **305**, 1923.

62. Malarkey, E. B., Reyes, R. C., Zhao, B., Haddon, R. C., and Parpura, V. (2008) Water soluble single-walled carbon nanotubes inhibit stimulated endocytosis in neurons, *Nano Lett.*, **8**, 3538.

63. Mazzatenta, A., Giugliano, M., Campidelli, S., Gambazzi, L., Businaro, L., Markram, H., Prato, M., and Ballerini, L. (2007) Interfacing neurons with carbon nanotubes: electrical signal transfer and synaptic stimulation in cultured brain circuits, *J. Neurosci.*, **27**, 6931.

64. (a) Cellot, G., Cilia, E., Cipollone, S., Rancic, V., Sucapane, A., Giordani, S., Gambazzi, L., Markram, H., Grandolfo, M., Scaini, D., Gelain, F., Casalis, L., Prato, M., Giugliano, M., and Ballerini, L. (2009) Carbon nanotubes might improve neuronal performance by favouring electrical shortcuts, *Nat. Nanotechnol.*, **4**, 126. (b) Silva, G. A. (2009) Shorting neurons with nanotubes, *Nat. Nanotechnol.*, **4**, 82.

65. Zakharenko, S., and Popov, S. (2000) Plasma membrane recycling and flow in growing neuritis, *Neuroscience*, **97**, 185.

66. Betz, W. J., and Bewick, G. S. (1992) Optical analysis of synaptic vesicle recycling at the frog neuromuscular junction, *Science*, **255**, 200.

67. (a) Haddon, R. C. (2002) Carbon nanotubes, *Acc. Chem. Res.*, **35**, 997. (b) Ouyang, M., Huang, J. L., and Lieber, C. M. (2002) Fundamental electronic properties and applications of single-walled carbon nanotubes, *Acc. Chem. Res.*, **35**, 1018. (c) Dai, H. (2002) Carbon nanotubes: synthesis, integration, and properties, *Acc. Chem. Res.*, **35**, 1035. (d) Zhou, O., Shimoda, H., Gao, B., Oh, S., Fleming, L., and Yue, G. (2002) Materials science of carbon nanotubes: fabrication, integration, and properties of macroscopic structures of carbon nanotubes, *Acc. Chem. Res.*, **35**, 1045. (e) Niyogi, S., Hamon, M. A., Hu, H., Zhao, B., Bhowmik, P., Sen, R., Itkis, M. E., and Haddon, R. C. (2002) Chemistry of single-walled carbon nanotubes, *Acc. Chem. Res.*, **35**, 1105.

Chapter 7

BIOMEDICAL APPLICATIONS VI: CARBON NANOTUBES AS BIOSENSING AND BIO-INTERFACIAL MATERIALS

Yupeng Ren

Shanghai Institute of Materia Medica, Chinese Academy of Sciences, Shanghai, China
renyunpeng@yahoo.com

7.1 INTRODUCTION

Since their discovery,[1] carbon nanotubes (CNTs) were widely studied for their application as materials for bio-engineering, which include two promising areas: biosensing and tissue engineering.[2] More and more research work showed that CNTs are novel materials with advanced properties.

First of all, the special structure of CNTs can be looked on as wrapped graphite layers. CNTs' electronic properties are dependent on the way of the wrapping. CNTs could be divided into two groups: metallic CNTs and semiconducting CNTs. Both of them have demonstrated potential for application in development of nano-electric devices. Moreover, CNTs possess advanced physical, chemical and biological properties, which make them possible candidates for tissue engineering.

This chapter focuses on the forefront research of utilising CNTs to produce medical and biological tools, including biosensing devices and tissue engineering materials. The feasibility of design of CNT-based biosensing devices, such as nano-transistors or nano-capacitors, is mainly due to the electric properties of CNTs. On the other hand, CNT-based materials for tissue engineering takes advantage of CNTs' physical, chemical and biological characteristics. Those novel materials include bone growth scaffold, neural cell grow matrix and so forth.

Carbon Nanotubes: From Bench Chemistry to Promising Biomedical Applications
Edited by Giorgia Pastorin
Copyright © 2011 Pan Stanford Publishing Pte. Ltd.
www.panstanford.com

7.2 BIOSENSOR

CNTs have many novel electronic properties, which have profound value for further developing them into functional devices. The applications of CNTs in the areas of chemical/biological analysis and sensing are strongly based on these properties.

7.2.1 Structure and Electric Properties of CNTs

The structure of CNTs can be thought of as seamless cylinders wrapped up from graphitic sheets with a hexagonal lattice.[1] As shown in Fig. 7.1, to make two equivalent sites of the hexagonal lattice coincide, the wrapping should follow the vector **C**, which defines the direction of the wrapping and relative position of the two sites. A pair of integers (n, m) is often used to describe the vector **C**. For example, Fig. 7.1 shows the formation of an (11, 7) CNT. With different values of n and m, the structure of CNTs can be grouped into three types. CNTs are "armchair" tubes when n equals m, and "zigzag" tubes in case n equals 0. Other CNTs belong to the "chiral" type, with the φ between 0° and 30°.[3] The electronic properties of CNTs are critically dependent on the tube structure indices (n, m). Wilder and collaborators[3] found that when $n = m$ or $n - m = 3l$ (where l is an integer), the CNTs are metallic. If $n - m \neq 3l$, the CNTs are semiconducting. This correlated with the study of Odom's group,[4] which described the providence of being semiconducting as $(2n + m)/3$ being an integer.

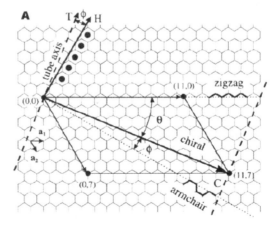

Figure 7.1 Wrapping of a graphite sheet to form a CNT. The wrapping vector **C** determines the structure of the formed CNT. The CNT is armchair tube when $\varphi = 0°$ $(n = m)$, zigzag tube when $\Phi = 30°$ $(n = 0)$, and chiral type when $0 < \varphi < 30°$. Redrawn from Wilder *et al.*[3].

The semiconducting CNTs have the chiral angle φ fall between 0° and 30° and $n - m \neq 3l$. The semiconductivity of CNTs was reported in some experiments.[3,4] When analysed by scanning tunneling spectroscopy, the chiral CNTs showed low conductance at low bias voltage but significantly reduced resistance at high voltage. Chiral CNTs showed several kinks at high bias voltage, while the metal CNTs showed a constant conductance.[3] The energy gap was measured of the order of ~0.5 eV. Furthermore, the Odom group's research clearly showed the conductance of the CNTs with indices of (12, 3) and (14, 3), following the model of metal and semiconducting CNTs, respectively.[4]

The metallic CNTs with ballistical electron transport are very good conductors. Because of the specific wrapping of CNTs, the electronic structure of CNTs is nearly one-dimensional, making the electronic transport in CNT ballistical.[5] This transport renders CNTs highly conductive with essentially no heating.[6,7] Previous studies showed that the room-temperature resistances of most metallic CNTs were below 100 kΩ and that some were around 15 kΩ, while the lowest resistance values were observed at 7 kΩ. These findings suggested metallic CNTs as excellent materials for producing electronic devices.

In addition to the transportation of electrons, propagation of phonons along the CNTs was also observed ballistically.[8] The phonon mean free path of the CNTs was estimated around hundreds of nanometres, comparable to the length of some CNTs, which may range from a few micrometres to hundreds of nanometres. This indicates that the phonons have only a few scattering events between the thermal reservoirs and that the phonon transport is "nearly" ballistic. Therefore CNTs could be good conductors for thermal transport. The theoretical prediction was supported by the data that the measured room temperature thermal conductivity for an individual multi-wall carbon nanotube (MWNT) was 3000 W/m K, greater than that of natural diamond and the basal plane of graphite (both 2000 W/m K).

With nanoscale dimension and special structure, CNTs often show quantum characteristics. The prediction that the conductance is quantised when electrons flow ballistically through CNTs[9] was proven by experiment designed by Frank,[7] who measured the conductance of CNTs dipped into liquid metal surface. The conductance of CNTs remained zero when the depth of the dipping was below 2 μm but jumped up and remained at a constant value when the insertion length was from 2 to 2.5 μm. The result was inconsistent with classical conductors and indicated that the CNTs were ballistic conductors and quantum resistors. Besides the electric conductance, the thermal conductance was also observed as quantised[10] and correlated with theoretical predictions for phonon transport in a ballistic, one-dimensional channel.[11]

In brief, carbon atoms arranged in condensed benzene rings such as in CNTs represent unique nanoscale tube structures. Because of their specific molecular structure, they show properties that distinguish them from other materials.

7.2.2 CNTs as Electric Sensors

7.2.2.1 CNT-based electric devices

Because of the novel properties described above, CNTs were used to produce many kinds of electric devices, such as transistors. Developments in these areas have critical meaning for utilising CNTs for chemical and biological sensing and analysis.

Semiconductive CNTs have been often used as the key component of field-effect transistors (FETs) at room temperature.[12,13] Transistors are semiconductor devices used to amplify or switch electronic signals. The electron transport in the transistors may be dominated by electrons or by holes. FETs are composed of three parts: source, drain and gate, which control the switch by electric field. CNT-based FETs were first prepared by linking the source and drain electrodes by a semiconductive CNT (Fig. 7.2), normally with Au[13] or Pt[12] used as the source and drain electrode materials. The gate voltage was used to control the conductivity of the CNTs. The source–drain current in a single-wall carbon nanotube (SWNT)-based FET decreased strongly with increasing gate voltage. Such transfer characteristics of current–gate voltage of a CNT-based transistor indicated that the transistor was a p-channel FET and the conductance was dominated by holes (positive carriers).[12] By modulating the gate voltage the conductance of the SWNT FET could be modulated by several orders of magnitude.

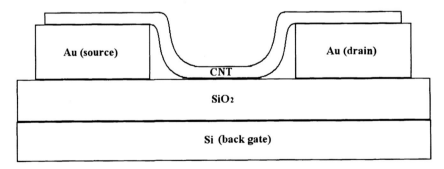

Figure 7.2 Scheme of a CNT-based field-effect transistor. The CNT bridges the drain and source electrodes. The gate voltage controls the "on" and "off" of the transistor.

Moreover, the materials used to prepare nano-transistors are not limited to semiconductive nanotubes. Because of their ultra-small and regular structure, metal CNTs could be used to produce single-electron transistors working at room temperature.[14] This was achieved by introducing two buckles in an individual metallic SWNT using AFM. The conductance of the resulting device was affected by the gate voltage and showed opening and closing effects in a periodic manner, thus demonstrating Coulomb blockade (single-electron tunneling) at room temperature.

However, it was a challenge to precisely connect CNTs to each part of the nano-devices and assemble them into nano-transistors. Therefore, Keren *et al.*[15] reported a novel method to precisely prepare those nano-transistors. In their study, they used DNA-RecA protein as the template. Then CNTs assembled with DNA by a lot of RecA/anti-RecA/biotin antimouse/ streptavidin linkers. The protruded terminals of DNA directed the deposition of gold, which formed the electrodes linking the CNTs. This study made it feasible to efficiently produce transistors with a precise structure.

In addition to the transistors, CNTs can be manipulated by various techniques to prepare many types of functional molecular devices. For example, CNTs can be used to produce wafers.[16] This was achieved by assembling individual CNTs hierarchically by two self-assembly stages and form closely packed and aligned nanotube films. Another example is the preparation of serpentines using CNTs (Fig. 7.3).[17] During the process of CNT synthesis, the CNTs grew standing up from the surface and then fell on the substrate surface. The structure of the serpentines could be controlled and were dramatically dependent on the gas flow: in fact, by changing the direction and rate of the gas flow, the structure of the serpentines could be controlled precisely.

Further applications of these CNT-based nano-devices resulted in inventions of several kinds of nano-machines, such as nano-sensors, nano-antennas, heating and cooling elements, optoelectronic devices and single-molecule dynamos. An interesting example of such applications is the nanotube radio.[18] The four critical radio components, antenna, tuner, amplifier and demodulator, can be implemented with a single CNT. Compared with the traditional radio, a special point of the CNT radio was that it sensed the electromagnetic waves and responded by physical vibration.

All of the studies and applications listed above suggested that, because of their novel properties, CNTs could be used to produce functional nano-electric devices in the areas of electronic transport as well as signal detection. Therefore, CNTs have become more and more attractive in the area of chemical/biological sensing and analysis.

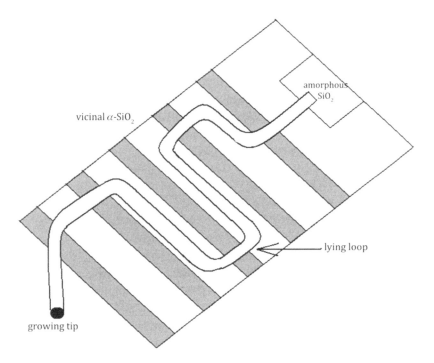

vicinal α-SiO$_2$

amorphous SiO$_2$

lying loop

growing tip

Figure 7.3 CNT falling on the substrate surface with gas flow forms a CNT serpentine. Reproduced from Geblinger *et al.*[17] with permission.

7.2.2.2 CNT-based sensors

Possessing novel properties, CNTs have been used as functional materials for chemical/biological analysis, especially in the area of nano-sensor production. On the basis of their intrinsic mechanisms, these sensors can be divided into many subtypes.

7.2.2.2.1 Mass/force sensor

Defect-free CNTs are the materials with the highest strength, low density and ultra-small cross-sections. Meanwhile, a nanotube can act as a transistor and thus may be able to sense its own motion. Therefore, CNTs were used to prepare nano-electromechanical systems proposed to detect mass or force with ultra-sensitivity. Sazonova *et al.*[19] reported of suspending CNTs over a trench (1.2–1.5 µm wide, 500 nm deep) between two metal electrodes to form a transistor structure. Then a gate voltage was used to induce an additional charge on the CNTs. The induced opposite charges on the gate caused an electrostatic force downward on the CNTs. This force controlled the tension of the CNTs and resulted in guitar-string-like oscillation. When a time-varying voltage was added, a driving frequency was produced. The transistor properties of semiconducting and small-band-gap semiconducting CNTs made it convenient to detect the driving force and driving frequency, which could be monitored by measuring the conductivity. Such a device can be used to transduce very small forces into electric signals.

7.2.2.2.2 *Chemical sensors*

The special atom structure and electronic structure make CNTs suitable to develop novel devices for the detection of various types of chemicals. Applications of CNT-based sensors on detection of both small molecules and macromolecules have been explored.

CNTs have been shown to be extremely sensitive to chemical gas. The electronic properties of nanotubes, including the electrical resistance R, thermoelectric power S (voltages between junctions caused by interjunction temperature differences) and local density of states $N(E)$, are exceedingly sensitive to environmental conditions, such as gas exposure. Collins *et al.*[20] reported that the exposure to air or oxygen dramatically influences the nanotubes' electrical resistance and thermoelectric power, which can be easily measured through the electric signal. In addition, these electronic parameters can be reversibly tuned. In this study, the resistance of an SWNT sample, with the surrounding medium cycled between vacuum and air, was detected. A rapid and reversible change in the SWNT resistance occurred in concomitance with the changing environment. The resistance of the CNT was higher in vacuum and decreased by 10–15% when it was surrounded by air. Moreover, the thermoelectric power S also showed a high sensitivity to chemical gas exposure. When CNTs were exposed to air, their thermoelectric power was a constant value and with a positive magnitude near 120 mV/K. However, when oxygen was gradually removed from the chamber containing the CNTs, the thermoelectric power changed continuously from positive to negative, with a final equilibrium value of ~210 mV/K. When oxygen was reintroduced into the chamber, the thermoelectric power again reversed sign and once again became positive. These results indicated CNTs as sensitive sensors for oxygen.

Kong *et al.*[21] reported using a single SWNT-based transistor to detect NO_2 and NH_3 gases. Gas-sensing experiments were carried out by exposing a semiconducting SWNT sample in flowing diluted NO_2 (2–200 parts per million) or NH_3 (0.1–1%) and monitoring the resistance of the SWNTs. Upon exposure to gaseous molecules NO_2 or NH_3, the electrical resistance of the semiconducting SWNT dramatically changed. The nanotube sensors exhibited a fast response and a substantially higher sensitivity than that of other existing solid-state sensors at room temperature. At a gate voltage of +4 V, the conductance of the SWNT sample increased sharply by about three orders of magnitude after 200 ppm of NO_2 was introduced. The response times of the resistance changed by one order of magnitude for 200 ppm of NO_2 in the range of 2–10 seconds. The conductance increase was reversible because of the NO_2 molecules' desorption from the nanotube sample. Moreover, after exposing the recovered SWNT sample to 1% NH_3 molecules, the current versus voltage curve showed a 100-fold conductance depletion, with a response time of about 1–2 minutes. Therefore, the electric conductance provided information about the target molecules. On the other hand, the back gate voltages, able to induce a high conductance for SWNTs exposed to NO_2 or NH_3 molecules, were about +4 V and 0 V, respectively, suggesting that the selectivity of different molecules can be achieved by adjusting the electrical gate to set the SWNT sample in an initial conducting or insulating state.

Interaction between CNTs and Br led to substantially change the conductance.[22] As a prototypical electron acceptor, doped Br decreased the resistivity of CNT at 300 K and enlarged the region where the temperature coefficient of resistance is positive (the signature of metallic behaviour). In addition, intercalation of charged polyiodide chains into the interstitial channels in an SWNT rope lattice resulted in a new carbon material with a significantly low resistance.[23]

The mechanism of chemical sensors lies on two main aspects: (1) the chemical nature of the molecules and (2) the CNT samples, which appear to be hole-doped (p-type). Oxidising molecules, including NO_2, Br, oxygen and air, have unpaired electrons, and there would be charge transfer from SWNTs to these molecules. Therefore, the conductivity of the p-type CNTs would increase. Removal of these molecules from CNT samples diminishes the charge transfer and causes a reverse effect. On the other hand, charge would be transferred to CNTs when Lewis base molecules, such as NH_3, are targeting compounds. The effect of the charge on the hole would result in increased resistance. These studies have indicated that utilising CNT-based transistors or conductors as chemical sensors is not limited to specific chemicals. The nanotube electronic transport may be sensitive to other substances as long as they can affect the amount of charge transfer between CNTs. In this way it is possible to design many CNT transistor-based nano-sensors to detect various types of molecules.

Besides transistors, a chemicapacitor was constructed from SWNT electrodes, and it showed the features of stability, high sensitivity to broad range of analytes and fast response time. Snow *et al.*[24] designed a nanoscale electrode by using an SWNT network. The electrode served as one plate of the capacitor, while Si was another electrode. Fringing electric fields radiated outward from the SWNTs under an applied bias voltage. The strongest fringing fields at the surface of the SWNTs produced a net polarisation of the adsorbates, which changed the capacitance of the system. The SWNT capacitor responded to a number of chemical molecules, including *N,N*-dimethylformamide (DMF), benzene, hexane, chloroform, isopropyl alcohol, tetrahydrofuran, acetone, dimethylmethylphosphonate and so on. Sensing chemicals by the capacitor was quick and reversible, and the response was proportional to the analyte concentration. The magnitude of the capacitance response depended on two factors. The first one was the dipole moment of the analytes. Non-polar molecules such as hexane produced small responses, whereas polar molecules like DMF produced a large capacitance response. The other factor was the surface interaction between the analytes and the capacitor. Because of the surface interactions between the analytes and the capacitor, chlorobenzene, 1,2-dichlorobenzene, 4-dinitrotoluene and water showed a low response. Covering the SWNT electrode dramatically increased the sensitivity for detecting chemicals, while introducing functional groups on the SWNTs greatly shortened the response time by hundreds of times and improved the selectivity of the sensor.

Further study on the mechanism of the CNT chemicapacitor was carried out by exploring the electronic response of SWNTs to trace levels of chemical vapours; Robinson *et al.*[25] found that adsorption at defect sites produced a large electronic response, which resulted from increased adsorbate binding energy and charge transfer and dominated the SWNTs' capacitance and conductance sensitivity. The sensitivity of an SWNT network sensor could be enhanced by controlled introduction of oxidation defects on the tubes.

Besides the pristine CNTs, modified CNTs appeared to be more functional in chemical and biological sensing. Functionalised CNTs could be used as electron conductors in enzyme-based electrochemical sensors.[26] The electrocatalytic oxidation o(1,4-dihydronicotinamide adenine dinucleotide [phosphate]) cofactors and the reduction/oxidation of H_2O_2 stimulated by CNTs are particularly important, since these electrocatalytic reactions may be easily coupled to enzymatic transformations.[27-28] Amperometric biosensors have been prepared by encapsulating alcohol dehydrogenase or glucose oxidase in CNT/Teflon composite material to analyse ethanol and glucose.

The enzymes yielded NADH or H_2O_2 in the presence of NAD+ and ethanol or O_2 and glucose, respectively. As the CNTs were linked with those enzymes, they could function as amperometric sensors for H_2O_2 or O_2 detection.

CNT-based nano-electric devices not only could be used to detect small chemical molecules, but they could be also applied in the area of biological macromolecule analysis. CNT FETs could monitor the change in surface charge when DNA was hybridised on the sensor surface. Star *et al.*[29] reported wrapping SWNTs using single-stranded DNA (ssDNA) by means of the aromatic interactions between nucleotide bases and SWNTs. The ssDNA 5-CCT AAT AAC AAT-3, when wrapped on an SWNT, could function as a capture probe to detect the complementary DNA sequences. The match and mismatch of the target DNA changed the transistor characteristics, such as G–Vg transfer curves, and gave out obvious electric signals for accurate monitoring.

Moreover, CNT-based nano-sensors have been explored for highly sensitive electronic detection of proteins, such as antibodies. Chen *et al.*[30]

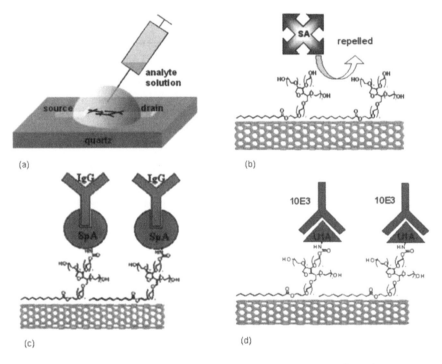

Figure 7.4 (a) A nano-protein detector composed of a CNT-based transistor. Functionalised CNT for specifically sensing (b) SA, (c) IgG and (d) mAbs. Reproduced from Chen *et al.*[30] with permission. Copyright (2003) National Academy of Sciences, U.S.A.

reported the study of developing highly specific electronic biomolecule detectors for proteins using CNTs as a platform. Nanotubes can be directly used as an electronic analytical tool to detect and monitor protein adsorption with high sensitivity. In Fig. 7.4a is shown a typical of such electronic devices, in which a semiconductive CNT-based transistor plays a crucial role. As the transistor could respond to non-specific bonded protein, some polymers, such as Tween 20, were used to cover the CNTs for protein resistance, while specific binding with proteins was re-enabled by covalent bonding to the polyethylene oxide (PEO)-functionalised nanotubes. Applications of such sensors included the detection of streptavidin (SA) (where biotin was conjugated on the tween terminal by 1,1-carbonyldiimidazole-mediated nucleophilic addition (Fig. 7.4b), the detection of IgG (with staphylococcal protein A [SpA]) conjugated (Fig. 7.4c) and the detection the mAbs (with human autoantigen immobilised [Fig. 7.4d]). Binding of the functional molecules resulted in an electric response in the CNTs and made the direct electronic readout possible, providing a more convenient technique in comparison with the traditional protein labelling method.

Star *et al.*[31] prepared a nano-FET for detection of proteins. The non-specific binding between CNTs and proteins was eliminated by polymer wrapped on CNTs. The biotin conjugated on the polymer could bind with streptavidin. The conductance, as well as the gate voltage of the nano-FET, changed significantly upon exposure to streptavidin.

Because of the ability to detect proteins as described above, CNTs were used for some medical purposes, such as cancer diagnosis. Li[32] used a CNT-based transistor to detect prostate-specific antigen (PSA), which is an oncological marker for prostate cancer. PSA monoclonal antibodies linked with CNTs were used to anchor PSA. The conductance and gate voltage of the transistor responded to PSA sensitively and specifically. Therefore it is plausible that more novel therapeutic and diagnostic devices will be developed using CNTs in future.

7.2.2.2.3 *Structure sensor*

The helix structure of CNTs might recognise special chemical molecules. Ju[33] reported the selective assemble of chiral CNTs with other compounds: for example, the flavin mononucleotide wrapping around specific SWNTs in a helical pattern. The selection was due to the structure of the assembled molecules. The chiral CNT has the helix structure, while the flavin mononucleotide consists of an aromatic isoalloxazine moiety and a chiral D-ribityl phosphate group. The isoalloxazine moiety of the flavin mononucleotide may form pi–pi interactions with SWNTs, and the uracil moiety of isoalloxazine can self-assemble into ribbons on the CNTs' surface. The assembly imparts efficient individualisation and chirality selection to CNTs. In other words, CNTs function as a template platform to select a specific isoalloxazine ribbon structure.

CNTs are also sensitive to the structure of lipid molecules. Different lipid derivatives may form ring, helix or double helix structures on CNTs.[34] Triton X-100, adsorbed onto the CNT surface through π-staking interactions, leads a full coating without any organisation, while octadecyltrimethylammonium bromide, with a long carbon atom aliphatic chain, may form a regular helix structure on CNTs, suggesting the CNTs' capability to distinguish different types of compounds.

7.2.2.2.4 Electric probes

As a sensitive nanoscale structure with outstanding electronic properties, CNTs have often been explored to be used as electric probes. Guo *et al.*[35] reported that CNTs can be used as probes to detect the mismatch of a DNA duplex. The CNTs provided two electric poles and a single DNA duplex bridging a CNT gap (Fig. 7.5). The nano-structure can function as a nano-transistor as well as a nano-resistor. How well the DNA duplex matched could be easily read through the source–drain current (I_{SD}) value. A well-matched DNA duplex showed low resistance, while a mismatch increased the resistance by hundreds of times. Therefore, such nano-devices could provide biochemical information for DNA reorganisation and detection. Similar results were reported by Roy and coworkers.[36] They used the carbon tips to detect the integrity of double-strain (ds)DNA. The dsDNA bridged the gap between two CNT tips and showed an obvious conductivity; on the contrary, conductivity decreased to a very low level after destroying of the double-strand structure. Again this study demonstrated that SWNTs can be employed as nano-electrodes to detect the conformation and structure of DNA chains.

Other studies on DNA detection using CNTs were performed by many groups. Cai and collaborators[37] reported the ability to detect DNA through a CNT-enhanced electrochemical DNA biosensor. They covalently bonded oligonucleotides with CNTs. When the CNT-ssDNA formed hybridisation with target DNA strands, daunomycin could intercalate in the DNA duplex and increased the electrochemical response. Then the CNTs functioned as molecular wires and accepted electrons from the redox intercalator through the dsDNA. Li *et al.*[38] produced a similar nano-device to detect targeting DNA chains, but they designed the detection method on the basis of Ru(bpy)$_3^{2+}$-mediated guanine oxidation. It seems that CNTs are particularly attractive for scientists involved in DNA detection and identification.

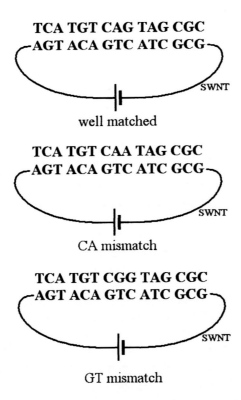

TCA TGT CAG TAG CGC
AGT ACA GTC ATC GCG
SWNT

well matched

TCA TGT CAA TAG CGC
AGT ACA GTC ATC GCG
SWNT

CA mismatch

TCA TGT CGG TAG CGC
AGT ACA GTC ATC GCG
SWNT

GT mismatch

Figure 7.5 Schematic illustration of well match (WM), C and A mismatch (CA) and G and T mismatch DNA duplex bridging between CNT poles. Conductance decreased when mismatch was introduced in the DNA duplex.

7.2.2.2.5 *Microscope sensors*

A potential important application of CNTs could be the materials for atomic force scanning microscope probes. The small diameters of CNTs make them ideal structures to be used as the tips for atomic force scanning. Both MWNTs[39] and SWNTs[40,41] could be used to produce tips of those microscopes. Hafner *et al.*[40] and Cheung *et al.*[41] reported that it is possible to grow individual CNT probe tips directly, with control over the orientation from the ends of silicon tips. Those based on SWNTs showed very high-resolution tips with end radii of 3–6 nm, in comparison with other commercial available tips, whose radii are generally greater than 5–10 nm.

Another advantage of CNT-based microscope tips is their versatility for chemical modification, which may give them the ability to discriminate between different chemical and biological objects. Wong *et al.*[42] prepared such a chemical force microscope by coupling the CNT tips with basic or hydrophobic functionalities. The microscope equipped with these tips showed images with phase contrast reversed for the same materials because the tips have different affinity towards the CH_3 and COOH area. Furthermore, the CNT tip functionalised with biotin was able to detect the adhesion with streptavidin protein receptor. Those studies suggested that it is feasible to modify CNT tips with different functional groups, active ligands or other macromolecules to create chemical and biological maps.

7.2.2.2.6 *Liquid flow sensor: transfer momentum to current*

Most interestingly, CNTs can be used as liquid flow sensors. Král and Shapiro[43] predicted theoretically that flowing a liquid may generate an electric current in a metallic CNT and suggested that the voltage may be caused by momentum transfer from the flowing liquid or by the scattering of free charge carriers in the flowing liquid. Ghosh *et al.*[44] observed that the flowing liquid induced a voltage on the CNT along the direction of the flow. The voltage response to flow velocity was almost logarithmic (Fig. 7.6), indicating that the current was dependent on a direct forcing of the free charge carriers in the nanotubes by the fluctuating Coulombic field of the flowing liquid. In addition, the voltage was also strongly correlated with the ionic and polar nature of the flowing liquid: liquids with higher ionic strength could induce a higher voltage along the flow direction. Meanwhile, a high viscosity did not result in a higher induced voltage. Although the voltage-inducing mechanisms suggested by different groups were not the same, the novel momentum–current change property hints CNTs as possible candidates to modulate the voltage and current source in a flowing liquid environment, which may have interesting biomedical applications.

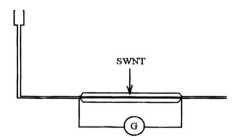

Figure 7.6 The scheme of the device to measure flowing-induced voltage.

7.2.3 Fluorescence Emission, Quenching and Detection

CNTs have shown many novel electroluminescence properties. They can be functionalised to become both fluorescence emitters and detectors. Light energy can be transferred into an electric signal in CNTs for convenient detection, and on the contrary, an electric current can produce light emission and make CNTs a possible light source for sensing.[45] On the other hand, the quenching of CNTs' fluorescence can be used as a tool for the detection of target materials.

7.2.3.1 Fluorescence emitter

The development of an ultrasonic dispersion method to wrap surfactants around nanotubes, thereby isolating them, is the major milestone in the optical characterisation of SWNTs. O'Connell and coworkers[46] reported a study in which they sonicated CNTs with long chain molecules and sodium dodecyl sulphate (SDS) in aqueous solution to disperse the CNTs and obtain single CNT suspension. The SDS could form micelles with individual CNTs as the cores. This method has profound meaning for understanding CNTs' fluorescence properties because these individual CNTs, differently from the CNTs encapsulated in bundles, were not subjected to perturbation by the surrounding tubes and showed much better resolved optical absorption spectra. Furthermore, because some long chain molecules can selectively wrap specific CNTs, it is possible to study their spectra separately.[47] For example, in a suspension of CNTs, only SWNTs (7, 5) were wrapped by poly(9,9-dioctyfluoreny1-2,7-diyl) (PFO) and gave out fluorescence at a correlated wavelength (Fig. 7.7). These dispersed CNTs represent potential fluorescence materials for biomarking and bioanalysis.

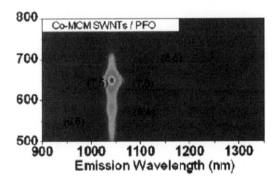

Figure 7.7 Poly(9,9-dioctyfluoreny1-2,7-diyl)-wrapped CNTs (7,5) showed fluorescence emission. Reproduced from Chen *et al.*[47].with permission. See also Colour Insert.

A detailed study on the CNTs' optical spectroscopy showed that the fluorescence is size-tunable and in close relation with the CNTs' atom structure.[48] Each CNT has a specific fluorescence spectroscopy. It was found that the emission–excitation spectrum of each type of CNT is specific. As described by the following equations, the wavelength $(\lambda_{11}, \lambda_{22})$ and frequency (v_{11}, v_{22}) of emission and excitation are closely related with the diameter and chiral angle of CNTs, where d_t is the tube diameter and α is the tube chiral angle. Other constants include Planck's constant h, speed of light c, C–C bond distance a_{cc} and the interaction energy between the neighbouring C atoms, γ_0.

$$\lambda_{11} = \frac{hcd_t}{2a_{cc}\gamma_0} \qquad \lambda_{22} = \frac{hcd_t}{4a_{cc}\gamma_0}$$

$$v_{11} = \frac{1 \times 10^7 \text{cm}^{-1}}{157.5 + 1066.9d_t} + \frac{A_1 \cos(3\alpha)}{d_t^2}$$

$$v_{22} = \frac{1 \times 10^7 \text{cm}^{-1}}{145.6 + 575.7d_t} + \frac{A_1 \cos(3\alpha)}{d_t^2}$$

Furthermore, by combining the two-dimensional excitation–emission fluorimetric spectrum and resonance Raman data, the optical transition can be attributed to a specific (n, m) nanotube structure (Table 7.1).

Table 7.1 Spectral data and assignments for SWNTs (modified from Kam *et al.*[49])

Assignment	λ_{11}	λ_{22}	Predicted* v_{RBM} (cm^{-1})	Observed v_{RBM} (cm^{-1})
(5, 4)	833	483	372.7	373
(6, 4)	873	581	335.2	
(9, 1)	912	693	307.4	
(8, 3)	952	663	298.1	297
(6, 5)	975	567	307.4	
(7, 5)	1023	644	281.9	283
(10, 2)	1053	734	265.1	264
(9, 4)	1101	720	256.4	
(8, 4)	1113	587	278.3	
(7, 6)	1122	647	262.1	264
(9, 2)	1139	551	289.7	
(12, 1)	1171	797	237.0	236
(8, 6)	1172	716	243.7	
(11, 3)	1197	792	232.8	233
(9, 5)	1244	671	241.4	
(10, 3)	1250	633	251.1	251
(10, 5)	1250	786	225.1	225
(11, 1)	1263	611	256.4	
(8, 7)	1267	728	228.9	

(13, 2)	1307	859	211.9	
(9, 7)	1323	790	214.9	215
(12, 4)	1342	857	207.5	
(11, 4)	1372	714	221.5	
(12, 2)	1376	685	227.0	
(10, 6)	1380	756	213.4	
(11, 6)	1397	858	200.8	
(9, 8)	1414	809	203.4	
(15, 1)	1425	927	193.6	
(10, 8)	1474	868	192.5	
(13, 5)	1485	928	187.2	
(12, 5)	1496	795	198.3	
(13, 3)	1497	760	203.4	
(10, 9)	1555	892	183.3	

* Using $\bar{v}_{RBM} = \dfrac{223.5}{d_t(nm)} + 12.5$ and assuming a C–C bond distance of 0.144 nm.

As shown in Table 7.1, most CNTs have their fluorescence emission wavelength in the near-infrared range (800–1600 nm).[48] Actually, the near-infrared (NIR) spectrum of CNTs has attracted a lot interest for bioanalysis, because this wavelength range is a special spectral window, which is transparent to biological systems. Kam et al.[49] reported solubilising CNTs in the aqueous phase by non-covalently adsorbing either 15-mer fluorescently Cy3-labelled ssDNA or polyethylene (PEG)-grafted phospholipids (PLs). The molar extinction coefficient of the solubilised SWNTs (length ~150 nm, diameter ~1.2 nm) measured at λ = 808 nm in the NIR was ε = 7.9 × 10[6] M^{-1}cm^{-1}. Therefore the light at this wavelength transferred energy to CNTs without disrupting the cells.

Meanwhile, the NIR fluorescence emission of CNTs has strong advantages over visible fluorescence for biological sensing and analysis. First, there is low noise from biological tissues and cells at this wavelength range. The blood and tissues often give light interference with the NIR from the dispersed CNTs. Second, the dispersed CNTs have a relatively high quantum yield of fluorescence, which was hinted by the high signal-to-noise ratio.[50] The quantum yield of fluorescence for semiconducting CNTs was initially estimated at 0.1 %.[47] It was believed that such low quantum yield of fluorescence was caused by the ensemble of tubes into bundles, since suspended SWNTs in density gradient fractions showed a quantum yield not less than 1%.[51] Carlson[52] dispersed CNTs with sodium cholate in D$_2$O and measured the fluorescence from SWNTs by epifluorescence microscopy on an inverted microscope. The fluorescence intensities from individual SWNTs showed the average quantum yield value of 3±1%, which is two orders greater than the value of CNTs in ensemble bundles, suggesting that dispersed CNTs are efficient NIR emitters. The high stability is another advantage of CNT-based fluorophors.[50] CNT fluorescence was extremely photostable, showing no evidence of blinking or photobleaching after prolonged exposure to excitation at high fluorescence. Figure 7.8 shows the rates of photobleaching of the NIR organic fluorophore 78-CA, NIR quantum dots consisting of a CdTe core and CdSe shell, and CNTs. The dispersed SWNTs did not show attenuation of fluorescence emission after exposure to excitation at a fluence of 6.17×10^6 mW/cm^2 for 10 hours. Because of these novel properties and advantages, the NIR fluorescence has attracted more and more attention in the area of biosensing and analysis.

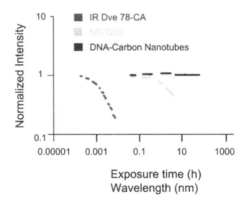

Figure 7.8 Fluorescence intensity–exposure time curves. CNTs showed higher photostability than organic fluorophore 78-CA and quantum dots. Reproduced from Cherukuri *et al.*[54] with permission.

Heller *et al.*[53] reported a study of incubated live murine 3T3 and myoblast stem cells with ssDNA-wrapped SWNTs. The excitation radiation passes through live cells without significant scattering, absorption, heating, or damage to tissues. The CNT fluorescence did not photobleach even under prolonged excitation and remained visible after a week in live cells. Cherukuri *et al.*[54] found that macrophage cells could ingest substantial amount of SWNTs. Those SWNTs emitted fluorescence around the wavelength of 1100 nm, making it possible to trace the interaction of CNTs with cells and give high-contrast imaging of the cells.

An advantage of CNT-based fluorescence emitters is that they can be easily modified with functional ligands for cell targeting and can act as cancer cells tracers. Folate moiety, which is a cancer-cell-targeting ligand, was linked with CNTs to selectively target cancer cells.[49] Conjugation of SWNTs with antibodies also made the system targeting fluorescence to specific cells. Rituxan, an antibody that recognizes the CD20 cell surface receptor, was coupled to an

SWNT (Fig. 7.9).[55] The measured quantum yield of SWNT photoluminescence is relatively low but can be up to 3–8%. After incubation of B cells and T cells in solutions of SWNT–Rituxan conjugates, bright NIR emission of SWNTs on B-cell lymphoma (CD20 positive) was observed when excited under 785 nm, suggesting SWNT–Rituxan binding to CD20 cell surface receptors. In contrast, NIR fluorescence images taken on SWNT–Rituxan incubated T cells (CD20 negative) showed little NIR photoluminescence, indicating the success of fluorescent targeting for detection of specific cells. Similarly, Herceptin conjugation to SWNTs recognised the HER2/neu cell surface receptors, which were overexpressed on various cancer cells. NIR fluorescence from the SWNTs clearly differentiated the positive and negative cell lines.

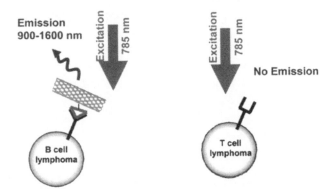

Figure 7.9 SWNT–Rituxan targeted B cell and labelled it with NIR fluorescence. T cell, without CD20 receptor, was not labelled. Reproduced from Leeuw *et al.*[56] with permission. See also Colour Insert.

In vivo detection of the CNTs confirmed CNTs' optical properties. Leeuw and collaborators[56] dispersed SWNTs by sonication with bovine serum albumin (BSA) solutions. The *in vivo* experiment was performed on *Drosophila melanogaster* (fruit flies). NIR fluorescence from SWNTs in the digestive tracts was clearly imaged, while a video showed the nanotubes' movements in the digestive system. Further scanning microscopy for NIR fluorescence of individual tissues showed SWNTs concentrated in the brain and dorsal vessel.

Detection of SWNT NIR fluorescence was also performed in mammalian animals for biological imaging.[57] Cherukuri and coworkers dispersed SWNTs using surfactant Pluronic F108 with the assistance of sonication and detected the blood concentration of SWNTs by monitoring the NIR fluorescence of the nanotubes. Therefore this procedure was convenient to study the pharmacokinetics of SWNTs. In addition NIR fluorescence also provided methods to detect the bio-distribution of the SWNTs. Clear NIR fluorescence was detected from tissue specimen slices. This provides a powerful tool to speed up the development of related medical applications.

7.2.3.2 Raman spectrum

Besides fluorescence, CNTs also give out Raman spectra.[50] CNTs often produce three typical strong Raman bands. The first band can be found within 150–300 cm^{-1}. It is caused by uniaxial vibrations and is called radial breathing mode (RBM). The second band locates at 1590–1600 cm^{-1} and is named tangential mode (or G band). The last one is called the disorder mode (D band, at 1250–1450 cm^{-1}) and is observed in all sp^2-hybridised disordered carbon materials. Both metallic and semiconducting SWNTs demonstrate intense Raman scattering.[58,59] Heller *et al.*[53] encapsulated SWNTs by DNA oligonucleotide. The dispersed SWNTs not only emitted fluorescence spectra but were also accessible for detection via Raman scattering in live cells. Raman scattering of CNTs in murine 3T3 cells and murine myoblast stem cells remained constant during continuous excitation, suggesting that CNTs may provide another feasible method for biological sensing and analysis.

7.2.3.3 Electric luminescence

Polarised infrared optical emission can be electrically induced from CNTs.[60] This was achieved on a CNT-based transistor by drawing electrons into the channel at the source and by producing holes into the channel at the drain. The mechanism of the IR emission is the radiative recombination of injected electrons and holes. Such an IR source may form the basis for ultra-small integrated photonic devices. It provides a possible light source for responsive sensors and analysis. Freitag and coworkers[61] studied the factors influencing the electric luminescence of CNTs and indicated that the environment defects, such as local dielectric changes, nanotube–nanotube contacts in looped tubes as well as nanotube segments close to the electronic contacts may change the electric luminescence. These results suggested that the electric luminescence is susceptible to environmental changes.

7.2.3.4 Fluorescence quenching

The fluorescence of CNTs can be strongly suppressed by specific chemicals, and sometimes such quench is reversible. Hence the change of the fluorescence may provide information for the sensing and analysis of the chemical environment.

Strano *et al.*[62] reported that when the pH was between 6 and 2.5, the photoluminescence of dispersed SWNTs quenched as long as the nanotubes were previously adsorbed with oxygen at the sidewall. The quench was reversible and fluorescence recovered when pre-adsorbed oxygen was diminished. In some other cases, the disturbance on the fluorescence caused by chemicals was not reversible. Usrey and coworkers[63] found that reaction with 4-chlorobenzene diazonium caused a prompt decrease in SWNTs' photoluminescence in a few minutes, proportionally to the addition of the reactant. The quenching of the fluorescence was caused by both non-covalent and covalent defects on the nanotubes and could be predicted as a linear combination of the covalent and non-covalent behaviours.

The mechanism of the fluorescence quench has been extensively studied. The reversible quenching of SWNTs' fluorescence by means of oxidative charge-transfer reactions with small redox-active organic dye molecules indicated that the chiral selective reactivity of compounds with CNTs is not only diameter dependent but also charge transfer dependent.[64] The quenching of fluorescence for SWNTs, which experience a transition between (6, 5) and (7, 5), can be used to detect the existence of 4-amino-1,1-azobenzene-3,4-disulphonic acid (AB), since AB can act as an electron acceptor and shows a band gap of ~1.2 eV. On the contrary, the effectiveness of bleaching occurs at a band gap of ~1.1 eV when mordant yellow 10 is the acceptor. Therefore the tubes with transitions between (10, 2) and (9, 4) can show sensitive fluorescence quenching upon the addition of mordant yellow 10. So SWNTs with a different chirality can be used as selective sensors for chemical analysis.

The NIR from semiconducting CNTs is a very sensitive tool and can even be used to detect a single molecule. Cognet and collaborators[65] reported that the excitation fluorescence of CNTs could be quenched by single-molecule reactions. They immobilised dispersed single nanotubes in agarose gel. Reactants, including sulphuric acid, approached the nanotube molecule from one edge of the gel. The fluorescence from the individual CNTs showed stepwise quenching by acid, rather than a continuous decrease. These discrete fluorescence intensity steps were attributed to individual protonation reactions at the nanotube surface. Moreover, upon addition of NaOH solution to the quenched nanotubes, the NIR fluorescence intensity recovered and showed a stepwise increase in intensity, suggesting individual un-protonation reaction on the SWNT sidewalls. However, although 4-chlorobenzenediazonium tetrafluoroborate, instead of sulphuric acid, could quench SWNTs, the fluorescence could not be re-established because of the irreversible covalent reaction. These research results indicated that SWNTs may be highly sensitive to local chemical and physical perturbations.

Reversible quenching of CNT fluorescence was applied also for macro-biomolecular sensing. Satishkumar and coworkers[66] designed a biosensing device, which takes advantage of the reversible fluorescence quenching. In their study, nanotube-based optical sensing was used to detect dye–ligand conjugates (DLCs) (Fig. 7.10). The DLCs used in this study consisted of biotinylated anthracene and biotinylated phenylazoaniline, respectively. The fluorescence spectra of doped samples showed that the nanotube fluorescence was quenched substantially relative to the undoped sample. The fluorescence quench caused by the DLCs may be due to the transfer of photo-induced excited-state electrons from the nanotube conduction band to the lowest unoccupied molecular orbital of the DLCs. This is supported by the evidence that B-nicotinamide adenine dinucleotide (NADH), which is a reducing agent and can inhibit the excited-state charge transfer through competitive reduction of DLCs, recovered the nanotube fluorescence. The target protein, avidine, served as the model analyte of the SWNT-DLC system. After successive addition of a nanomolar amount avidin, significant recovery of fluorescence was observed, suggesting a very sensitive response to the target protein. The recovery was complete, and the fluorescence almost reached the original level. The possible mechanism was that the binding affinity between biotin and avidin disrupted the weak non-covalent interaction between DLC and SWNT and diminished the quenching effect of the DLC. Although BSA could also react with biotin and repristinate the fluorescence, the recovery efficiency was two to three orders lower than that of avidin, indicating the system's high selectivity.

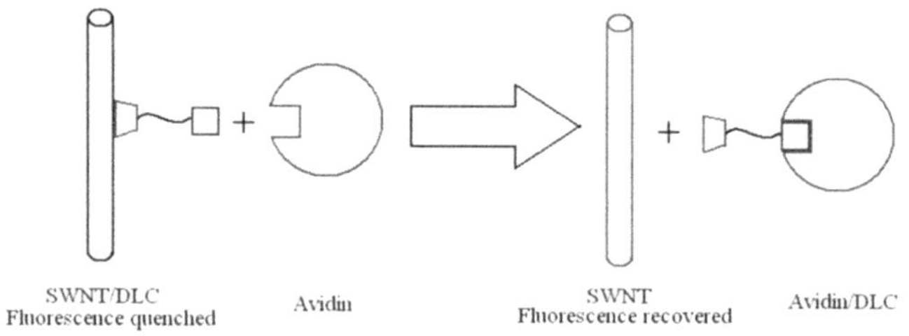

SWNT/DLC Fluorescence quenched   Avidin   SWNT Fluorescence recovered   Avidin/DLC

Figure 7.10 The dye–ligand conjugates (DLCs) quenched the fluorescence of CNTs. Upon addition of avidin, the DLC detached from CNTs and the fluorescence was reversed. Redrawn from Barone *et al.*[67]

Barone *et al.*[67,68] explored a method to detect the glucose concentration by NIR of CNTs. Glucose oxidase was immobilised on CNTs. The enzyme catalyses β-D-glucose to the D-glucono-1,5-lactone with H_2O_2 as co-product. The subsequent reduction of $Fe(CN)_6^{3-}$ resulted in a significant change in the fluorescence strength, making the system a sensitive detection device.

7.2.3.5 Photoconductivity

Semiconducting CNTs are possible photo-sensors because their conductivity was reported to be sensitive to photoradiation. Freitag and coworkers[69] prepared single nanotube ambipolar FETs and then excited them by infrared light. The incident infrared light that was used excites the second exciton state of the semiconducting nanotube (i.e., the $n = 2 \rightarrow n' = 2$ ($E_{2\text{-}2}$) transition). For SWNTs with diameters centred around 1.3 nm, the energy of $E_{2\text{-}2}$ is centred at about 1.35 eV. This excited state decayed and gave an energetic electron–hole pair within the nanotube molecule[70,71] and the charge carriers could diffuse within the channel or be separated by an applied electric field between the source and drain contacts. With a diameter 1000 times smaller than the wavelength of the light, the nanotube FET had an estimated quantum efficiency of >10%. Upon irradiation, the current was increased by an order of magnitude and the gate voltage shifted significantly. Therefore the semiconductor-based FET could act as a photodetector.

Qiu *et al.*[72] obtained correlated results on CNT-based transistors but found a weaker absorption band, which correlated with an energy 200 meV higher than that of the $E_{2\text{-}2}$ resonance band. That sideband was attributed to the C–C bond stretching phonon. Qiu and coworkers also showed clear evidence of the dependence of current quantum yield on the polarisation of the laser. There was no photocurrent generated when the polarisation was perpendicular to the carbon axis.

7.3 BIO-INTERFACE

CNTs can be used to prepared bio-compatible materials, which may possess advantages such as high tensile strength and low cytotoxicity. Therefore CNTs can be used for many kinds of biomaterials for tissue/cell engineering.

7.3.1 The Fundamental Properties for Bio-interface Application

Before the application of CNTs for bio-interface purposes, cytotoxicity is the first issue to be considered. Studies on CNTs' cytotoxicity showed controversial results.[73-74] However some recent evidence showed that the CNTs are versatile for chemical modifications, which can introduce some functional groups, ligands and even biomolecules.[75] The functionalised CNTs might display a very low cytotoxicity,[76] suggesting that they are possible bio-engineering materials for bio-interface applications. More details will be described in the chapter for cytotoxicity (Chapter 8).

Another special aspect of CNTs is their novel physical properties. Some materials with novel and extensively strong properties have been produced using CNTs. It was even believed that CNTs might be suitable for building space elevators and other robust constructions.[77]

Vigolo *et al.*[78] found CNTs' high flexibility and resistance to torsion. They designed a method to prepare carbon fibres using CNTs. By injecting CNTs' suspension to polyvinylalcohol (PVA) solution flow, the CNTs underwent alignment followed by aggregation and resulted in long CNT fibres. Those CNT fibres were excellent in flexibility and resistance to torsion and could be curved through 360° in a few micrometres without breaking. Dalton and coworkers[79] produced super-tough CNT composite fibres of 100 m length by a coagulation-based spinning method. The fibres were many times stronger than steel wires. Walters and collaborators[80] reported that the SWNTs' strength is very high (37 GPa) and that the CNT ropes could be extended elastically by 5.8% before breaking.

Furthermore, the density-normalised strength of SWNTs is more impressive and excellent for applications in which light structural materials are needed. Wong *et al.*[81] indicated that the density-normalised strength of SWNTs were 56 times that of steel wires. Such outstanding mechanical characteristics suggest CNTs to be good candidates for applications as biomaterials, such as artificial bone scaffold materials.

At last, CNTs are known to orient under the influence of an alternating current (AC) electric field and form aligned patterns on a flat surface.[82] Chung *et al.*[83] demonstrated a scalable and highly parallel process of electric-field-induced CNT positioning and alignment that can lead to mass assembly of nanostructures and nanoscale devices. Chen and Zhang[84] were recently able to align SWNTs avoiding CNT entanglement between 2D co-planar parallel electrodes for potential applications in high-performance nano-devices. Chen and coworkers[85] aligned CNTs between electrode gaps with a high degree of alignment and the appearance of uniform coverage. In addition, the application of dielectrophoretically aligned CNTs as an ordered substrate for tissue engineering was explored.[86] These studies made it convenient to manipulate CNTs for bio-engineering materials.

In addition, CNTs possess some novel advantages which make them suitable for next-generation scaffolds. For instance, CNTs may carry a neutral electric charge, which sustains the highest cell growth and production of plate-shaped crystals.[87-88] Meanwhile, CNTs may be metallic and show good conductivity. An electric field may stimulate the healing of various tissues. A mechanical bone injury is often accompanied by an electrical signal, and therefore a conductive scaffold is very promising for stimulating cell growth and tissue regeneration by facilitating physioelectrical signal transfer.[89,90]

7.3.2 Applications

Because of the excellent chemical, physical, mechanical, biological and electrical properties, CNTs have been widely studied for their application as biomaterials.

7.3.2.1 Applications for bone tissue engineering

As CNTs are the strongest materials in the world, they can be used as agents to obtain extremely high mechanical properties.[91,92] For example, 5% wt of SWNTs significantly enhanced poly(vinyl alcohol)/ poly(vinyl pyrrolidone)/ sodium dodecyl sulphate (PVA/PVP/SDS) tensile strength,[93] and Shi *et al.*[94] reported that the SWNTs, functionalised by covalent attachment with 4-*tert*-butylphenylene groups were much better than pristine SWNTs in enhancing the mechanical strength for poly(propylene fumarate)–CNT composite. Wang and coworkers[95] obtained the biopolymer chitosan–MWNT nanocomposite by a simple solution evaporation method. With the MWNTs homogeneously dispersed throughout the chitosan matrix, the tensile modulus and strength of the nanocomposites were improved by more than 90%. Therefore, CNTs were often considered as reinforced bone tissue engineering scaffolds.

Adhesion of cells such as osteoblasts is a crucial prerequisite to subsequent cell functions such as synthesis of extracellular matrix proteins and formation of mineral deposits. Some research groups showed that CNT-based materials appeared to be efficient as a nano-matrix for the nucleation and growth of hydroxyapatite (HA), which is important for bone tissue engineering. Thus, it is possible to design CNT-based materials as scaffolds with suitable surface morphology for specific cell adhesion and proliferation via the formed CNT/ HA layers.

Zanello *et al.*[88] and Zhao *et al.*[96] explored the application of CNTs as scaffold materials for osteoblast proliferation and bone formation. SWNTs showed that they could mineralise with hydroxyapatite. Biocompatible studies indicated that, compared with other functionalised CNTs, pristine CNTs and PEG-linked CNTs, which were neutral in charge, resulted in a higher proliferation of osteoblast ROS 17/2.8 cells. Those cells were active in producing HA crystals and responding to electric signals, thus suggesting the biocompatibility of CNTs. These results, combined with CNTs' excellent physical properties, promote CNTs as promising candidates to improve the mechanical properties of damaged bone tissue.

Ultra-short CNTs (US tubes) were used to produce porous nanocomposite scaffolds for bone tissue engineering.[97,98] Both original US tubes and dodecylated US tubes could form porous structure with cavities interconnected. *In vitro* studies showed that the porous materials provided substrates for marrow stromal cells' attachment and proliferation, suggesting good osteoconductivity.[97] When the US tubes were used as reinforcing scaffold and tested in an *in vivo* rabbit model, they exhibited favourable hard and soft tissue responses at different time points.[98] Bone tissue growth in US-tube-based materials was many times greater than that in poly(propylene fumarate), which was used as control. Moreover, US tubes caused fewer inflammatory effects on the tissue.

CNTs can serve as a template for the ordered formation of a nanostructured HA layers for tissue engineering. Boccaccini and coworkers[99] prepared bioglass-based tissue scaffolds for bone tissue engineering. Electrophoretic deposition (EPD)[100] was used to deposit CNTs on planar substrates and form porous layers. The CNT scaffolds were bioactive for the formation of HA. CNTs in these scaffolds functioned as templates for nucleation and growth of the HA nanosize crystals and could completely be embedded in the HA layer. Therefore, by coating CNTs it was possible to improve the bioglass-based scaffolds' performance in bone tissue engineering applications.

CNTs were often associated with other polymers to achieve novel properties and overcome several disadvantages intrinsically present in polymers. Mei *et al.*[101] reported the synthesis of poly(L-lactic acid) MWNTs and hydroxyapatite (PLLA/MWNT/HA) composite membrane by the electrospinning technique. The materials showed selectivity for adhesion and proliferation of different cells. The PLLA/MWNT/HA membrane enhanced the adhesion and proliferation of periodontal ligament cells (PDLCs) by 30% and inhibited the adhesion and proliferation of gingival epithelial cells by 30%. The selectivity may be due to the higher surface energy of the nanostructured surface, which would enhance the adhesion of osteoblast cells and inhibit the adhesion of osteoblast competitive cells.[102,103] Bacakova *et al.*[104] designed a terpolymer–CNT composite. The substrate composed of polytetrafluoroethylene, polyvinyldifluoride, polypropylene and CNTs showed enhanced compatibility for cell spreading. Cells grown on the terpolymer–CNT substrate showed higher activity and proliferation.

Bone morphogenetic protein could assist the growth of bone on CNTs. Usui and coworkers[105] incubated recombinant human bone morphogenetic protein-2 (rhBMP-2) with collagen and MWNTs. The MWNTs may accelerate bone formation, and histological studies found that immature bone conspicuously surrounded MWNTs at an early stage. The bone mineral content (BMC) in rh-BMP-2/MWNT/collagen sample was higher than in the control, which did not contain MWNTs. Although the difference was not significant, the result suggested that MWNTs are biocompatible for bone formation.

Abarrategi *et al.*[106] combined the idea of polymer–CNT composite and CNT-BMP complex and studied the co-effect of CNT/chitosan/rhBMP-2. The growth of bone cells C2C12 was examined after histological stains. Observation of the cross-section of the microchannelled structure of CNT/chitosan showed that cells could penetrate into the scaffold cavities and attach on its walls, suggesting that cellular infiltration into the porous structure was possible (Fig. 7.11a). The depth of the cell penetration might be limited by nutrient diffusion. Furthermore, rhBMP-2 greatly enhanced the formation of bone. Alkaline phosphatase (ALP) activity increased by many times in the presence of rhBMP-2 (Fig. 7.11b).

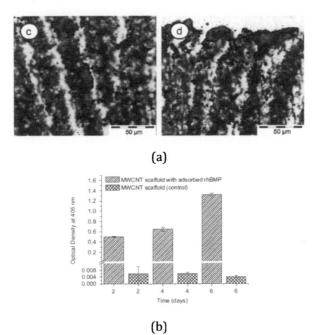

(a)

(b)

Figure 7.11 (a) Microscope images of SMA immunohistochemical stained C2C12 cells penetrating into the scaffold structure. (b) rhBMP-2 significantly enhanced the proliferation of C2C12 cells in the scaffold. Reproduced from Gabay *et al.*[111] with permission.

Wang *et al.*[107] proved the *in vivo* activity of CNT-based materials for bone cell proliferation. The CNT–polycarbosilane composite implanted under subcutaneous tissue were covered by surrounding fibrous connective tissue after 1 week. Some inflammation appeared but did not worsen after 4 weeks. Those composite implants in hard tissue simulated callus formation. After 4 weeks some parts of the newly formed bone attached to the composite.

7.3.2.2 Applications for neural tissue engineering

As strong, flexible and conductive materials, CNTs were explored for their potential for neural tissue engineering. Mattson *et al.*[108] applied CNTs for neural cell culture. The embryonic rat-brain neurons were cultured on unmodified CNTs and CNTs coated with bioactive 4-hydroxynonenal elaborate multiple neuritis. In unmodified CNTs, the growth of neurons was limited to only one or two neurite extensions. However functionalized CNTs are good candidates and enable extensive branching of neural cells..

Gheith's group[109] demonstrated that SWNT layer-by-layer film could support the growth, viability and differentiation of neuronal NG108-15 neuroblastoma/glioma hybrid cells. Furthermore, the electrical conductivity of the SWNTs made the layer-by-layer film substrate, which could transfer electric signals to the neural cell NG108-15.[110] Those results indicated CNTs as novel biomaterials for neuron tissue.

Gabay and coworkers[111] established a method for the self-organisation of neural networks on CNTs. CNT islands on SiO_2 were used as template. Cells proliferated on the CNT clusters and formed connections between the islands. The cells were active in generating action potential with normal shapes in response to electrical stimulation, demonstrating that the CNT template did not change the functionality of the neurons. The growth of neurons could be directed by various patterned CNTs and for 2D neural networks. Zhang and collaborators[112] prepared different CNT templates, including lines and circles. The growth of neurite cells was observed to be restricted along the sidewalls of the MWNTs.

Hu *et al.*[113] studied the effect from the charge of functionalised CNTs on the morphology of neuron cells. A positive charge resulted in longer-than-average length of neuritis and more branches of neurites, suggesting the possibility of controlling the cell outgrowth by chemical modification.

Further information can be found in the chapter dedicated to CNTs for neuronal living networks (Chapter 6).

7.3.2.3 Application for other cells and tissues engineering

Besides bone cells and neural cells, many kinds of other cells were also found to be compatible with CNTs.

Ago and coworkers[114] developed a simple but effective method to immobilize Hela cells on a CNT array. Hela cells were centrifuged under 100–200 g for 5 minutes in a centrifuge tube containing aligned nanotubes covering a Si substrate. The cells were then pressed onto the nanotube array.

The interaction between CNTs and cells might include mechanical sticking as well as insertion of nanotube tips into the cells.

MacDonald *et al.*[115] investigated CNTs' compatibility with rat aortic smooth muscle cells. CNTs formed composites with collagen. Rat aortic smooth muscle cells were cultured inside the CNT/collagen scaffolds. Cell viability, proliferation and morphology were not significantly changed.

Non-woven SWNTs appeared to be a good substrate for the growth of mouse fibroblast cell 3T3-L1.[116] Compared with blank, polyether polyurethane and carbon fibres, the existence of SWNTs significantly enhanced the cell adhesion and proliferation. Observation of the cytoskeleton through fluorescence labelling of actin revealed that cells sensed better growing environments and produced stronger responses to SWNTs than to the plain polyether polyurethane film.

Correa-Duarte's study[117] suggested that 3D CNT networks are ideal scaffolds for murine fibrosarcoma L929 cell growth. SEM showed that after incubation for 1 day, most of L929 cells attached on the CNT network, while 7 days' growth resulted in most CNT surfaces' being covered by the cell layer. The extensive adhesion, growth and spreading of the fibroblast cells suggested the 3D network's potential application.

7.4 CONCLUSIONS

CNTs are novel materials with special properties, and they are becoming more and more attractive in the areas of chemistry, physics, biomedicine, biosensing as well as bio-interfacial engineering.

CNTs have been widely studied for biosensing applications. Because of their electronic and structural properties, they can be used to prepare various kinds of sensors, such as mass/force sensors, chemical sensors, structure sensors, electric probes, microscope sensors as well as liquid flow sensors. Moreover, CNTs can emit fluorescence upon excitation by light or current, and the spectra can reflect the changes of chemical and physical environments. It is easy to translate the changes to electric signals, because CNTs can act as nano-electric transistors, resistors, capacitors, and so on.

In addition to biosensing, CNTs have also been widely studied for their possible application in bio-interfacial materials. CNTs were found to represent good materials for growth of bone, neural, stem and some other cells.

On the basis of these studies, we believe that CNTs are promising materials in biosensing and bio-interfacial areas.

References

1. Iijima, S. (1991) Helical microtubules of graphitic carbon, *Nature*, **354**, 56–58.

2. Harrison, B. S., and Atala, A. (2007) Carbon nanotube applications for tissue engineering, *Biomaterials*, **28**, 344–353.

3. Wilder, J. W. G., Venema, L. C., Andrew, G. R., Smalley, R. E., and Dekker, C. (1998) Electronic structure of atomically resolved carbon nanotubes, *Nature*, **391**, 59–62.

4. Odom, T. W., Huang, J. L., Kim, P., and Lieber, C. M. (1998) Atomic structure and electronic properties of single-walled carbon nanotubes, *Nature*, **391**, 62–64.

5. White, C. T., and Todorov, T. N. (1998) Carbon nanotubes as long ballistic conductors, *Nature*, **393**, 240–242.

6. Liang, W., Bockrath, M., Bozovic, D., Hafner, J. H., Tinkham, M., and Park, H. (2001) Fabry-Perot interference in a nanotube electron waveguide, *Nature*, **411**, 665–669.

7. Frank, S., Poncharal, P., Wang, Z. L., and de Heer W. A. (1998) Carbon nanotube quantum resistors, *Science*, **280**, 1744–1746.

8. Kim, P., Shi, L., Majumdar, A., and McEuen, P. L. (2001) Thermal transport measurements of individual multiwalled nanotubes, *Phys. Rev. Lett.*, **87**, 215502.

9. Chico, L., Benedict, L. X., Louie, S. G., and Cohen, M. L. (1996) Qujantum conductance of carbon nanotubes with defects, *Phys. Rev. B*, **54**, 2600.

10. Schwab, K., Henriksen, E. A., Worlock, J. M., and Roukes, M. L. (2000) Measurement of the quantum of thermal conductance, *Nature*, **404**, 974–977.

11. Rego, L. G. C., and Kirczenow,G. (1998) Quantized thermal conductance of dielectric quantum wires, *Phys. Rev. Lett.*, **81**, 232–235.

12. Martel, R., Schmidt, T., Shea, H. R., Hertel, T., and Avourisa, Ph. (2002) Single- and multi-wall carbon nanotube field-effect transistors, *Appl. Phys. Lett.*, **73**, 2247–2249.

13. Sander, J. T., Alwin, R. M., Verschueren, and Dekker, C. (1998) Room-temperature transistor based on a single carbon nanotube, *Nature*, **393**, 49–52

14. Postma, H. W. C., Teepen, T., Yao, Z., Grifoni, M., and Dekker, C. (2001) Carbon nanotube single-electron transistors at room temperature, *Science*, **293**, 76–79.

15. Keren, K., Berman, R. S., Buchstab, E., Sivan, U., and Braun, E. (2003) DNA-templated carbon nanotube field-effect transistor, *Science*, **302**, 1380–1382.

16. Hayamizu, Y., Yamada, T., Mizuno, K., Davis, R. C., Futaba, D. N., Yumura, M., and Hata, K. (2008) Integrated three-dimensional microelectromechanical devices from processable carbon nanotube wafers, *Nat. Nanotechnol.*, **3**, 289–294.

17. Geblinger, N., Ismach, A., and Joselevich, E. (2008) Self-organized nanotube serpentines, *Nat. Nanotechnol.*, **3**, 195–200.

18. Jensen, K., Weldon, J., Garcia, H., and Zettl, A. (2007) Nanotube radio, *Nano Lett.*, **7**, 3508–3511.

19. Sazonova, V., Yaish, Y., Űstűnel, H., Roundy, D., Arias, T. A., and McEuen, P. L. (2004) A tunable carbon nanotube electromechanical oscillator, *Nature*, **431**, 284–287.

20. Collins, P. G., Bradley, K., Ishigami, M., and Zettl, A. (2008) Extreme oxygen sensitivity of electronic properties of carbon nanotubes, *Science*, **287**, 1801–1804.

21. Kong, J., Franklin, N. R., Zhou, C., Chapline, M. G., Peng, S., Cho, K., and Dai, H. (2000) Nanotube molecular wires as chemical sensors, *Science*, **287**, 622–625.

22. Lee, R. S., Kim, H. J., Fischer, J. E., Thess, A., and Smalley, R. E. (1997) Conductivity enhancement in single-walled carbon nanotube bundles doped with K and Br, *Nature*, **388**, 255.

23. Grigorian, I., Williams, K. A., Fang, S., Sumanasekera, G. U., Loper, A. L., Dickey, E. C., Pennycook, S. J., and Ecklund, P. C. (1998) Reversible intercalation of charged iodine chains into carbon nanotube ropes, *Phys. Rev. Lett.*, **80**, 5560.

24. Snow, E. S., Perkins, F. K., Houser, E. J., Badescu, S. C., and Reinecke, T. L. (2005) Chemical detection with a single-walled carbon nanotube capacitor, *Science*, **307**, 1942–1945.

25. Robinson, J. A., Snow, E. S., Bădescu, S. C., Reinecke, T. L., and Perkins, F. K. (2006) Role of efects in single-walled carbon nanotube chemical sensors, *Nano Lett.*, **6**, 1747–1751.

26. Katz, E., and Willner, I. (2004) Biomolecule-functionalized carbon nanotubes: applications in nanobioelectronics, *Chem. Phys. Chem.*, **5**, 1084–1104.

27. Musameh, M., Wang, J., Merkoci, A., and Lin, Y. (2002) Low-potential stable NADH detection at carbon-nanotube modified glassy carbon electrodes, *Electrochem. Commun.*, **4**, 743–746.

28. Wang, J., Musameh, M., and Lin, Y. (2003) Solubilization of carbon nanotubes by nafion toward the preparation of amperometric biosensors, *J. Am. Chem. Soc.*, **125**, 2408–2409.

29. Wang, J., and Musameh, M. (2003) Carbon nanotube/teflon composite electrochemical sensors and biosensors, *Anal. Chem.*, **75**, 2075–2079.

30. Star, A., Tu, E., Niemann, J., Gabriel, J. C. P., Joiner, C. S., and Valcke, C. (2006) Label-free detection of DNA hybridization using carbon nanotube network field-effect transistors, *PNAS*, **103**, 921–926.

31. Chen, R. J., Bangsaruntip, S., Drouvalakis, K. A., Kam, N. W. S., Shim, M., Li, Y., Kim, W., Utz, P. J., and Dai, H. (2003) Noncovalent functionalization of carbon nanotubes for highly specific electronic biosensors, *PNAS*, **100**, 4984–4989.

32. Star, A., Gabriel, J.-C. P., Bradley, K., and Grüner, G. (2003) Electronic detection of specific protein binding using nanotube FET devices, *Nano Lett.*, **3**, 459–463.

33. Li, C., Curreli, M., Lin, H., Lei, B., Ishikawa, F. N., Datar, R., Cote, R., Thompson, M., and Zhou, C. J. (2005) Complementary detection of prostate-specific antigen using In2O3 nanowires and carbon nanotubes, *J. Am. Chem. Soc.,* **127,** 12484–12485.

34. Ju, S. Y., Doll, J., Sharma, I., and Papadimitrakopoulos, F. (2008) Selection of carbon nanotubes with specific chiralities using helical assemblies of flavin mononucleotide, *Nat. Nanotechnol.,* **3,** 356–362.

35. Richard, C., Balavoine, F., Schultz, P., Ebbesen, T. W., and Mioskowski, C. (2003) Supramolecular self-assembly of lipid derivatives on carbon nanotubes, *Science,* **300,** 775–778.

36. Guo, X., Gorodetsky, A. A., Hone, J. Barton, J. K., and Nuckolls, C. (2008) Conductivity of a single DNA duplex bridging a carbon nanotube gap, *Nat. Nanotechnol.,* **3,** 163–167.

37. Roy, S., Vedala, H., Roy, A. D., Kim, D. H., Doud, M., Mathee, K., Shin, H. K., Shimamoto, N., Prasad, V., and Choi, W. (2008) Direct electrical measurements on single-molecule genomic DNA using single-walled carbon nanotubes, *Nano Lett.,* **8,** 26–30.

38. Cai, H., Cao, X., Jiang, Y., He, P., and Fang, Y. (2003) Carbon nanotube-enhanced electrochemical, DNA biosensor for DNA hybridization detection, *Anal. Bioanal. Chem.,* **375,** 287–293.

39. Li, J., Ng, H. T., Cassell, A., Fan, W., Chen, H., Ye, Q., Koehne, J., Han, J., and Meyyappan, M. (2003) Carbon nanotube nanoelectrode array for ultrasensitive DNA detection, *Nano Lett.,* **3,** 597–602.

40. Dai, H., Hafner, J. H., Rinzler, A. G., Colbert, D. T., and Smalley, R. E. (1996) Nanotubes as nanoprobes in scanning probe micrscopy, *Nature,* **384,** 147–150.

41. Hafner, J. H., Cheung, C. L., and Leiber, C. M. (1999) Growth of nanotubes for probe microscopy tips, *Nature,* **398,** 761–762.

42. Cheung, C. L., Hafner, J. H., and Lieber, C. M. (2000) Carbon nanotube atomic force microscopy tips: Direct growth by chemical vapor deposition and application to high-resolution imaging, *PNAS,* **97,** 3809–3813.

43. Wong, S. S., Joselevich, E., Woolley, A. T., Cheung, C. L., and Lieber, C. M. (1998) Covalently functionalized nanotubes as nanometresized probes in chemistry and biology, *Nature,* **394,** 52–55.

44. Král, P., and Shapiro, M. (2001) Nanotube electron drag in fowing liquids, *Phys. Rev. Lett.,* **86,** 131–134.

45. Ghosh, S., Sood, A. K., and Kumar, N. (2003) Carbon nanotube flow sensors, *Science,* **299,** 1042–1044.

46. Avouris, P., Chen, Z., and Perebeinos, V. (2008) Carbon-based electronics, *Nat. Nanotechnol.,* **2,** 605–615.

47. O'Connell, M. J., Bachilo, S. M., Huffman, C. B., Moore, V. C., Strano, M. S., Haroz, E. H., Rialon, K. L., Boul, P. J., Noon, W. H., Kitrell, C., Ma, J., Hauge, R. H., Weisman, R. B., and Smalley, R. E. (2002) Band gap fluorescence from individual single-walled carbon nanotubes, *Science,* **297,** 593–596.

48. Chen, F., Wang, B., Chen, Y., Li, and L. J. (2007) Toward the extraction of single species of single-walled carbon nanotubes using fluorene-based polymers, *Nano Lett.*, **7**, 3013–3017.

49. Bachilo, S. M., Strano, M. S., Kittrell, C., Hauge, R. H., Smalley, R. E., and Weisman, R. B. (2002) Structure-assigned optical spectra of single-walled carbon nanotubes, *Science*, **298**, 2361–2366.

50. Kam, N. W., O'Connell, M., Wisdom, J. A., and Dai, H. (2005) Carbon nanotubes as multifunctional biological transporters and near-infrared agents for selective cancer cell destruction, *PNAS*, **102**, 11600–11605.

51. Hartschuh, A., Pedrosa, H. N., Novotny, L., and Krauss, T. D. (2003) Simultaneous fluorescence and Raman scattering from single carbon nanotubes, *Science*, **301**, 1354–1356.

52. Arnold, M. S., Stupp, S. I., Hersam, and M. C. (2005) Enrichment of single-walled carbon anotubes by diameter in density gradients, *Nano Lett.*, **5**, 713–718.

53. Carlson, L. J., Maccagnano, S. E., Zheng, M., Silcox, J., and Krauss, T. D. (2007) Fluorescence efficiency of individual carbon nanotubes, *Nano Lett.*, **7**, 3698–3703.

54. Heller, D. A., Baik, S., Eurell, T. E., and Strano, M. S. (2005) Single-walled carbon nanotube spectroscopy in live cells: towards long-term labels and optical sensors, *Adv. Mater.*, **17**, 2793–2799.

55. Cherukuri, P., Bachilo, S. M., Liovsky, S. H., and Weisman, R. B. (2004) Near-infrared fluorescence microscopy of single-walled carbon nanotubes in phagocytic cells, *J. Am. Chem. Soc.*, **126**, 15638–15639.

56. Welsher, K., Liu, Z., Daranciang, D. and, Dai, H. (2008) Selective probing and imaging of cells with single walled carbon nanotubes as near-infrared fluorescent molecules, *Nano Lett.*, **8**, 586–590.

57. Leeuw, T. K., Reith, R. M., Simonette, R. A., Harden, M. E., Cherukuri, P., Tsyboulski, D. A., Beckingham, K. M., and Weisman, R. B. (2007) Single-walled carbon nanotubes in the intact organism: Near-IR imaging and biocompatibility studies in drosophila, *Nano Lett.*, **7**, 2650–2654.

58. Cherukuri, P., Gannon, C. J., Leeuw, T. K., Schmidt, H. K., Smalley, R. E., Curley, S. A., and Weisman, R. B (2006). Mammalian pharmacokinetics of carbon nanotubes using intrinsic near-infrared fluorescence, *PNAS*, **103**, 18882–18886.

59. Strano, M. S., Doorn, S. K., Haroz, E. H., Kittrell, C., Hauge, R. H., and Smalley, R. E. (2003) Assignment of (n, m) Raman and optical features of metallic single-walled carbon nanotubes, *Nano Lett.*, **3**, 1091–1096.

60. Doorn, S. K., Heller, D. A., Barone, P. W., Usrey, M. L., and Strano, M. S. (2004) Resonant Raman excitation profiles of individually dispersed single walled carbon nanotubes in solution, *Appl. Phys. A: Mater. Sci. Process.*, **78**, 1147–1155.

61. Misewich, J. A., Martel, R., Avouris, Ph., Tsang, J. C., Heinze, S., and Tersoff, J. (2003) Electrically induced optical emission from a carbon nanotube FET, *Science*, **300**, 783–786.

62. Freitag, M., Tsang, J. C., Kirtley, J., Carlsen, A., Chen, J., Troeman, A., Hilgenkamp, H., and Avouris, P. (2006) Electrically excited, localized infrared emission from single carbon nanotubes, *Nano Lett.*, **6**, 1425–1433.

63. Strano, M. S., Huffman, C. B., Moore, V. C., O'Connell, M. J., Haroz, E. H., Hubbard, J., Miller, M., Rialon, K., Kittrell, C., Ramesh, S., Hauge, R. H., and Smalley, R. E. (2003) Reversible, band-gap-selective protonation of single-walled carbon nanotubes in solution, *J. Phys. Chem. B*, **107**, 6979–6985.

64. Usrey, M. L., Lippmann, E. S., and Strano, M. S. (2005) Evidence for a two-step mechanism in electronically selective single-walled carbon nanotube reactions. *J. Am. Chem. Soc.*, **127**, 16129–35.

65. O'Connell, M. J., Eibergen, E. E., and Doorn, S. K. (2005) Chiral selectivity in the charge-transfer bleaching of single-walled carbon-nanotube spectra, *Nat. Mater.*, **4**, 412–418.

66. Cognet, L., Tsyboulski, D. A., Rocha, J. D. R., Doyle, C. D., Tour, J. M., and Weisman, R. B. (2007) Stepwise quenching of exciton fluorescence in carbon nanotubes by single-molecule reactions, *Science*, **316**, 1465–1468.

67. Satishkumar, B. C., Brown, L. O., Gao, Y., Wang, C. C., Wang, H. L., and Doorn, S. K. (2008) Reversible fluorescence quenching in carbon nanotubes for biomolecular sensing, *Nat. Nanotechnol.*, **2**, 560–564.

68. Barone, P. W., Baik, S., Heller, D. A., and Strano, M. S. (2005) Near-infrared optical sensors based on singlewalled carbon nanotubes, *Nat. Mater.*, **4**, 86–92.

69. Song, C., Pehrsson, P. E., and Zhao, W. (2005) Recoverable solution reaction of hipco carbon nanotubes with hydrogen peroxide, *J. Phys. Chem. B*, **109**, 21634–21639.

70. Freitag, M., Martin, Y., Misewich, J. A., Martel, R., and Avouris, Ph. (2003) Photoconductivity of single carbon nanotubes, *Nano Lett.*, **3**, 1067–1071.

71. Pedersen, T. G. (2003) Variational approach to excitons in carbon nanotubes, *Phys. Rev. B*, **67**, 073401.

72. Kane, C. L., and Mele, E. J. (2003) The ratio prolem in single carbon nanotube fluorescence spectroscopy, *Phys. Rev. Lett.*, **90**, 207401.

73. Qiu, X., Freitag, M., Perebeinos, V., and Avouris, Ph. (2005) Photoconductivity spectra of single-walled carbon nanotubes: implications on the nature of their excited states, *Nano Lett.*, **5**, 749–752.

74. Shvedova, A. A., Castranova, V., Kisin, E. R., Schwegler-Berry, D., Murray, A. R., Gandelsman, V., Maynard, A., and Baron, P. (2003) Exposure to carbon nanotube material, assessment of nanotube cytotoxicity using human keratinocyte cells, *J. Toxicol. Environ. Health A*, **66**, 1909–1926.

75. Zhang, Z., Yang, X., Zhang, Y., Zeng, B., Wang, S., Zhu, T., Roden, R. B. S., Chen, Y., and Yang, R. (2006) Delivery of telomerase reverse transcriptase small interfering RNA in complex with positively charged single-walled carbon nanotubes suppresses tumor growth, *Clin. Cancer Res.*, **12**, 4933–4939.

76. Bianco, A., Kostarelos, K., Partidos, C. D., and Prato, M. (2005) Biomedical applications of functionalised carbon nanotubes, *Chem. Commun.*, 571–577

77. Dumortier, H., Lacotte, S., Pastorin, G., Marega, R., Wu, W., Bonifazi, D., Briand, J. P., Prato, M., Muller, S., and Bianco, A. (2006) Functionalized carbon nanotubes are non-cytotoxic and preserve the functionality of primary immune cells, *Nano Lett.*, **6**, 522–1528.

78. Yakobson, B. I., and Smalley, R. E. (1997) Fullerene nanotubes: C-1000000 and beyond, *Am. Sci.*, **85**, 324.

79. Vigolo, B., Pénicaud, A., Coulon, C., Sauder, C., Pailler, R., journet, C., Bernier, P., and Poulin, P. (2000) Macroscopic fibers and ribbons of oriented carbon nanotubes, *Science*, **290**, 1331–1334.

80. Dalton, A. B., Collins, S., Muñoz, E., Razal, J. M., Ebron, V. H., Ferraris, J. P., Coleman, J. N., Kim, B. G., and Baughman, R. H. (2003) Super-tough carbon-nanotube fibres, *Nature*, **423**, 703.

81. Walters, D. A., Ericson, L. M., Casavant, M. J., Liu, J., Colbert, D. T., Smith, K. A., and Smalley, R. E. (1999) Elastic strain of freely suspended single-wall carbon nanotube ropes, *Appl. Phys. Lett.*, **74**, 3803.

82. Wong, E. W., Sheehan, P. E., and Lieber, C. M. (1997) Nanobeam mechanics: elasticity, strength, and toughness of nanorods and nanotubes, *Science*, **277**, 1971–1975.

83. Yamamoto, K., Akita, S., and Nakayama, Y. (1998) Orientation and purification of carbon nanotubes using ac electrophoresis, *J. Phys. D. Appl. Phys.*, **31**, L34–L36.

84. Chung, J., Lee, K. H., Lee, J., and Ruoff, R. S. (2004) Toward largescale integration of carbon nanotubes, *Langmuir*, **20**, 3011–3017.

85. Chen, C., and Zhang, Y. (2006) Manipulation of single-wall carbon nanotubes into dispersively aligned arrays between metal electrodes, *J. Phys. D. Appl. Phys.*, **39**, 172–176.

86. Chen, X. Q., Saito, T., Yamada, H., and Matsushige, K. (2001) Aligning single-wall carbon nanotubes with an alternating-current electric field, *Appl. Phys. Lett.*, **78**, 3714–3716.

87. Yuen, F. L. Y., _Zak, G., Waldman, S. D., and Docoslis, A. (2008) Morphology of fibroblasts grown on substrates formed by dielectrophoretically aligned carbon nanotubes, *Cytotechnology*, **56**, 9–17.

88. Sinnott, S. B. J. (2002) Bone cell proliferation on carbon nanotubes, *Nanosci. Nanotechnol.*, **2**, 113.

89. Zanello, L. P., Zhao, B., Hu, H., and Haddon, R. C. (2006) Bone cell proliferation on carbon nanotubes, *Nano Lett.*, **6**, 3562–3567.

90. Supronowicz, P. R., Ajayan, P. M., Ullmann, K. R., Arulanandam, B. P., Metzger, D. W., and Bizios, R. (2002), *J. Biomed. Mater. Res.*, **59**, 499–506.

91. Zanello, L. P. (2006) Electrical properties of osteoblasts cultured on carbon nanotubes, *Micro Nano Lett.*, **1**, 19–22.

92. Calvert, P. (1999) A recipe for strength, *Nature*, **399**, 210–211.

93. Ajayan, P. M., Schadler, L. S., Giannaris, C., and Rubio, A. (2000) Single-walled carbon nanotube±polymer composites: strength and weakness, *Adv. Mater.*, **12**, 750–753.

94. Zhang, X. F., Liu, T., Sreekumar, T. V., Kumar, S., Moore, V. C., Hauge, R. H., and Smalley, R. E. (2003) Poly (vinyl alcohol)/SWNT composite film, *Nano Lett.*, **3**, 1285–1288.

95. Shi, X., Hudson, J. L., Spicer, P. P., Tour, J. M., Krishnamoorti, R., and Mikos, A. G. (2006) Injectable nanocomposites of single-walled carbon nanotubes and biodegradable polymers for bone tissue engineering, *Biomacromolecules*, **7**, 2237–2242.

96. Wang, S. F., Shen, L., Zhang, W. D., and Tong, Y. J. (2005) Preparation and mechanical properties of chitosan/carbon nanotubes composites, *Biomacromolecules*, **6**, 3067–3072.

97. Zhao, B., Hu, H., Mandal, S. K., and Haddon, R. C. (2005) A bone mimic based on the self-assembly of hydroxyapatite on chemically functionalized single-walled carbon nanotubes, *Chem. Mater.*, **17**, 3235–3241.

98. Shi, X., Sitharamana, B., Phama, Q. P., Liang, F., Wu, K., Billups, W. E., Wilson, L. J., and Mikos, A. G. (2007) Fabrication of porous ultra-short single-walled carbon nanotube nanocomposite scaffolds for bone tissue engineering, *Biomaterials*, **28**, 4078–4090.

99. Sitharaman, B., Shi, X., Walboomers, X. F., Liao, H., Cuijpers, V., Wilson, L. J., Mikos, A. G., and Jansen, J. A. (2008) In vivo biocompatibility of ultra-short single-walled carbon nanotube/biodegradable polymer nanocomposites for bone tissue engineering, *Bone*, **43**, 362–370.

100. Boccaccini, A. R., Chicatun, F., Cho, J., Bretcanu, O., Roether, J. A., Novak, S., and Chen, Q. (2007) Carbon nanotube coatings on bioglass-based tissue engineering scaffolds, *Adv. Funct. Mater.*, **17**, 2815–2822.

101. Du, C., Heldbrandt, D., and Pan, N. (2002) Preparation and preliminary property study of carbon nanotubes films by electrophoretic deposition, *Mater. Lett.*, **57**, 434–438.

102. Mei, F., Zhong, J., Yang, X., Ouyang, X., Zhang, S., Hu, X., Ma, Q., Lu, J., Ryu, S., and Deng, X. (2007) Improved biological characteristics of poly(L-lactic acid) electrospun membrane by incorporation of multiwalled carbon nanotubes/ hydroxyapatite nanoparticles, *Biomacromolecules*, **8**, 3729–3735.

103. Price, R. L., Waid, M. C., Haberstroh, K. M., and Webster, T. J. (2003) Selective bone cell adhesion on formulations containing carbon nanofibers, *Biomaterials*, **24**, 1877–1887.

104. Webster, T. J., Siegel, R. W., and Bizios, R. (1999) Osteoblast adhesion on nanophase ceramics, *Biomaterials*, **20**, 1221–1227.

105. Bacakova, L., Grausova, L., Vacik, J., Fraczek, A., Blazewicz, S., Kromka, A., Vanecek, M., and Svorcik, V. (2007) Improved adhesion and growth of human osteoblast-like MG 63 cells on biomaterials modified with carbon nanoparticles, *Diamond Rel. Mater.*, **16**, 2133–2140.

106. Usui, Y., Aoki, K., Narita, N., Murakami, N., Nakamura, I., Nakamura, K., Ishigaki, N., Yamazaki, H., Horiuchi, H., Kato, H., Taruta, S., Kim, Y. A., Endo, M., and Saito, N. (2008) Carbon nanotubes with high bone-tissue compatibility and bone-formation acceleration effects, *Small*, **4**, 240–246.

107. Abarrategi, A., Gutiérrez, M. C., Moreno-Vicente, C., Hortigüela, M. J., Ramos, V., Lòpez-Lacomba, J. L., Ferrer, M. L., and Monte, F. (2008) Multiwall carbon nanotube scaffolds for tissue engineering purposes, *Biomaterials*, **29**, 94–102.

108. Wang, W., Watari, F., Omori, M., Liao, S., Zhu, Y., Yokoyama, A., Uo, M., Kimura, H., and Ohkubo, A. (2006) Mechanical properties and biological behavior of carbon nanotube/polycarbosilane composites for implant materials, *J. Biomed. Mat. Res. B: Appl. Biomater.*, **82B**, 223–230.

109. Mattson, M. P., Haddon, R. C., and Rao, A. M. (2000) Molecular functionalization of carbon nanotubes and use as substrates for neuronal growth, *J. Mol. Neurosci.*, **14**, 175–182.

110. Gheith, M. K., Sinani, V. A., Wicksted, J. P., Matts, R. L., and Kotov, N. A. (2005) Single-walled carbon nanotube polyelectrolyte multilayers and freestanding films as a biocompatible platform for neuroprosthetic implants, *Adv. Mater.*, **17**, 2663–2670.

111. Gheith, M. K., Pappas, T. C., Liopo, A. V., Sinani, V. A., Shim, B. S., Motamedi, M., Wicksted, J. P., and Kotov, N. A. (2006) Stimulation of neural cells by lateral currents in conductive layer-by-layer films of single-walled carbon nanotubes, *Adv. Mater.*, **18**, 2975–2979.

112. Gabay, T., Jakobs, E., Ben-Jacob, E., and Hanein, Y. (2005) Electrochemical properties of carbon nanotube based multielectrode arrays, *Physica A*, **350**, 611–621.

113. Zhang, X., Prasad, S., Niyogi, S., Morgan, A., Ozkan, M., and Ozkan, C. S. (2005) Guided neurite growth on patterned carbon nanotubes, *Sens. Actuators B*, **106**, 843–850.

114. Hu, H., Ni, Y., Montana, V., Haddon, R. C., and Parpura, V. (2004) Chemically functionalized carbon nanotubes as substrates for neuronal growth, *Nano Lett.*, **4**, 3507–511.

115. Ago, H., Uchimura, E., Saito, T., Ohshima, S., Ishigami, N., Tsuji, M., Yumura, M., and Miyake, M. (2006) Mechanical immobilization of Hela cells on aligned carbon nanotube array, *Mater. Lett.*, **60**, 3851–3854.

116. MacDonald, R. A., Laurenzi, B. F., Viswanathan, G., Ajayan, P. M., and Stegemann, J. P. (2009) Collagen–carbon nanotube composite materials as scaffolds in tissue engineering, *J. Biomed. Mater. Res. A*, **74A**, 489–496.

117. Meng, J., Song, L., Meng, J., Kong, H., Zhu, G., Wang, C., Xu, L., Xie, S., and Xu, H. (2009) Using single-walled carbon nanotubes nonwoven films as scaffolds to enhance long-term cell proliferation in vitro, *J. Biomed. Mater. Res. A*, **74A**, 298–306.

118. Correa-Duarte, M. A., Wagner, N., Rojas-Chapana, J., Morsczeck, C., Thie, M., and Giersig, M. (2004). Fabrication and biocompatibility of carbon nanotube-based 3D networks as scaffolds for cell seeding and growth, *Nano Lett.*, **4**, 2233–2236.

Chapter 8

TOXICITY OF CARBON NANOTUBES

Tapas Ranjan Nayak and Giorgia Pastorin

Department of Pharmacy, Faculty of Science, National University of Singapore, Singapore

phapg@nus.edu.sg

8.1 INTRODUCTION

Until very recently, the development of nanotechnology was mainly constrained to electronics and engineering devices, representing an apparently inoffensive phenomenon, but lately it has envisaged its potential application into medicine and biology as well. Therefore, two major concerns have emerged, one focused on the effects of these materials on the environment, and the other one more vigilant on their role in the improvement of health conditions. Due to the novelty of the topic, current available information concerning the relative environmental and health risks of manufactured nanoparticles or nanomaterials is severely absent and defective. For example, the first generation of nanomedicines (liposomal preparations) were approved much before a real awareness existed about safety of nanomaterials.

Amid all the materials reported under the common umbrella of nanotechnology, carbon nanotubes (CNTs) have attracted the attention of numerous scientists, due to the possibility of being applied in engineering, physics, chemistry, materials science and biology. CNTs, both in single- (SWCNTs) and multi-walled (MWCNTs) forms, are classified as "synthetic graphite" by the National Occupational Safety and Health Administration, on the basis of the same honeycomb pattern (http://www.osha.gov/dts/chemicalsampling/data/CH_244000.html). However, such definition might not be exhaustive for the exposure to CNTs, since they present physicochemical

Carbon Nanotubes: From Bench Chemistry to Promising Biomedical Applications
Edited by Giorgia Pastorin
Copyright © 2011 Pan Stanford Publishing Pte. Ltd.
www.panstanford.com

properties often influenced by several variables, including size, chemical composition, solubility and aggregation.[1] These parameters could affect cellular uptake and protein binding and cause tissue damage. At a more general level, one severe drawback is represented by the inability to fabricate structurally and chemically controlled CNTs with identical characteristics in terms of properties and contaminants amount, and this has also limited their clinical and pharmacological applications. Moreover, differently from mono-dispersed particles, their shape belongs to both fibres and nanoparticles, thus triggering some unpredictable effects. The potential risk associated with CNTs includes the toxicity of eventual impurities (e.g., metal catalysts) in the samples, and the possibility, due to their small dimensions, that CNTs escape from the normal phagocytic defences and deposit into organ and tissues, with hazardous effects on the body. CNTs commonly show different levels of purity, which are strictly dependent on the methods employed for their production. The impurities are essentially made up of residual catalysts and amorphous carbon. If present in high amount, they might enhance the toxicity[2] and determine unwanted effects.

Many articles have indicated that CNTs are toxic, proposing some valuable reasons, which range from reduced dimensions (which enable them to be subtracted from common scavengers in our body) and hydrophobic nature (incompatible with physiological fluids) to impurities (difficult to be eliminated even after CNTs manipulation), problems of dispersion (which might form sediments on the cell culture and therefore reduce cell viability) and functionalisations.[3-6] Nevertheless, in the most of the cases, data showing toxicity had been obtained from non-functionalised samples, rather than ultrapure or chemically modified nanotubes; therefore, we have recently reported on the investigation of crucial parameters that might be responsible for their cytotoxicity.[7] These factors have been summarised in the following paragraphs, and they include surface functionalisation, concentration, purity and so forth.

8.2 PARAMETERS RESPONSIBLE FOR THE TOXICITY OF CNTS

8.2.1 Surface of CNTs

Up to now, the size cut-off below which particles are surely toxic has not been defined. However, it has been observed that the smaller the particles, the more toxic they become. This is due to the fact that there is more surface area per mass unit. As a consequence, any intrinsic toxicity of the surface will deeply influence the toxicological profile of the samples. Accordingly, the tubular structure of CNTs provides a large and reactive surface area,[8,9] which might render CNTs accountable for the observed cytotoxicity and thus unsuitable for biomedical applications. An inverse correlation between toxicity of CNTs and extent of their functionalisation has been suggested[10]; in other words, the higher the functionalisation at the tips and sidewalls of CNTs, the lower is their influence on cell viability. As support of such evidence, it has also emerged that several alarming data were mainly due to studies performed only on pristine, non-functionalised CNTs. For example, Sayes and Ausman studied the effect of some water-dispersible SWCNTs on human fibroblasts (HDF).[11] It was found that cytotoxicity of compounds decreased significantly with the increased degree of functionalisation on the surface.[1] The effect of functionalised, water-soluble CNTs on cell viability was also analysed by Pantarotto *et al.* during the study of translocation of bioactive peptides across the cell membrane.[12] 3T6 and 3T3 fibroblasts were exposed to 1–10 µM concentration of fluorescent SWCNTs. It was noticed that below 5 µM almost 90% of the cell population remained alive, suggesting a non-toxic behaviour of functionalised nanotubes (*f*-CNTs). Alternatively, HeLa cells, incubated for several hours with about 1 mg/mL of CNTs mixed with plasmid DNA in different charge ratios, did not demonstrate signs of apoptosis.[13] In addition, *f*-CNTs complexed to different types of nucleic acids, including plasmid DNA, RNA and oligodeoxynucleotides CpG sequences were not toxic for cells such as breast cancer cells (MCF7) or splenocytes.[14–16] This was confirmed by other groups that investigated the effects of both *f*-SWCNTs and *f*-MWCNTs in different cell subtypes.[17,18]

In parallel investigations, SWCNTs were evaluated in terms of their ability to activate mouse spleen cells.[19] Since different cell types (monocytic leukemia THP-1 and spleen cells) and incubation times were employed, it was difficult to compare the obtained results. It was only possible to confirm that SWCNTs induced an immune response, although the stimulating activity resulted to be lower than that of microbial systems.

In an *in vivo* study, SWCNTs were investigated to verify their influence on the immune system through the activation of complement.[20] For this purpose, SWCNTs in concentrations between 0.62 and 2.5 mg were tested in rabbit red blood cells. They displayed a dose-dependent potency in complement activation comparable to that of the positive control, the potent activator Zymosan. In particular, it seemed that such activation followed the classical pathway, but with high selectivity, as confirmed by the direct binding of CNTs to the main complement subunit C1q. On the contrary, chemical modifications at the surface of SWCNTs reduced or even eliminated the complement activation, but further investigations are necessary to confirm this observation.

In our case, both surface charges and functionalisation on the tubes have been evaluated on the basis of the evidence that surface charges might alter blood–brain–barrier integrity, with concomitant modulation of toxic effects,[21] although some cationic liposomes exhibited improved cell uptake while lowering toxicity.[22] No remarkable effects were observed for the presence of charges on the surface in comparison with uncharged CNTs. Conversely, in agreement with earlier studies done in this regard,[23] we demonstrated that CNTs, once functionalised, caused less cell damage. Moreover, it was observed that toxicity inversely correlated with the length of water-soluble chains (e.g., polyethylene glycol [PEG]) attached (Fig. 8.1), since PEG decreases the hydrophobic character of CNTs and promotes their dispersion in aqueous solvents. Therefore, as the number of ethylene oxide groups increased, cell viability showed a better correlation with water dispersibility. As expected, longer PEG-CNTs showed a better profile than shorter tri-ethylene glycol ones (TEG-CNTs) (this was statistically significant, with $p < 0.05$). Interestingly, raw (unfunctionalised) SWCNTs displayed higher cell viability than MWCNTs, but such difference became less obvious after oxidation (not statistically significant, with $p > 0.05$), so that PEG-MWCNTs displayed similar cell loss to PEG-SWCNTs ($p > 0.05$). Conversely, pristine CNTs, being the most hydrophobic, had the worst cytotoxic profile. This result is in accordance with the previous result pertaining to more hydrophobic molecules leading to greater cell toxicity due to their higher propensity to interact with the hydrophobic membrane.[24]

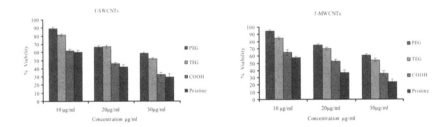

Figure 8.1 Percentage of cell viability of MCF-7 cells after 24 hours exposure to Pristine, oxidised, TEG and PEG functionalised CNTs (*f*-SWCNTs on the left and *f*-MWCNTs on the right) at three concentrations: 10 µg/mL, 20 µg/mL and 30µg/mL. Dose-dependent cytotoxic effect was observed for all the samples ($p < 0.05$). Reproduced from Nayak *et al.*[7] with permission.

An exception to this common trend is represented by the work by Monteiro-Riviere and collaborators, who investigated whether MWCNTs were able to cross the external membrane of human embryo kidney cells (HEK293) and affect the cell functions. It was shown that pristine MWCNTs were able to enter HEK293 cells and to induce the release of the proinflammatory cytokine IL-8.[25] In addition, it was observed that they were less harmful towards T lymphocytes than to chemically functionalised tubes.[6] The authors speculated that such result was due to the fact that oxidised MWCNTs were better dispersed in aqueous solution, determining higher weight/volume concentrations and, thus, a deeper impact on toxicity. At the dose of 400 μg/mL, oxidised tubes killed more than 80% of cells in 5 days, while pristine MWCNTs decreased the cell viability of less than 40%. However, although concentration could not be ascertained as the only parameter involved in the cytotoxic effect, the dose used in this study was very high, and the same experiments with concentrations inferior to 40 μg/mL did not affect the function of T cells.[6]

8.2.2 CNTs' Concentration

CNTs' cytotoxic profile is not yet well elucidated, since different studies have shown contradictory results.[26,27] The disparity could be ascribed to different cell viability methods, cell lines and sources of CNTs. Despite this degree of uncertainty, there is a common agreement regarding a concentration-dependent toxicity for all nanomaterials, in the form of a direct correlation between cell loss and doses used. More precisely, the higher the concentration of the nanosystems incubated with cells, the more remarkable the cell loss.[11,24,28] In the study by Sayes and Ausman, although a dose-response relationship of toxicity in the considered range of concentrations (0.003–30 mg/mL) was observed, cell death did not exceed 50% in all cases, apart from one in which 1% surfactant was employed. This could be attributed to the surfactant that was coated on the surface of the nanotubes in a non-covalent, reversible way; hence it could have been removed in the conditions of the biological tests.

Table 8.1 summarises the most recent *in vitro* studies and the concentrations used.

Table 8.1 Effect of CNTs at different concentrations in several cell lines

No.	Article title	Journal/date of article	Type of carbon nanotubes		Cell line used	Viability	Concentration (if applicable)
			Single-walled (SWCNTs)	Multi-walled (MWCNTs)			
1	In vitro toxicity evaluation of single walled carbon nanotubes on human A549 lung cells	*Toxicology in Vitro*, 2007, **21**(3), 438–448	Non-functionalised		Human A549 lung cells	Low acute toxicity	Concentration range used: 1.56–800µg/mL
2	Multi-walled carbon nanotubes induce T lymphocyte apoptosis	*Toxicology Letters*, 2006, **160**, 121–126		1. Oxidised 2. Pristine	Jurkat T-leukemia cells	Oxidised: Loss of >80%	400µg/mL
						Pristine: Loss of less than 50%	400 µg/mL
3	Cellular toxicity of carbon based nanomaterials	*Nano Letters*, 2006, **6**, 1121–1125		1. Pristine 2. Oxidised	Initial: Three different human lung-tumour cell lines, H596, H446 and Calu-1. Principal experiment: H596	Pristine: Cell viability decreased but less pronounced than that of the other CBN materials used in experiment	0.02 µg/mL
						Acid treated: Significant increase in toxicity compared to the pristine MWCNTs.	
4	Functionalised carbon nanotubes are non-cytotoxic and preserve the functionality of primary immune cells	*Nano Letters*, 2006, **6**, 1522–1528	1. 1,3-dipolar cycloaddition reaction 2. Oxidation/ amidation methodology		1. Purified B 2. T lymphocytes 3. Macrophages (obtained from the spleen, lymph nodes, and peritoneal cavity of BALB/c mice)	No significant toxicity for all three cell lines	10 µg/mL

No.	Title	Reference	CNT type		Cell type	Results	Concentration range
5	Biological interactions of functionalised single-wall carbon nanotubes in human epidermal keratinocytes	*International Journal of Toxicology*, 2007, **26**(2), 103–113	Purified CNTs + AHA (6-aminohexanoic acid) in DMF (covalent bond formation)		Neonatal human epidermal keratinocytes	Significant decrease in viability from 0.00005 to 0.05mg/mL	Concentration range used: 0.00000005–0.05mg/mL
6	Chemical modification of SWNT alters in vitro cell-SWNT interactions	*Journal of Biomedical Research Part A*, 2005, **76**(3), 614–625	1. Pristine 2. Purified 3. Functionalised with glucosamine (amide bond formation between glucosamine and SWCNTs)		3T3 mouse fibroblasts	1. AP NT: At the lowest tested concentration, 55% cell viability 2. Purified and glucosamine functionalised: Dose-dependent decrease in viability but less toxic than AP-NTs	Concentration range used: 0.001–1.0% (wt/vol)
7	Cytotoxicity of single-wall carbon nanotubes on human fibroblasts	*Toxicology In Vitro*, 2006, **20**, 1202–1212	Refined SWCNTs	Refined MWCNTs	Human dermis fibroblasts	Refined SWCNTs exhibited the most toxic effect (at 25 µg/mL)	Concentration range used: 0.8–100 µg/mL
8	Impact of carbon nanotube exposure, dosage and aggregation on smooth muscle cells	*Toxicology Letters*, 2007, **169**(1), 51–63	Purified and acid treated		Rat aortic smooth muscle cells	Significant effect on cell viability	Concentration range used: 0.18–0.22 mg/mL
9	Effect of single wall carbon nanotubes on human HEK293 cells	*Toxicology Letters*, 2005, **155**(1), 73–85	Not stated whether it is pristine or purified		Human embryo kidney cells (HEK293)	SWCNTs inhibit the proliferation of HEK293 cells.	Concentration range used: 0.78125–200 µg/mL

No	Description	Reference	Material/fractions		Cell line	Result	Concentration
10	The degree and kind of agglomeration affect carbon nanotube cytotoxicity	*Toxicology Letters*, 2007, **168**, 121–131	Four different fractions: 1. CNT-raw material 2. CNT-agglomerates 3. CNT-bundles 4. CNT-pellet		Mesothelioma cell line (MSTO-211H)	All CNTs fractions were able to significantly decrease cell activity and proliferation in a dose-dependent way	Concentration range used: 7.5–30 g/mL
11	Investigation of the cytotoxicity of CCVD (catalytic chemical vapour deposition) carbon nanotubes towards human umbilical vein endothelial cells	*Carbon*, 2006, **44**, 1093–1099	Three different samples, each synthesised by different catalytic chemical vapour deposition		Human umbilical vein endothelial cells (HUVEC)	Non-toxic	Concentration range: 100–1% (vol/vol) Volume used: 100 μL
12	In-vitro studies of carbon nanotubes biocompatibility	*Carbon*, 2006, **44**, 1106–1111		High purity MWNTs coated on polysulfone films	1. Human osteoblastic line hFOB 1.19 ATCC CRL-11372 2. Human fibroblastic line HS-5 ATCC CRL-11882	Good biocompatibility	Not applicable
13	Spectroscopic analysis confirms the interaction between single walled carbon nanotubes and various dyes commonly used to assess cytotoxicity	Carbon 2007, **45**, 1425-1432	Pristine SWCNTs		Human alveolar carcinoma epithelial cell line (A549)	Toxicity was observed	Concentration range used: 0.00156–0.8 mg/mL

14	Effects of fullerenes and single-wall carbon nanotubes on murine and human macrophages	*Carbon*, 2006, **44**, 1100–1105	Purified SWCNTs		1. Murine macrophage cells (J 774 cell line) 2. Human monocytes derived macrophages	Low cytotoxicity	Concentration range: 30–60 µg/mL
15	Functionalization density dependence of single-walled carbon nanotubes cytotoxicity in vitro	*Toxicology Letters*, 2006, **161**(2), 135–142	Four water-dispersible SWNT samples: 1. SWCNT-phenyl-SO$_3$H 2. SWCNT-phenyl-(COOH)$_2$ 3. SWCNT in 1% pluronic F108 4. SWCNT-phenyl-SO$_3$Na		Human dermal fibroblasts (HDF)	Covalent functionalisation reduced HDF cytotoxic response Overall, limited impact on cell viability.	Concentration range: 3–30 mg/mL
16	Cytotoxicity assessment of some carbon nanotubes and related carbon nanoparticle aggregates and the implications for anthropogenic carbon nanotube aggregates in the environment	*International Journal of Environment Research and Public Health*, 2005, **2**(1), 31–42	Pristine SWNT sample	Two different MWCNT samples from two companies.	Murine alveolar macrophages (RAW267.9 cells)	Significant cytotoxic effect. Induction of cellular death at a threshold of 2.5 µg/mL	Concentration range used: 10 µg/mL with 11 doubling concentrations subsequently

No	Title	Reference	Material	CNT type	Cell line	Findings	Concentration range
17	Influence of length on cytotoxicity of multi-walled carbon nanotubes against human acute monocytic leukaemia cell line THP-1 in vitro and subcutaneous tissue of rats in vivo	*Molecular Biosystems*, 2005, **1**, 176–182		Purified, acid-treated MWCNTs of two different lengths: 1. 220 nm 2. 825 nm	Human acute monocytic leukaemia cell line (THP-1)	Cytotoxic effects was based on the production of TNF-α Both 220-CNTs and 825-CNTs possess induction activity towards macrophages.	Concentration range used: 5–50 ng/mL
18	Multi-walled carbon nanotube interactions with human epidermal keratinocytes	*Toxicology Letters*, 2005, 155(3), 377–384		Self-prepared MWCNTs	Human Epidermal Keratinocytes (HEK)	HEK viability assessed by the NR assay slightly decreased in a dose-dependent manner at 24 and 48 hours (data not shown) after exposure to the nanotubes	Concentration range used: 0.1, 0.2, and 0.4 mg/mL
19	Carbon nanotube biocompatibility with cardiac muscle cells	*Institute of Physics Publishing Nanotechnology*, 2006, **17**, 391–397	Highly purified SWCNTs.		Cardiac muscle cell lines (rat cell line H9c2)	SWCNTs can be seeded with cardiomyocytes without affecting cell viability	Concentration range used: 0.2 mg/mL
20	In-vitro toxicity assessment of single and multi-walled carbon nanotubes in human astrocytoma and lung carcinoma cells	*Toxicology Letters*, 2007, **172S**, S1–S240	SWCNTs	1. MWCNTs 2. MWCNT-COOH 3. MWCNT-NH$_2$	1. Human astrocytoma D384-cells 2. Lung carcinoma A549-cells	MTT results revealed a strong cytotoxicity Calcein/PI staining did not confirm MTT cytotoxicity in both cell lines	Concentration range used: 0.1–100 µg/mL

8.2.3 CNTs' Dispersibility

Due to their hydrophobic nature, CNTs tend to aggregate in large bundles especially in media commonly used for toxicological testing.[29] Dispersion in aqueous solvents can be improved,[3,30] for example, through sonication, stabilisation with surfactants and CNTs' functionalisation. In previous studies it has been shown that functionalisation is more effective than surfactants in reducing CNTs' cytotoxicity,[11] while prolonged sonication can cut the tubes excessively and thus decrease their aspect ratio, which is a desirable physicochemical property of CNTs.[31] Functionalisation of CNTs, besides improving water dispersibility compared with their pristine counterparts,[24] offers the advantage of incorporating biologically active moieties such as antibodies and drug molecules.[32] Physiological solutions are the most appropriate, but sometimes the low solubility of carbon material requires organic solvents. It has already been reported that human keratinocytes, once exposed to nanotubes, displayed cell death.[33] This was attributed to oxidative stress within cells and to activation of NF-kB factor. However, dimethylformamide (DMF) was used to dissolve the samples, and it might have induced some alterations, since it is a toxic solvent, which should be avoided in cell manipulation. Similarly, in our investigations we have maintained di-methyl sulfoxide (DMSO) in concentrations as low as 1% (vol/vol) of the final volume, to avoid any deceptive result. DMSO, together with CNTs' functionalisation, improved the tubes' dispersibility, indicating the maximum weight of nanotubes that could be added with minimal formation of aggregates. To reiterate, aggregates of f-CNTs should be kept to a minimum as they could cause cell death.[34]

Moreover, we observed better dispersibility and loading for f-MWCNTs in comparison with f-SWCNTs (this was statistically significant, with $p < 0.05$); this could be attributed to the wider surface exposed in the case of multi-walled tubes, suggesting that further sidewall functionalisation caused an increase in f-CNTs concentration that could form stable suspensions.

8.2.4 Length and Diameter

Besides the above-mentioned factors, length could also be involved in the intrinsic toxicity of CNTs. It has been demonstrated that shorter CNTs (220 nm) displayed lower inflammatory effects compared with longer CNTs (825 nm).[4] Clusters of both samples were surrounded by macrophages, due to activation of innate immunity.[4] The shorter tubes displayed a lower inflammatory response, suggesting that tubes' length plays a remarkable role in toxicity. In both cases, no severe effects, such as necrosis or degeneration, were observed around CNT throughout the experimental period of 4 weeks.

In general, shorter tubes can be easily obtained from controlled acidic conditions, which are known to cut the tubes mainly in correspondence of the tips and sidewall defects. The tubes collected at the end of the process present lengths that are inversely proportional to the oxidising times (Table 8.2). We confirmed previous data indicating that long sonication time in strong acidic environment results in excessive cutting, leading to loss of tubular structure of SWCNTs.[35,36] In fact SWCNTs treated beyond 9 hours resulted in too small fragments, while MWCNTs preserved their structure for much longer time (beyond 24 hours).

Table 8.2 Length of different CNTs oxidised by treatment with strong acids for different time intervals

Time of Sonication (h)	Length of MWCNTs (nm)	Length of SWCNTs (nm)
0	1,000–2,000	1,000–5,000
3	600–800	800–1,000
6	250–350	150–200
9	150–250	50–100
12	100–150	–
24	50–100	–

Surprisingly, there was no remarkable difference between toxicity of SWCNTs and MWCNTs, even though their structure was much different after 9 hours. This result suggests that high aspect ratio is not a favourable parameter in enhancing cell viability, since cells were shown to prefer fragmented tubes (Fig. 8.2).

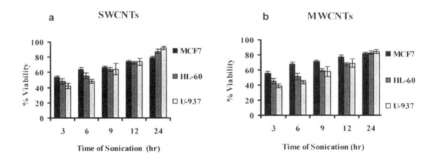

Figure 8.2 Percentage of cell viability of MCF7, HL-60 and U937 cells after 24 hours exposure to oxidised CNTs (20 µg/mL) for different time periods: 3, 6, 9, 12 and 24 hours. Length-dependent cytotoxic effect was observed for both SWCNTs (A) and MWCNTs (B). Reproduced from Nayak *et al.*[7] with permission.

Moreover, the influence of diameter on cell viability is not clear. In previous studies it was reported that MWCNTs with diameters of 60–100 nm induced more serious cytotoxicity on alveolar macrophages in comparison with smaller-diameter MWCNTs (<60 nm).[37] Conversely, in another study, toxicity of SWCNTs with a mean diameter of 1.4 nm was much higher than MWCNTs of 10–20 nm, with serious damage of alveolar macrophages at doses as low as 0.38 μg/mL.[38] In our experiments, the effect was strictly dependent on the cells under investigation (Fig. 8.2), although it was not particularly determinant ($p > 0.05$): HL60 and U937 cells displayed a slightly better cell viability than MCF7, especially in case of SWCNTs. One explanation could be that HL60 and U937 cells are suspended cells of smaller dimensions (2–10 μm) than MCF7 cells (30–40 μm), and thus they could accommodate SWCNTs much better than MWCNTs. On the other hand, the adherence of MCF7 cells and their bigger dimensions seemed more influenced by the better dispersion of MWCNTs. Another possibility could be that MWCNTs, being more easily suspended, have a more prolonged contact with suspended cells (HL60 and U937) than with adherent cells (MCF7). Hence, we could conclude that the tubes' diameter does not play a crucial role in the cytotoxic profile.

Such difference could also be attributed to the method employed for the cell detection; that is, MTT assay could be more sensitive towards HL60 cells. However, cell viability evaluation through an alternative Cyquant assay demonstrated the same profile ($p > 0.05$),[7] thus excluding any significant interference deriving from the method of analysis employed.

8.2.5 Purity

Several methods adopted for the production of CNTs involve the use of metals as catalysts[39] which, if present in high amount, enhance cytotoxicity.[38] In fact, free iron or nickel and transition-metal complexes are known to originate reactive radicals, which could induce radical oxidation and enhance oxidative stress to the cells.40

In order to investigate this aspect, Shvedova *et al.* identified and quantified the reactive species involved during the manipulation of carbon-based nanomaterials; more precisely, they evaluated the effects of pristine SWCNTs on human epidermal keratinocytes (HaCaT).[41] Images under scanning and transmission electron microscopy clearly indicated that oxidative stress, induced by treatment with SWCNTs, caused a change in the morphology of cells and altered surface integrity. The same hazardous effects were reduced by the subsequent use of an iron chelator that displayed a protecting role towards HaCaT cells and, thus, confirmed that cytotoxicity of SWCNTs was mainly correlated with iron catalytic effects.

Jia *et al.* investigated the cytotoxicity determined by some nanostructures such as SWCNTs, MWCNTs and fullerenes on alveolar macrophages (AM).[38] The different carbon particles were stably suspended in RPMI medium and sonicated for 20 minutes with a Dounce homogenizer before usage. Although all the samples had a purity superior than 90%, they showed remarkably different effects on the phagocytic ability of AM after 6 hours' exposure to carbon materials. More precisely, at low doses (1.41 µg/cm²) SWCNTs displayed a high cytotoxic effect, corresponding to >20% inhibition of cell growth in a dose-dependent trend. MWCNTs and fullerenes were instead much less toxic. One possible explanation for such a deep variation could be attributed to the purity level of SWCNTs that was lower than the other samples. Indeed, residual amorphous carbon and trace amount of metallic catalysts, such as Fe, Ni and Y, were present in SWCNT samples. Similarly, another investigation demonstrated that lower purified CNTs induced cytotoxicity, while homogeneously suspended CNTs with high purity did not alter cell growth.[42]

Interestingly, one of our samples, devoid of functional groups, without catalysts (the initial $MmNi_3$ was not detected once subjected to thermogravimetric analysis) and negligible amorphous carbon,[43] showed no sign of toxicity (Fig. 8.3) in the concentrations normally used for cell viability assays (10–50 µg/mL); for that reason, doses were increased, and the first evidence of toxicity was detected only at concentrations above 150 µg/mL. To our knowledge, very few articles have ever reported cytotoxic profiles testing such high doses, analogue compatibility with cells *in vitro* and high purity by energy-dispersive X-ray spectroscopy (EDS) (further data available in Ref. 7), suggesting that further functionalisations could be introduced in this last sample for a better cell targeting but without harmful consequences. This result not only allowed us to identify the sample's purity as the most crucial parameter that is responsible for the toxicity of CNTs, but it also encourages further application of these non-toxic nanomaterials in the field of nanomedicine.

8.3 ENVIRONMENTAL EXPOSURE

Carbon nanotubes' needle-like fibre shape has been recently compared to asbestos[44]: previous studies in populations exposed to this material showed that the main body of the lung was a target for asbestos fibres, resulting in both lung cancer and scarring of the lungs (asbestosis). Therefore, the analogy between CNTs and asbestos has resulted in huge concerns since CNTs' widespread use may lead to analogue inflammation and formation of lesions known as granulomas.

Lam *et al.*[45] and Warheit *et al.*[46] were among the first to investigate the pulmonary toxicity of CNTs (specifically SWCNTs) in rodents. In particular, Lam *et al.* evaluated histopathological alterations in mice at 7 and 90 days after exposure to three differently manufactured SWCNTs that contained varying amounts of residual catalytic metals: (i) raw nanotubes (RNTs) and (ii) purified nanotubes (PNTs), both iron-containing HiPCO products from Rice University, and (iii) CarboLex's nickel-containing electric-arc nanotubes (CNTs). Carbon black and quartz were employed as low and high pulmonary toxicity controls, respectively. The experiments indicated that the three types of tubes induced dose-dependent lung lesions, consisting on interstitial granulomas, regardless of the amount of metal impurities inside the samples.[45] Even in case of low quantities of iron (2% by weight) granulomas were formed, suggesting that CNTs themselves are toxic. However, the methods employed for this study were somehow unclear, since the suspension was obtained by briefly shearing and short sonication to avoid any alteration of the nature of SWCNTs, but the instillation was performed through a plastic catheter placed in the trachea. So further studies would be needed to elucidate the cause of death. However, it was clear that mice treated with 0.5 mg of nanotubes showed the formation of granulomas, containing macrophages laden with black particles, a few lymphocytes, neutrophils, eosinophils and other inflammatory cells. Interestingly, the lowest dose (0.1 mg/30 kg) determined no evident clinical signs. The same (5 mg/kg), or even a lower (1 mg/kg), amount of SWCNTs was employed in the study by Warheit *et al.* in which histopathological evaluation of lung tissue was conducted after 24 hours, 1 week and 3 months post-instillation in rats.[46] In these experiments, SWCNTs and graphite particles, together with the corresponding controls (carbonyl iron and quartz particles), were intratracheally instilled in the animals. In the case of carbonyl iron or graphite, no adverse effects were observed, while non-dose-dependent multifocal granulomas were visible after exposure to SWCNTs, even if they did not progress beyond 1 month. In 15% of the rats, the highest dose induced mortality, but the main reason of that was imputable to the mechanical blockage of the upper airways, and not to the inner toxicity of SWCNT particulate. The death index was somehow incorrect, because the nanotubes tended to form nanoropes instead of being individually dispersed. A proof was the observation that the survived animals appeared normal through the whole 3 months. In addition, studies on chemotaxis concerning quartz treatment showed a reduced motility as a consequence of a deficiency in macrophages after 1 week. This phenomenon was not observed in the case of SWCNTs, which differed from quartz also in the formation of granulomas in a non-dose-dependent manner.

An interesting investigation has been reported on the adverse effects determined by aerosol release obtained through the mechanical collisions of as produced SWCNTs and bronze beads.[47] Even though the article provided additional insights on the toxicity of CNTs, the production of ultra-fine powders was not very efficient, and it was not possible to discriminate if such particles were made up mainly of nanotubes or of catalyst. As a consequence, most probably the respirable fraction of the nanotube aerosol was smaller than the estimated mass concentrations indicated above. Even more important, in none of the cases reported there was any indication that handling the nanotube material led to an increased concentration of fine particles, suggesting that eventual released particles tended to be larger than 1 μm in diameter. Anyway, it is worth mentioning that, although nanotubes tended to agglomerate into nanoropes, thus reducing the formation of an appreciable respirable aerosol (with estimated airborne concentrations of nanotube material lower than 53 μg/m³), it is also possible that such nanoparticles remain in the mouth and nasopharyngeal regions, still causing a potential health risk.

In the same manuscript, the authors also reported a preliminary quantification of the amount of nanoparticles deposited on the gloves of normal workers: potential dermal loading of SWCNTs was estimated by placing cotton gloves over the rubber gloves generally used by the workers. Glove deposits of SWCNTs during handling were estimated at between 0.2 mg and 6 mg per hand. Even though the maximum concentration of 6 mg is quite alarming, it is very likely that it is overestimated to a certain extent. In fact the cotton gloves used to collect the hand samples are likely to retain more material than latex gloves (or similar) or nude skin, so these results are useful to encourage protective measures, but they also require additional investigations.

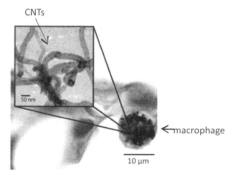

Figure 8.4 Inhaled carbon nanotubes accumulate within cells at the pleural lining of the lung as visualised by light microscopy. Reproduced from www.nanotech-now.com/news_images/35119.jpg with permission of Dr James Bonner, North Carolina State University. See also Colour Insert.

In a very recent article, Ryman-Rasmussen and collaborators[48] showed that MWCNTs reached the subpleura in mice after a single inhalation exposure of 30 mg/m^3 for 6 hours and that subpleural fibrosis increased after 2 weeks following inhalation (Fig. 8.4).

As emerges also by comparison with the data reported in all the articles mentioned in the present chapter, it is evident that the exposure of 30 mg/m^3 is a huge dose which is unluckily to occur if nanometerial is handled cautiously. Even in the case of aerosols, typical drug concentrations by weight may range up to 1% so that fluid metering, of say 25–100 μL of suspension, commonly releases the drug in around 100 μg quantities.[49] Moreover, the authors admitted that none of these effects was seen in mice that inhaled lower doses of nanotubes (1 mg/m^3), thus implying an overestimation of the risks and a general scepticism until further long-term assessments are conducted.

Additional studies by Huczko *et al.* have contributed to a further understanding of the impact of nanomaterials on health. The researchers tested the effect of fullerenes and CNTs on skin irritation and allergen risks.[50,51] Before these studies, there were only a few evidences of contact dermatitis caused by carbon fibre exposure.[52,53] The authors applied two protocols usually adopted for testing skin irritation, namely a patch and a Draize rabbit eye test, respectively. In the first case, 40 people showing predisposition to irritation and allergy were treated with a patch of filter paper that Whatman saturated with a water suspension of CNTs, and then they were examined for 96 hours. In the second method, four albino rabbits were instilled with 0.2 mL of CNT aqueous suspension, and observed after 24, 48 and 72 hours. The absence of adverse effects seemed to be reliant neither on the time of exposure nor on the type of material used, since there were no differences in comparison with the reference material (which did not contain CNTs).

In a different study, aimed to evaluate the effects of ingested CNTs, SWCNTs or MWCNTs were delivered through the food to the larval stage. They both had no detectable consequence on egg to adult survivorship, despite evidence that the nanomaterials are taken up and become sequestered in tissue.[54] However, when these same nanocarbons were exposed in dry form to adults, some materials (such as SWCNTs) adhered extensively to fly surfaces, overwhelmed natural grooming mechanisms and led to impaired locomotor function and mortality. Conversely, MWCNTs adhered weakly, could be removed by grooming and did not reduce locomotor function or survivorship. Therefore, it seems more likely that these discrepancies are primarily due to differences in nanomaterial superstructure, or aggregation state, and that the combination of adhesion and grooming can lead to active fly-borne nanoparticle transport. This also confirms our findings concerning the MWCNTs' better dispersibility profile and the lower cell loss in comparison with SWCNTs.

Interesting preliminary epidemiological data were also provided by Murr and collaborators,[55] who indicated a correlation between asthma (or related lung inflammations) incidence and exposure to "gas stoves", especially because indoor number concentrations for MWCNT aggregates (of about 1–2 μm in dimension) is at least 10 times the outdoor concentration, and virtually all gas combustion processes are variously effective sources. These results also raise concerns for manufactured CNT aggregates and related fullerene nanoparticles.

A crucial aspect to consider is that in most of the cases just reported, the observations do not correspond to samples chemically modified with different chains and molecules, since they mainly consist of unrefined or as-produced material. Therefore, as we have already mentioned, the tubes that have undergone further functionalisations display remarkable differences in comparison with raw material, but their toxicological impact has been poorly investigated. As a confirmation of this, Kostarelos and collaborators[56] have recently shown that intravenously administered pristine SWCNTs accumulated mainly in lung, liver and spleen, while the functionalised tubes tended to persist much less in tissues and organs (Fig. 8.5 and Table 8.3), with an accumulation proportional to the degree of functionalisation but independent of the characteristic of the attached groups.[57] For the purpose, water-soluble SWCNTs were functionalised with the chelating agent diethylenetriaminepentaacetic acid (DTPA) and labelled with indium (^{111}In) for imaging purposes. Radioactivity tracing indicated that the tubes were not retained in any of the reticuloendothelial system organs (liver or spleen) of the mice. Nearly all the tubes were excreted, through the normal renal pathway, as intact tubes in the urine, without any remarkable tissue damage even at high concentrations (max 400 μg/mouse) of tubes. The results could be explained by the fact that unfunctionalised CNTs are extremely hydrophobic and difficult to disperse in aqueous milieu owing to the van der Waals forces leading to aggregation in bundles, while functionalisation allows to obtain more individual tubes.

On the whole, despite some encouraging results and the lack of absolute substantiation of toxicity directly associated with CNTs, specific precautions should always be adopted while handling such nanomaterial.

Table 8.3 [¹¹¹In]DTPA–CNT 5 and [¹¹¹In]DTPA–CNT 3 percentage of injected dose per organ after intravenous administration[a]

	30 min		3 h		24 h	
Organ	Mean % dose per organ	SD	Mean % dose per organ	SD	Mean % dose per organ	SD
[¹¹¹In]DTPA–CNT 5						
Blood	4.5182	2.7105	0.2699	0.2069	0.0442	0.0314
Bone	0.2215	0.2354	0.0166	0.0147	0.0030	0.0027
Heart	0.0275	0.0230	0.0039	0.0032	0.0002	0.0004
Kidney	2.4806	3.1174	0.1840	0.1431	0.1417	0.0459
Liver	0.1784	0.0686	0.1655	0.0237	0.1040	0.0544
Lung	0.0897	0.0620	0.0036	0.0018	0.0016	0.0012
Muscle	49.2787	71.9382	0.6740	0.2814	0.0022	0.0038
Skin	7.6407	6.8005	0.6903	0.1996	0.6572	0.6642
Spleen	0.0375	0.0280	0.0037	0.0002	—	—
[¹¹¹In]DTPA–CNT 3						
Blood	5.3802	3.3357	0.1130	0.0026	0.0233	0.0227
Bone	0.3124	0.4894	0.0021	0.0030	—	—
Heart	0.0483	0.0221	0.0006	0.0002	0.0001	0.0001
Kidney	4.7936	5.5527	0.2262	0.0577	0.1303	0.0260
Liver	0.1693	0.0319	0.1144	0.0396	0.0773	0.0117
Lung	0.1326	0.0428	0.0002	0.0003	—	—
Muscle	68.4333	113.1819	0.0528	0.0747	—	—
Skin	36.3054	41.1663	0.3930	0.0620	0.4495	0.0277
Spleen	0.0327	0.0034	0.0064	0.0038	0.0001	0.0001

Note: CNTs were covalently attached to an indium chelator (diethylenetriaminepentaacetic dianhydride, DTPA), which was added in amounts able to saturate partially (CNT5) or completely (CNT3) the available functionalities.

[a] Reproduced from Singh *et al.*[56] with permission.

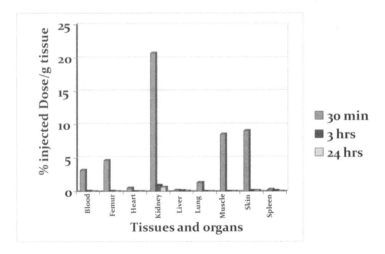

Figure 8.5 Biodistribution per collected gram of tissue of [¹¹¹In]DTPA–SWCNTs after intravenous administration of 60 μg of SWCNTs in 200 μL of phosphate buffered saline (PBS). Reproduced from Singh *et al.*[56] with permission. See also Colour Insert.

8.4 CONCLUSION

Even though in the last decade carbon nanotubes have shown exquisite properties and encouraging results, they still present a fundamental aspect that has limited their widespread applications: none among their characteristics in term of structure, size and composition can be ascertained with unquestionable certainty. Manufactured samples present huge differences and their further modification, through chemical procedures, enhances their diversity. Therefore, it becomes extremely important to identify the dependence from a particular property, in order to obtain more reproducible results. It is imperative to obtain a clear understanding of what really happens to nanoparticles months and years after their release, since some encouraging evidences are not exhaustive in guaranteeing complete innocuous behaviour. For all these very important reasons, further studies are required to investigate whether CNTs can be widely used for their useful characteristics, and their impact on both health and environment can be monitored in a meticulous and completely safe manner.

References

1. Donaldson, K., Aitken, R., Tran, L., Stone, V., Duffin, R., Forrest, G., and Alexander, A. (2006) Carbon nanotubes: a review of their properties in relation to pulmonary toxicology and workplace safety, *Toxicol. Sci.*, **92**, 5–22.

2. Chen, R. J., Bangsaruntip, S., Drouvalakis, K. A., Kam, N. W., Shim, M., Li, Y., Kim, W., Utz, P. J., and Dai, H. (2003) Noncovalent functionalization of carbon nanotubes for highly specific electronic biosensors, *Proc. Nat.l Acad. Sci. USA*, **100**, 4984–4989.

3. Lacerda, L., Bianco, A., Prato, M., and Kostarelos, K. (2006) Carbon nanotubes as nanomedicines: from toxicology to pharmacology, *Adv. Drug. Deliv. Rev.*, **58**, 1460–1470.

4. Sato, Y., Yokoyama, A., Shibata, K., Akimoto, Y., Ogino, S., Nodasaka, Y., Kohgo, T., Tamura, K., Akasaka, T., and Uo, M., Motomiya, K., Jeyadevan, B., Ishiguro, M., Hatakeyama, R., Watari, F., and Tohji, K. (2005) Influence of length on cytotoxicity of multi-walled carbon nanotubes against human acute monocytic leukemia cell line THP-1 in vitro and subcutaneous tissue of rats in vivo, *Mol. Biosyst.*, **1**, 176–182.

5. Garibaldi, S., Brunelli, C., Bavestrello, V., Ghigliotti, G., and Nicolini, C. (2006) Carbon nanotube biocompatibility with cardiac muscle cells, *Nanotecnology*, **17**, 391–397.

6. Bottini, M., Bruckner, S., Nika, K., Bottini, N., Bellucci, S., Magrini, A., Bergamaschi, A., and Mustelin, T. (2006) Multi-walled carbon nanotubes induce T lymphocyte apoptosis, *Toxicol. Lett.*, **160**, 121–126.

7. Nayak, T. R., Pay Chin Leow, Pui-Lai Rachel Ee, Arockiadoss, T., Ramaprabhu, S., and Pastorin, G. (2010) Crucial parameters responsible for carbon nanotubes toxicity, *Curr. Nanosci.*, **6**, 141–154.

8. Donaldson, K., Stone, V., Tran, C. L., Kreyling, W., and Borm, P. J. (2004) Nanotoxicology, *Occup. Environ. Med.*, **61**, 727–728.

9. Tran, C. L., Buchanan, D., Cullen, R. T., Searl, A., Jones, A. D., and Donaldson, K. (2000) Inhalation of poorly soluble particles. II. Influence of particle surface area on inflammation and clearance, *Inhal. Toxicol.*, **12**, 1113–1126.

10. Lacerda, L., Ali-Boucetta, H., Herrero, M. A., Pastorin, G., Bianco, A., Prato, M., and Kostarelos, K. (2008) Tissue histology and physiology following intravenous administration of different types of functionalized multiwalled carbon nanotubes, *Nanomedicine*, **3**, 149–161.

11. Sayes, C. M., Liang, F., Hudson, J. L., Mendez, J., Guo, W., Beach, J. M., Moore, V. C., Doyle, C. D., West, J. L., and Billups, W. E., Ausman, K. D., and Colvin, V. L. (2006) Functionalization density dependence of single-walled carbon nanotubes cytotoxicity in vitro, *Toxicol. Lett.*, **161**, 135–142.

12. Pantarotto, D., Briand, J. P., Prato, M., and Bianco, A. (2004) Translocation of bioactive peptides across cell membranes by carbon nanotubes, *Chem. Commun. (Camb).*, 16–17.

13. Pantarotto, D., Singh, R., McCarthy, D., Erhardt, M., Briand, J. P., Prato, M., Kostarelos, K., and Bianco, A. (2004) Functionalized carbon nanotubes for plasmid DNA gene delivery, *Angew. Chem. Int. Ed. Engl.*, **43**, 5242–5246.

14. Bianco, A., Hoebeke, J., Godefroy, S., Chaloin, O., Pantarotto, D., Briand, J. P., Muller, S., Prato, M., and Partidos, C.D. (2005) Cationic carbon nanotubes bind to CpG oligodeoxynucleotides and enhance their immunostimulatory properties, *J. Am. Chem. Soc.*, **127**, 58–59.

15. Liu, Y., Wu, D. C., Zhang, W. D., Jiang, X., He, C. B., Chung, T. S., Goh, S. H., and Leong, K. W. (2005) Polyethylenimine-grafted multiwalled carbon nanotubes for secure noncovalent immobilization and efficient delivery of DNA, *Angew. Chem. Int. Ed. Engl.*, **44**, 4782–4785.

16. Lu, Q., Moore, J. M., Huang, G., Mount, A. S., Rao, A. M., Larcom, L. L., and Ke, P. C. (2004) RNA polymer translocation with single-walled carbon nanotubes, *Nano Lett.*, **4**, 2473–2477.

17. Kam, N. W., and Dai, H. (2005) Carbon nanotubes as intracellular protein transporters: generality and biological functionality, *J. Am. Chem. Soc.*, **127**, 6021–6026.

18. Ajima, K., Yudasaka, M., Murakami, T., Maigne, A., Shiba, K., and Iijima, S. (2005) Carbon nanohorns as anticancer drug carriers, *Mol. Pharm.*, **2**, 475–480.

19. Kiura, K., Sato, Y., Yasuda, M., Fugetsu, B., Watari, F., Tohji, K., and Shibata, K. (2005) Activation of human monocytes and mouse splenocytes by single-walled carbon nanotubes, *J. Biomed. Nanotechnol.*, **1**, 359–364.

20. Salvador-Morales, C., Flahaut, E., Sim, E., Sloan, J., Green, M. L., and Sim, R. B. (2006) Complement activation and protein adsorption by carbon nanotubes, *Mol. Immunol.*, **43**, 193–201.

21. Lockman, P. R., Koziara, J. M., Mumper, R. J., and Allen, D. D. (2004) Nanoparticle surface charges alter blood-brain barrier integrity and permeability, *J. Drug. Target*, **12**, 635–641.

22. De Rosa, G., De Stefano, D., Laguardia, V., Arpicco, S., Simeon, V., Carnuccio, R., and Fattal, E. (2008) Novel cationic liposome formulation for the delivery of an oligonucleotide decoy to NF-kappa B into activated macrophages, *Eur. J. Pharm. Biopharm.*, **70**, 7–18.

23. Dumortier, H., Lacotte, S., Pastorin, G., Marega, R., Wu, W., Bonifazi, D., Briand, J. P., Prato, M., Muller, S., and Bianco, A. (2006) Functionalized carbon nanotubes are non-cytotoxic and preserve the functionality of primary immune cells, *Nano Lett.*, **6**, 1522–1528.

24. Nimmagadda, A., Thurston, K., Nollert, M. U., and McFetridge, P. S. (2006) Chemical modification of SWNT alters in vitro cell-SWNT interactions, *J. Biomed. Mater. Res. A*, **76**, 614–625.

25. Monteiro-Riviere, N. A., Nemanich, R. J., Inman, A. O., Wang, Y. Y., and Riviere, J. E. (2005) Multi-walled carbon nanotube interactions with human epidermal keratinocytes, *Toxicol. Lett.*, **155**, 377–384.

26. Cui, D., Tian, F., Ozkan, C. S., Wang, M., and Gao, H. (2005) Effect of single wall carbon nanotubes on human HEK293 cells, *Toxicol. Lett.*, **155**, 73–85.

27. Davoren, M., Herzog, E., Casey, A., Cottineau, B., Chambers, G., Byrne, H. J., and Lyng, F. M. (2007) In vitro toxicity evaluation of single walled carbon nanotubes on human A549 lung cells, *Toxicol. in Vitro*, **21**, 438–448.

28. Magrez, A., Kasas, S., Salicio, V., Pasquier, N., Seo, J. W., Celio, M., Catsicas, S., Schwaller, B., and Forro, L. (2006) Cellular toxicity of carbon-based nanomaterials, *Nano Lett.*, **6**, 1121–1125.

29. Smart, S. K., Cassady, A. I., Lu, G. Q., and Martin, D. J. (2006) The biocompatibility of carbon nanotubes, *Carbon*, **44**, 1034–1047.

30. Piret, J. P., Detriche, S., Vigneron, R., Vankoningsloo, S., Rolin, S., Mendoza, J. H. M., Masereel, B., Lucas, S., Delhalle, J., Luizi, F., Saout, C., and Toussaint, O. (2009) Dispersion of multi-walled carbon nanotubes in biocompatible dispersants, *J. Nanopart. Res.*, **12**, 75–82, DOI 10.1007/s11051-009-9697-8.

31. Hilding, J., Grulke, E. A., Zhang, Z. G., and Lockwood, F. (2003) Dispersion of carbon nanotubes in liquids, *J. Dispersion. Sci. Tech.*, **24**, 1–41.

32. Bianco, A., Kostarelos, K., and Prato, M. (2005) Applications of carbon nanotubes in drug delivery, *Curr. Opin. Chem. Biol.*, **9**, 674–679.

33. Manna, S. K., Sarkar, S., Barr, J., Wise, K., Barrera, E. V., Jejelowo, O., Rice-Ficht, A. C., and Ramesh, G. T. (2005) Single-walled carbon nanotube induces oxidative stress and activates nuclear transcription factor-kappaB in human keratinocytes, *Nano Lett.*, **5**, 1676–1684.

34. Raja, P. M., Connolley, J., Ganesan, G. P., Ci, L., Ajayan, P. M., Nalamasu, O., and Thompson, D. M. (2007) Impact of carbon nanotube exposure, dosage and aggregation on smooth muscle cells, *Toxicol. Lett.*, **169**, 51–63.

35. Mawhinney, D. B., Naumenko, V., Kuznetsova, A., Yates, J. T., Liu, J., and Smalley, R. E. (2000) Surface defect site density on single walled carbon nanotubes by titration, *Chem. Phys. Lett.*, **324**, 213–216.

36. Hu, H., Bhowmik, P., Zhao, B., Hamon, M. A., Itkis, M. E., and Haddon, R. C. (2001) Determination of the acidic sites of purified single-walled carbon nanotubes by acid-base titration, *Chem. Phys. Lett.*, **345**, 25–28.

37. Bai, R., Wang, W., Jin, X. L., and Song, W. H. (2007) Review on biological security of nanomaterials, *J. Environ. Health*, **24**, 59–61.

38. Jia, G., Wang, H., Yan, L., Wang, X., Pei, R., Yan, T., Zhao, Y., and Guo, X. (2005) Cytotoxicity of carbon nanomaterials: single-wall nanotube, multi-wall nanotube, and fullerene, *Environ. Sci. Technol.*, **39**, 1378–1383.

39. Flahaut, E., Durrieu, M. C., Remy-Zolghadri, M., Bareille, R., and Baquey, C. (2006) Investigation of the cytotoxicity of CCVD carbon nanotubes towards human umbilical vein endothelial cells, *Carbon*, **44**, 1093–1099.

40. Kagan, V. E., Tyurina, Y. Y., Tyurin, V. A., Konduru, N. V., Potapovich, A. I., Osipov, A. N., Kisin, E. R., Schwegler-Berry, D., Mercer, R., Castranova, V., and Shvedova, A. A. (2006) Direct and indirect effects of single walled carbon nanotubes on RAW 264.7 macrophages: role of iron, *Toxicol. Lett.*, **165**, 88–100.

41. Shvedova, A. A., Castranova, V., Kisin, E.R., Schwegler-Berry, D., Murray, A. R., Gandelsman, V. Z., Maynard, A., and Baron, P. (2003) Exposure to carbon nanotube material: assessment of nanotube cytotoxicity using human keratinocyte cells, *J. Toxicol. Environ. Health A*, **66**, 1909–1926.

42. Pensabene, V., Vittorio, O., Raffa, V., Menciassi, A., and Dario, P. (2007) Investigation of CNTs interaction with fibroblast cells, *Conf. Proc. IEEE Eng. Med. Biol. Soc.*, 6621–6624.

43. Rakhi, R. B., Sethupathi, K., Ramaprabhu, S. (2008) Synthesis and hydrogen storage properties of carbon nanotubes, *Int. J. Hydro. Energy*, **33**, 381–386.

44. Poland, C. A., Duffin, R., Kinloch, I., Maynard, A., Wallace, W. A., Seaton, A., Stone, V., Brown, S., Macnee, W., and Donaldson, K. (2008) Carbon nanotubes introduced into the abdominal cavity of mice show asbestos-like pathogenicity in a pilot study, *Nat. Nanotechnol.*, **3**, 423–428.

45. Lam, C. W., James, J. T., McCluskey, R., and Hunter, R. L. (2004) Pulmonary toxicity of single-wall carbon nanotubes in mice 7 and 90 days after intratracheal instillation, *Toxicol. Sci.*, **77**, 126–134.

46. Warheit, D. B., Laurence, B. R., Reed, K. L., Roach, D. H., Reynolds, G. A., and Webb, T. R. (2004) Comparative pulmonary toxicity assessment of single-wall carbon nanotubes in rats, *Toxicol. Sci.*, **77**, 117–125.

47. Maynard, A. D., Baron, P. A., Foley, M., Shvedova, A. A., Kisin, E. R., and Castranova, V. (2004) Exposure to carbon nanotube material: aerosol release during the handling of unrefined single-walled carbon nanotube material, *J. Toxicol. Environ. Health A*, **67**, 87–107.

48. Ryman-Rasmussen, J. P., Cesta, M. F., Brody, A. R., Shipley-Phillips, J. K., Everitt, J. I., Tewksbury, E. W., Moss, O. R., Wong, B. A., Dodd, D. E., Andersen, M. E., and Bonner, J. C. (2009) Inhaled carbon nanotubes reach the subpleural tissue in mice, *Nat. Nanotechnol.*, **4**, 747–751.

49. Byron, P. R. (1986) Some future perspectives for unit dose inhalation aerosols, *Drug. Dev. Ind. Pharm.*, **12**, 993–1015.

50. Huczko, A., Lange, H., and Calko, E. (1999) Short communication: fullerenes: experimental evidence for a null risk of skin irritation and allergy, *Fullerene Sci. Technol.*, **7**, 935–939.

51. Huczko, A., and Lange, H. (2001) Carbon nanotubes: experimental evidence for a null risk of skin irritation and allergy, *Fullerene Sci. Technol.*, **9**, 247–250.

52. Eedy, D.J. (1996) Carbon-fibre-induced airborne irritant contact dermatitis, *Cont. Dermatitis*, **35**, 362–363.

53. Kasparov, A. A., Popova, T. B., Lebedeva, N. V., Gladkova, E. V., and Gurvich, E. B. (1989) Evaluation of the carcinogenic hazard in the manufacture of graphite articles, *Vopr. Onkol.*, **35**, 445–450.

54. Liu, X., Vinson, D., Abt, D., Hurt, R. H., and Rand, D. M. (2009) Differential toxicity of carbon nanomaterials in Drosophila: larval dietary uptake is benign, but adult exposure causes locomotor impairment and mortality, *Environ. Sci. Technol.*, **43**, 6357–6363.

55. Murr, L. E., Garza, K. M., Soto, K. F., Carrasco, A., Powell, T. G., Ramirez, D. A., Guerrero, P. A., Lopez, D. A., and Venzor, J., III (2005) Cytotoxicity assessment of some carbon nanotubes and related carbon nanoparticle aggregates and the implications for anthropogenic carbon nanotube aggregates in the environment, *Int. J. Environ. Res. Public Health*, **2**, 31–42.

56. Singh, R., Pantarotto, D., Lacerda, L., Pastorin, G., Klumpp, C., Prato, M., Bianco, A., and Kostarelos, K. (2006) Tissue biodistribution and blood clearance rates of intravenously administered carbon nanotube radiotracers, *Proc. Natl. Acad. Sci. USA*, **103**, 3357–3362.

57. Kostarelos, K., Lacerda, L., Pastorin, G., Wu, W., Wieckowski, S., Luangsivilay, J., Godefroy, S., Pantarotto, D., Briand, J. P., Muller, S., Prato, M., and Bianco, A. (2007) Cellular uptake of functionalized carbon nanotubes is independent of functional group and cell type, *Nat. Nanotechnol.*, **2**, 108–113.

Chapter 9

OVERVIEW ON THE MAJOR RESEARCH ACTIVITIES ON CARBON NANOTUBES BEING DONE IN AMERICA, EUROPE AND ASIA

Cécilia Ménard-Moyon[a] and Giorgia Pastorin[b]

[a] *Institut de Biologie Moléculaire et Cellulaire, UPR 9021, CNRS, Immunologie et Chimie Thérapeutiques, 67084 Strasbourg Cedex, France*
[b] *Department of Pharmacy, Faculty of Science, National University of Singapore, Singapore*
c.menard@ibmc-cnrs.unistra.fr

9.1 INTRODUCTION

In the last few decades, several research groups around the world have focused their attention on the field of nanotechnology, dealing with advanced techniques and applied science through the use of nanomaterials. Such systems in nanoscale dimensions have been shown to revolutionise the way scientists approach conventional research and to combine diverse expertise towards exciting discoveries. Among the materials that have integrated different disciplines to the highest extent, carbon nanotubes (CNTs) represent a fascinating example. The reason at the basis of the high significance of CNTs is associated with their unique physicochemical properties, including not only high thermal and mechanical stability or nanometric dimensions (which enable the production of durable and compact systems), but also their ability to share electronic properties of both metals and semiconductors[1,2] (which can be applied in the fabrication of ultra-sensitive devices). In addition, they present a large inner volume that could be filled with several biomolecules,[3,4] while imparting chemical properties through the functionalisation of their external walls (which render CNTs water soluble and biocompatible).[5-9] As a

Carbon Nanotubes: From Bench Chemistry to Promising Biomedical Applications
Edited by Giorgia Pastorin
Copyright © 2011 Pan Stanford Publishing Pte. Ltd.
www.panstanford.com

consequence, several applications have been proposed for this nanomaterial, such as drug delivery, tissue engineering, biosensing and electronics.

In this chapter, we would like to offer an interesting overview on the major research activities on CNTs being developed in America, Europe and Asia. The purpose is to provide useful information about the most advanced discoveries concerning this novel material and to stimulate the readers' interest in pursuing their vocation on the line of excellent scientists and outstanding experts.

Data were cross-linked from SciFinder and high-impact factor journals over the period 2007–2009.

Overall, we selected nine groups or research institutes (Avouris, Hongjie Dai, Hersam, Iijima, Kataura, Prato, Smalley Institute at Rice, Strano, and Terrones) on the basis of the number of publications and their citation indices derived from journal articles with impact factors higher than 5. Where possible, the website of the groups or institutes was included, since it contains additional information on their latest research projects.

9.2 AMERICA

9.2.1 USA

> **Phaedon AVOURIS**
> @ IBM, T. J. Watson Research Center, Yorktown Heights, New York, 10598, USA
> Website: http://www.research.ibm.com/nanoscience/
> Expertise: Nanoelectronics

The expertise of this research group deals mainly with nanoelectronics where CNTs are incorporated in advanced systems for their potential use in nanoelectronic and nanomechanical devices. More precisely, the research scientists working at IBM in New York are interested in two main aspects of CNTs: (i) their manipulation into different shapes and orientations and (ii) their integration into field-effect transistors (FETs), to monitor the changes in both the nanotube phonon spectrum and its electronic resonances while passing an electrical current through it.

Avouris and collaborators are more focused on the electronic aspects associated with this material, as demonstrated by their recent publications. In their studies, they maintain the operating principles of the currently used FETs, but they replace the conducting channel with carbon nanomaterials such as one-dimensional (1D) CNTs or two-dimensional (2D) graphene layers. These materials, in fact, have demonstrated enhanced electrical properties.[10]

Moreover, owing to the excellent optical properties of semiconducting nanotubes, both electronic and optoelectronic devices from the same material seem feasible.

9.2.1.1 Electronic properties of CNTs

Although electrical properties of CNTs derive, to a large extent, from the unusual electronic structure of graphene, the additional rolling process forming the nanotube structure is further described by a pair of integers (n, m), defining the chiral vector $Ch = na_1 + ma_2$, where a_1 and a_2 are the unit vectors of the graphene honeycomb lattice. Depending on how the graphene sheet is rolled up, nanotubes show either metallic or semiconducting behaviour. The main electronic properties of CNTs can be summarised as resistance, capacitance and inductance.[11]

Electrical transport inside the CNTs is affected by scattering in correspondence with defects and by lattice vibrations that lead to a new type of quantised resistance (R_Q) related to their contacts with three-dimensional (3D) macroscopic objects, such as the metal electrodes.[12,13] When the only resistance present is the quantum resistance, transport in the CNT is ballistic, which means no carrier scattering or energy dissipation takes place in the body of the CNTs, and this is generally achieved with tubes' length ≤100 nm. Another electronic characteristic of CNTs is the capacitance that depends on how their energy states are distributed. This quantum capacitance, C_Q, is small (of the order of 10^{-16} F/μm).[14] In addition, a CNT incorporated in a structure has an electrostatic capacitance, C_G, which arises from its coupling to surrounding conductors; in a single-CNT FET, geometry leads to $C_G \approx 1/\ln(t_{ins})$, where t_{ins} is the thickness of the gate insulator. Finally, CNTs have inductance, which is a resistance to any changes in the current flowing through them.[15] The quantum inductance is usually called kinetic inductance, L_K, and it is the resistance to the change in the kinetic energy of the electrons of the CNT. It leads to electron velocities in phase with the external driving field.

9.2.1.2 CNT-FETs

Semiconducting CNTs present the interesting characteristic of an electronic transport that can be switched "on" and "off", thus paving the way for their use as novel FETs. In a typical FET, the basic features include a channel (made up of a semiconductor, e.g., silicon) connected to two electrodes, namely a source (S) and a drain (D). An insulating thin film, generally SiO_2, separates this part from a third electrode called the gate (G). By applying a voltage at this gate electrode (V_g), it is possible to modulate (switch) the conductance of the semiconducting channel (Fig. 9.1).

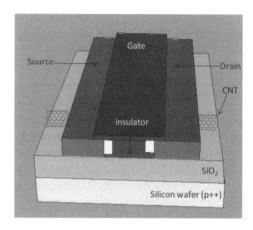

Figure 9.1 Schematic representation of a FET. Figure redrawn from Avouris *et al.*[11]

In the case of CNT-based FETs, the role of the channel is played by one or more semiconducting nanotubes. The first CNT-FETs were reported in 1998.[16,17] The key advantage is the low scattering in the CNT and the high mobility of the FET channel. In addition, as reported before, there are two contributions to the capacitance, C_G and C_Q, and in conventional devices, $C_G <$ C_Q, while in CNT-FETs $C_G \approx C_Q$.

Therefore, in a conventional FET the ability of the gate to control the potential in the channel is limited once V_g exceeds the threshold voltage, V_{th}.[18] As a consequence, increasing the gate voltage above this value does not change the bands anymore. Conversely, in a CNT-FET the gate can still retain control of the potential, thus contributing to the current.[19] In general, the main advantages of CNT-FETs over conventional systems are their much lower capacitance and their lower operating voltage. Furthermore, the small size of the CNTs allows for the fabrication of aligned arrays with a high packing density. However, careful design and engineering are needed to make sure that the system is efficient.

In an electronic device, energy can be released as heat (phonons), and a small fraction may involve the emission of a photon. This light emission process is called electroluminescence (EL) and is extensively used to produce solid-state light sources such as light emitting diodes (LEDs). Although the mechanism of this emission is similar to that of a LED, in CNTs the light does not originate from a fixed point along the tube, but its origin can be translated by simply changing V_g. As a consequence, more intense and brighter light sources can be produced with CNTs. Some properties of single-walled CNT (SWCNT) films, including their conductivity, their optical transparency and their flexibility, render them ideal anodes for organic LEDs (OLEDs).[20,21]

The advantage over the traditionally used indium thin oxide films is their cheaper price, flexibility and resilience to corrosion. For example, a CNT-FET has been modified as shown in Fig. 9.2 to induce a sharp discontinuity in the potential; that is, to produce a drop of energy along the CNT channel. Electrons that are accelerated at the discontinuity can acquire enough energy to generate a bright light emitter with a yield that is about 1,000 times higher than that produced by recombination in a conventional device (Fig. 9.2).[22] The excitation of CNTs can be used to create an efficient EL source.[22-25] At the same time, monitoring localised EL provides a tool for detecting defects in CNT devices. In future, additional integration to include optics could lead to a unified electronic-optoelectronic technology.

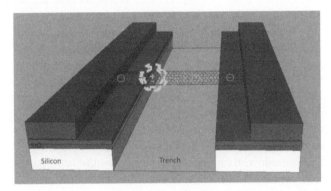

Figure 9.2 Schematic representation of a CNT as LED. Figure redrawn from Avouris *et al.*[11]

Unlike many other materials, CNTs have demonstrated very weak electron–acoustic–phonon coupling and a very large optical phonon energy of about 200 meV. Excitation of CNTs can be achieved through energetic ("hot") carriers flowing through the CNT. The carriers are accelerated by the applied electric field, and they gain energy and then lose some of it to phonons, primarily optical phonons. The energy of the optical phonons is usually efficiently dissipated into the heat bath provided by the substrate of the device. However, in the case of suspended nanotubes, evidence for the existence of a non-equilibrium optical phonon excitation has been reported, and it seems to degrade the carrier mobility in CNTs.[26,27]

Therefore, in a recent article,[28] the authors have investigated the whole electrical power dissipation in a CNT-based FET. The results from Raman analysis regarding G band showed both G^+ and G^- transitions clearly resolved. As the electrical power increases, both G^+ and G^- bands shifted and broadened, with a trend corresponding to the electric power. Similarly, the corresponding slopes matched with thermal heating results, although the widths of the bands showed activated behaviour, as expected for a decay into optical phonons.[29] On the whole, the reported data accounted for an anharmonic phonon decay, i.e., a non-equilibrium phonon distribution, in order to justify the observed electronic transport behaviour of CNTs.

Once exposed to photo-excitation, CNTs reach high-energy E_{22} state; subsequently they relax (decay) by means of phonons (fluorescence) to lower-energy E_{11} states, with a radiative lifetime at room temperature of about 1–10 ns.[30] Recent work has suggested that only about 10% of the E_{22} excitons decay frees electrons.[31] It must therefore be concluded that a non-radiative-decay process controls the E_{11} exciton lifetime. In other words, another intrinsic energy dissipation channel must influence the relaxation of the lowest exciton state. To explain this phenomenon, the authors have reported a theoretical study on the efficiency of decay pathways of nanotubes involving purely multiphonon decay (MPD) as well as other electronic decay mechanisms.[32] The calculations indicated that the MPD of free excitons is too slow to be responsible for the experimentally observed lifetimes; conversely, the combination of localised exciton MPD and the phonon-assisted indirect exciton ionization (PAIEI) allow explaining the range of available experimental data. The results suggested a mechanism in which PAIEI involved exciton decay into both an optical phonon and an *intraband* electron–hole pair. In fact, in the presence of the free carriers in CNTs, an exciton could decay fast by creating a phonon and an intraband electron–hole (*e-h*) pair, as shown in Fig. 9.3. The resulting lifetime was in the range of 5 to 100 ps.

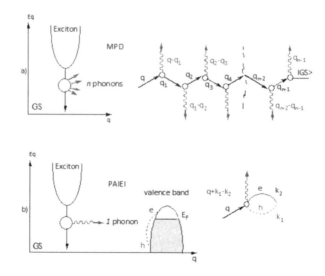

Figure 9.3 Schematic representation of the exciton decay mechanisms: (a) multiphonon decay (MPD) and (b) phonon-assisted indirect exciton ionisation (PAIEI) in CNTs. Figure redrawn from Perebeinos and Avouris.[32] See also Colour Insert.

On the other hand, the group has also considered the intersubband decay of E_{22} excitons in semiconducting CNTs, through both theoretical and experimental evaluations.[31,33] Photoluminescence spectra showed that the global behaviour of E_{22} line widths is strongly associated with E_{11} and E_{12} exciton bands, while it is independent of their chiral indices (n, m) normally used to describe the nanotube structure. Conversely, the tubes' chirality affected the optical properties of CNTs exposed to an external static electric field applied along the tubes' axis.[34] Theoretical calculations predicted different effects, including characteristic Franz–Keldysh oscillations, quadratic Stark effects and field dependence of the bound exciton ionisation rate.

In a recent investigation,[35] the group showed that the photoluminescence of a partially suspended, semiconducting CNT-based FET was quenched and red-shifted upon application of a longitudinal electric field. The authors explained the quenching by a loss of oscillator strength and by an increase in the non-radiative decay of the E_{11} exciton; the shifts towards lower frequencies were instead attributable to field-induced doping (i.e., change of electronic properties), most probably associated with the screening and heating of CNTs.

9.2.1.3 CNT nanophotonics

In addition to electronics, semiconducting SWCNTs have shown a promising future in nanophotonics. To that purpose, Avouris and collaborators have integrated a single, electrically excited, semiconducting nanotube transistor with a planar $\lambda/2$ microcavity towards nanotube-based nanophotonic devices.[36]

The device was based on a single semiconducting single-walled carbon nanotube-based FET combined with a planar optical $\lambda/2$ microcavity.[37-40] A three-dimensional schematic representation of the device is depicted in Fig. 9.4. The photonic microcavity consisted of three dielectric layers: 250 nm of polymethyl methacrylate (PMMA), 22 nm of aluminium oxide (Al_2O_3) and 250 nm of silicon oxide (SiO_2) sandwiched between a top gold mirror (20 nm) and a parallel bottom silver mirror (100 nm). Moreover, an individual nanotube was placed above the Al_2O_3 layer, parallel to the silver and the gold mirrors, to provide optimum coupling conditions. The applied current typically ranged between 5 and 10 µA, and electrons represented the main carrier injected. Subsequently, the radiative decay of the produced excitons determined light emission. Interestingly, the electroluminescence spectra obtained from this nanotube-based system involved the same E_{11} excitonic state as observed in the photoluminescence of nanotubes in suspension. In other words, little difference in the effective dielectric constants for a nanotube on SiO_2 and embedded in PMMA was observed in comparison with tubes suspended in solution. Therefore, it was suggested that similar devices might find applications in integrated nanophotonic circuits, quantum optics and on-chip optical interconnects.

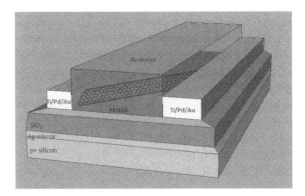

Figure 9.4 3D schematic representation of the device comprising a CNT-based FET integrated with a planar $\lambda/2$ microcavity. Figure redrawn from Xia *et al.*[36]

One main hurdle in the widespread application of CNTs in electronics is still represented by the current inability to produce large amounts of identical nanostructures. A solution to this problem has been proposed, which suggests the use of uniform CNTs as key components in thin-film transistors (TFTs). In this context, CNTs were self-assembled into micrometre-wide strips that formed regular arrays of dense and highly aligned CNTs covering the entire chip.[41] The initial limitation, due to the presence of heterogeneous mixtures of metallic tubes with concomitant low on/off currents, was overcome by pre-selection of the tubes, thus obtaining purely semiconducting CNT-based TFTs which, as a result, exhibited high current flows. This work was realised between two main groups of research, namely, Avouris's group and Hersam's (reported in another session), thus merging excellent skills towards common ambitious goals.

Another successful array has been reported through the assembly of aligned SWCNTs via lithography techniques.[42] More precisely, CNTs were initially derivatised with hydroxamic acid groups, which presented the double advantage of enhancing SWCNTs' water dispersibility while increasing their interaction with metal oxide surfaces (i.e., hafnium oxide [HfO_2]) (Fig. 9.5). Subsequently, the tubes were arranged and aligned into trenches of SiO_2 films with HfO_2 at the bottom. Thermal treatment at 600°C reconstituted pristine, pure SWCNTs, to which palladium source–drain electrodes were connected perpendicularly to the trench length. The array exhibited electrical properties comparable to those reported for suspended CNT-based devices, without deterioration following the alignment of the tubes, thus representing valuable starting points for large-scale arrays.

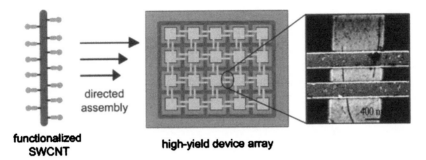

functionalized
SWCNT

high-yield device array

Figure 9.5 Assembly of functionalised SWCNTs through lithography techniques. Reproduced from *Tulevski et al.*[42] with permission.

9.2.2 USA

> **Hongjie DAI**
> @ Stanford University-California
> Website (research activities): http://www.stanford.edu/dept/chemistry/faculty/dai/group/research.htm
> Expertise: Nanoscale transistors and biomedical applications of CNTs

This group includes about 20 researchers working in the areas of material chemistry, inorganic synthesis, solid state physics, electron transport, scanning probe microscopy and bio-nanotechnology. The ongoing projects can be subdivided into three main areas: the first one includes bio-nanotechnology and nanomedicine, through the application of nanomaterials for drug/gene delivery and targeting. Both *in vitro* and *in vivo* experiments are performed to elucidate the mechanism of cellular uptake, as well as tissue accumulation and toxicity after *in vivo* administration. The group has already obtained promising results by improving cancer treatment efficacy through functionalised SWCNTs. Moreover, because of their unique optical properties, including bright Raman scattering and near-infrared (NIR) photoluminescence, SWCNTs can also be used for biological imaging.

Another area involves surface functionalisation, mainly focused on the incorporation of several molecules onto functionalised or pristine CNTs, thus originating attractive supramolecular complexes. In particular, the researchers have obtained a reversible hydrogenation of SWCNTs by creating C–H bonds over 65% of the tubes, with an overall 5% of hydrogen storage.

The functionalisation is also applied to semiconducting nanowires, essentially with long alkanethiol chains ($\geq$12 C), which impart high resistance against oxidation and hence a better durability of electronic devices.

The last research focus deals with the fabrication and characterisation of ordered nanomaterial architectures, a specific research program involving the development of new synthetic procedures to obtain well-defined CNT or graphene architectures on surfaces for future device applications. To that purpose, the group combines inorganic synthesis of mesoporous catalytic materials and chemical vapour deposition with microfabrication techniques, followed by electrical and electromechanical measurements of individual structures, in order to understand the properties of quasi-1D solids and explore their applications in future miniaturised devices. Therefore, the main objective is to realise nanoscale transistors through the fabrication of ordered substrates and semiconducting nanowires.

9.2.2.1 Functionalisation of CNTs for biomedical applications

An important aspect concerning CNTs is that the raw material needs to undergo an extensive surface functionalisation in order to become suitable tool for biomedical applications. Functionalisation of the tubes not only enhances water solubility and CNTs' biocompatibility, but it has also demonstrated to influence biological systems both *in vitro* (e.g., in the mechanism of cellular uptake) and *in vivo* (in terms of binding with blood proteins, tissue accumulation and CNT elimination).[43-45]

There are two major types of functionalisation protocols for CNTs, namely covalent reactions and non-covalent coating by amphiphilic molecules onto nanotubes. It is important to note that, although both are extremely useful, they are not interchangeable: covalent chemical reactions have been shown to be useful in drug and gene delivery,[46,47] where solubility in aqueous media and stability of the complexes between CNTs and bioactive molecules represent crucial aspects. However, through this procedure, the properties of CNTs are modified when the nanotube sidewall is functionalised, dramatically decreasing the Raman scattering and NIR fluorescence signals of SWCNTs, thus rendering this functionalisation not ideal for sensing and imaging applications.[48] Conversely, the structure and optical properties of the tubes are preserved when they are non-covalently functionalised, although the stability can be hampered by such weaker interactions. The research group has therefore developed a few protocols (summarised in a recent review)[49] that help maintain both chemical stability and bio-conjugation with drugs and molecules: an important example is represented by the incorporation of poly(ethylene glycol) (PEG) on the nanotube sidewalls, which has led to a remarkable follow-up on "supramolecular chemistry" with nanotubes.[50] This strategy has shown extremely high degrees of π-stacking of aromatic molecules, including some fluorescent dyes and even drug molecules (e.g., doxorubicin), which are commonly associated with efficacy but also unwanted side effects.[51,52] As regards the therapeutic efficacy of CNTs conjugated to doxorubicin (SWCNT–DOX), some *in vivo* experiments were performed in SCID mice bearing Raji lymphoma xenograft tumours;[53] these animals were treated weekly with SWCNT–DOX, free DOX or DOXIL (formulation of DOX using liposomes). Results showed that the size of the tumours in untreated controls rapidly increased by 7.5 fold. The SWCNT–DOX (5 mg/kg) treated group showed a greater inhibition of tumour growth than free DOX at the equivalent dose, while tumours in mice treated with plain SWCNTs increased by 6.4 fold, thus indicating that SWCNTs did not affect tumour growth. Finally, mice treated with DOXIL (5 mg/kg) developed severe toxicity and died. Therefore, loading of DOX onto SWCNTs increased its therapeutic efficacy compared with free DOX and decreased side effects of DOXIL. It was also noticed that such π-stacking interaction was dependent on the nanotube's diameter, thus suggesting a controlled release rate of molecules from CNTs on the basis of their diameter. In addition, the concept of "functionalisation partitioning" of CNTs was anticipated, meaning that it is possible to incorporate several molecules (e.g., drugs, fluorescent dyes and targeting agents) on the same tubes. As regards this last concept, non-covalent functionalisation was exploited to tether fluorescein and polyethylene glycol moieties onto CNTs and thus forming a supramolecular complex.[54] Fluorescein was adsorbed onto the tubes through π-stacking interaction, while PEG chains guaranteed water solubility.

Interestingly, through optical absorbance and fluorescence measurements, it was observed that the interaction between fluorescein and SWCNTs was pH dependent; in fact it diminished as the pH increased, with concomitant lower stability at high pHs. It is important to note that fluorescence emission from fluorescein adsorbed onto the tubes was quenched to a considerable extent (~67%), but it was still sufficient to act as both fluorescent label and intracellular transporter, as demonstrated by enhanced uptake of SWCNT-PEG-fluorescein by mammalian cells. In this case, Raman analysis was used to confirm the co-localisation of fluorescein and SWCNTs and disprove its detachment from the tubes.

A further development of this work is provided by the non-covalent functionalisation of raw, hydrophobic nanotubes with amphiphilic polymers (e.g., phospholipid-poly(ethylene glycol) [PL-PEG]). The hydrophobic lipid chains of PL tend to anchor onto the nanotube surface strongly, while the hydrophilic PEG chains afford CNT water solubility and biocompatibility. Recently, the research group exploited this functionalisation through the incorporation of SWCNTs with branched polymers based on poly(γ-glutamic acid) (γPGA) and poly(maleic anhydride-*alt*-1-octadecene) (PMHC18) (Fig. 9.6). More precisely, they took advantage of the carboxylate function of γPGA to incorporate lipophilic moieties (e.g., PL or pyrene [Py]) that can easily physisorb not only onto CNTs but also on the surface of nanoparticles (NPs) and nanorods (NRs) (Fig. 9.7).[55] On the other hand, they coupled the remaining COOH groups of γPGA and the anhydrides of PMHC18 with hydrophilic PEG chains, thus forming stable amphiphilic complexes. Moreover, because of the presence of PEG, a prolonged blood circulation of more than 20 hours after intravenous (i.v., through tail vein) injection in mice was observed in comparison with 5.4 hours for previous samples (Fig. 9.8).[56] As a consequence, there was also a delayed clearance of the nanomaterials by the reticuloendothelial system (RES) of the animals, thus improving the imaging quality. However, it remains to be considered whether a prolonged circulation of such complexes is associated with a higher toxicity for the treated mice.

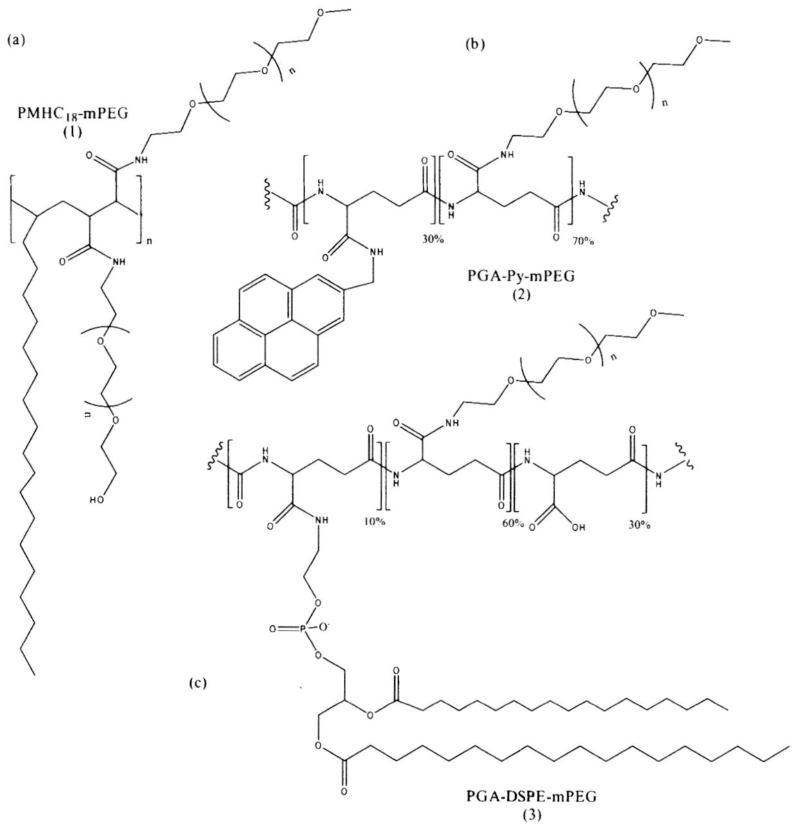

Figure 9.6 (a) PMHC18-mPEG (**1**): poly(maleic anhydride-*alt*-1-octadecene)-poly(ethylene glycol) methyl ethers. (b) γPGA-Py-mPEG (**2**): poly-(γ-glutamic acid)-pyrene(30%)-poly(ethylene glycol) methyl ethers (70%). (c) γPGA-DSPE-mPEG (**3**): poly-(γ-glutamic acid)-phospholipid 1,2-distearoyl*sn*-glycero-3-phosphoethanolamine (10%)-poly(ethylene glycol) methyl ethers (60%).

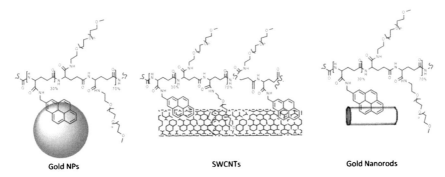

Gold NPs   SWCNTs   Gold Nanorods

Figure 9.7 Schematic representation of non-covalent functionalisations onto gold nanoparticles (NPs), single-walled carbon nanotubes (SWCNTs) and gold nanorods (NRs). Figure modified from Prencipe *et al.*[55] with permission. See also Colour Insert.

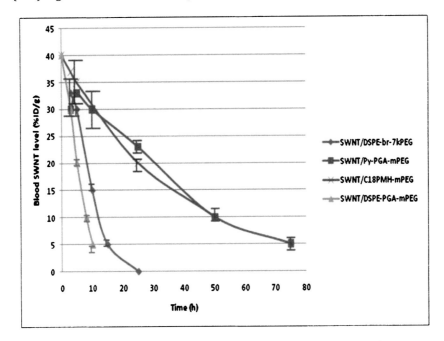

Figure 9.8 Blood circulation data of SWCNTs with different functionalisations in BALB/c mice. Blood circulation curves of **1-3** coated nanotubes compared with DSPE-branched-7kPEG coated nanotubes. The latter one was previously reported by the same research group. Graph modified from Prencipe *et al.*[55] with permission. See also Colour Insert.

The enhanced quality of CNTs wrapped with PL-PEG complexes has been confirmed through an investigation in the NIR region.[57] SWCNTs have the intrinsic ability to display NIR photoluminescence, but first they need to be dispersed in an aqueous medium and become biocompatible. To that purpose, the authors found better results with tubes previously sonicated in sodium cholate, which was subsequently replaced by the polymeric PL-PEG complex; although this step was not necessary for the functionalisation, it proved particularly useful at providing less damage to the nanotubes, a better quantum yield and sharp E_{11} absorption peaks (1,000–1,300 nm) in the UV-vis-NIR spectra in comparison with SWCNTs directly suspended in PL-PEG. In fact, despite the fact that the exchange SWCNTs were injected at a dose (17 mg/L) 15 times lower, they still showed a stronger *in vivo* NIR photoluminescence signal (red) than the direct-SWCNT-treated mice. This fact, in combination with the good tissue penetration and the ultra-low background autofluorescence in the emission range, envisages the use of biocompatible nanotube fluorophores as promising biological imaging agents.

Another non-covalent functionalisation consists of the recently published conjugation of dextran and a phospholipid 1,2-distearoyl-*sn*-glycero-3-phosphoethanolamine (dextran-DSPE), which was subsequently adsorbed onto nanomaterials (SWCNTs, NPs and NRs).[58] Dextran is a naturally occurring polymer made up of $\alpha(1\rightarrow6)$-linked glucose monomers with 5% branching at the 4 position. The hyperbranched structure and packed interior result in a globular shape, thus rendering it similar to dendrimers and allowing a coating more resistant to protein adhesion than a linear polymer. Moreover, its polyhydroxylated structure makes dextran a very hydrophilic polymer, while maintaining several hydroxyl groups, which may be further functionalised with PEG units to originate hyperbranched polymeric structures with longer circulation *in vivo*. The authors proposed a phospholipid-dextran conjugate, in which the DSPE was bound at a single point to the reducing end of the dextran (Fig. 9.9). This was possible through the anomeric end of dextran, which differs in reactivity from the hydroxyl groups on the rest of the polymer.[59] The complex was subsequently adsorbed onto nanomaterials, showing improved photophysical properties in comparison with commonly used PEGylated structures.

The authors also demonstrated that such conjugation maintained the key requirements necessary for an optimal coating in view of biomedical applications, such as (i) a stable coating even under high dilutions, as would be expected *in vivo*; (ii) avoidance of non-specific binding of proteins that might determine low blood circulation time; (iii) present functional groups for a selective targeting; (iv) absence of aggregation of nanomaterials under various conditions and preservation of the intrinsic physical properties of the material under investigation.

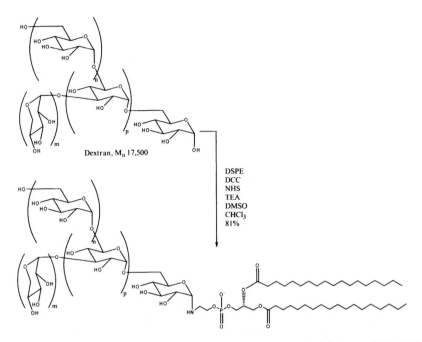

Figure 9.9. Synthesis of dextran-DSPE at the anomeric end of dextran. DCC: *N,N'*-Dicyclohexylcarbodiimide. NHS: *N*-Hydroxysuccinimide. TEA: Triethylamine.

A different strategy has been adopted for the delivery of paclitaxel (PTX) towards xenograft tumours in *in vivo* experiments:[60] SWCNTs, previously sonicated in the presence of PL-PEG, were subsequently conjugated with succinic anhydride-modified PTX, which formed cleavable ester bonds able to subsequently react with free H_2N-PEG groups via amide bonds. Results showed that delivery of PTX from SWCNTs afforded an improved treatment efficacy over the clinical drug formulation Taxol, as evidenced by its ability of slowing down tumour growth at a low PTX dose (5 mg/kg). The main reason for such a higher tumour suppression of SWCNT-PTX conjugates in comparison with Taxol alone or PEG-PTX is the up to 10-fold higher tumour uptake of PTX afforded by SWCNT carriers (Fig. 9.10). In fact, PEG-PTX remains a relatively small molecule that tends to be rapidly excreted via the renal route, thus providing short blood circulation as well as uptake and treatment efficacy similar to free Taxol. It has been already reported that the maximum tolerable dose of Taxol for BALB/c mice is between 20 and 50 mg/kg.[61-63] In this study, the achieved tumour growth suppression by SWCNT-PTX at 5 mg/kg dose once every 6 days suggests the promising application of SWCNT drug delivery for effective cancer treatment, also because SWCNTs have been shown to be safe at least in mouse models.

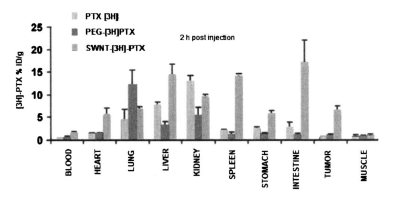

Figure 9.10 ³H-PTX biodistribution in 4T1 tumour-bearing mice injected with ³H-labelled Taxol, PEG-PTX and SWCNT-PTX at 2 hours after injection. Reproduced from Liu *et al.*[60] with permission. See also Colour Insert.

As a proof of that, a pilot toxicological study of CNTs was done on a small number of mice:[64] functionalised SWCNTs were injected into the bloodstream of mice, and no evidence of acute or chronic toxicity was observed over a period of 4 months. For the same period, functionalised SWCNTs apparently accumulated in liver and spleen macrophages, without affecting tissue composition, as demonstrated by necropsy and histological evaluations. These encouraging results were confirmed by biodistribution studies on intravenously injected SWCNT-PEG complexes in various organs and tissues of mice *ex vivo* over a period of 3 months.[65]

Functionalisation of SWCNTs by branched PEG chains enabled thus far the longest SWCNT blood circulation up to 1 day, with low uptake in the RES, and near-complete clearance from the main organs in ~2 months. Raman spectroscopy detected SWCNTs in the intestine, feces, kidney and bladder of mice, suggesting excretion via the biliary and renal pathways. No toxic side effect of SWCNTs to mice was observed, thus paving the way for future biomedical applications of CNTs.

In another study on biodistribution, the authors investigated the effects of radio-labelled SWCNTs in mice by *in vivo* positron emission tomography (PET), *ex vivo* biodistribution and Raman spectroscopy [66]. In particular SWCNTs, functionalised with phospholipids (PL) bearing PL-PEG chains were used, determining a long blood circulation and a selective targeting towards integrin-positive tumours, providing that SWCNTs were doped with PEG chains linked to a cyclic arginine–glycine–aspartic acid (RGD) peptide in the form of a multivalent system. The same complex, made up of SWCNTs combined with RGD peptide, has shown the incredible property of acting as a photoacoustic contrast agent, thus offering 8 times higher signal than non-targeted CNTs and a better spatial resolution in comparison with conventional optical techniques.[67]

The successful application of functionalised CNTs, for the delivery of otherwise toxic or insoluble drugs, has encouraged the exploitation of additional procedures. Among them, it is worth mentioning the use of SWCNTs as carrier of Pt-based pro-drugs:[68] in this study, a Pt(IV) complex, containing folic acid (FA) as targeting agent at an axial position, was attached to the free amino groups of the SWCNT-PL-PEG adduct (Fig. 9.11). As regards the mechanism involved, the SWCNTs were shown to have been taken into cancer cells by endocytosis, while the entrapment of the SWCNTs within the endosomes was confirmed by fluorescence microscopy of SWCNTs.[69] Once selectively inside cancer cells, which tend to overexpress the target folic receptor (FR), the reducing environment converted the pro-drug into Pt(II) derivative, which subsequently translocated into the nucleus and interacted with the DNA. These results confirmed the ability of CNT-Pt conjugates to trigger specific destruction of cancer cells.

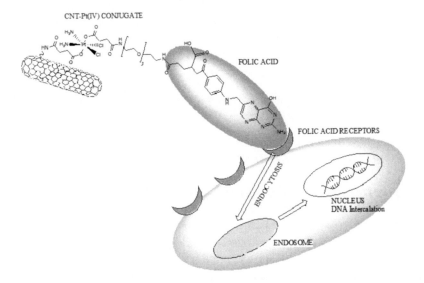

Figure 9.11 Cancer cell uptake of SWCNT-Pt(IV)-FA complexes and subsequent reduction and release of Pt(II) derivative into the nuclear DNA. Figure modified from Dhar *et al.*[68] with permission. See also Colour Insert.

Apart from drug delivery, the research group has also been involved in gene transfection of small interfering RNA (siRNA).[70] It has been already reported that the delivery of siRNA to human T cells to silence the expression of the human immunodeficiency virus (HIV) specific cell surface receptors

CD4 and/or coreceptors CXCR4/CCR5 can block HIV entry and thus reduce infection. To that purpose, SWCNTs were first reacted with a bifunctional linker, sulphosuccinimidyl 6-(3'-[2-pyridyldithio]-propionamido) hexanoate (sulpho-LC-SPDP) and then conjugated with thiolated siRNA$_{anti\text{-}CXCR4}$ through a cleavable disulphide bond (Fig. 9.12). Once transported into cells via endocytosis, siRNA was released from SWCNTs by sulphide cleavage and then bound to CXCR4 mRNA to induce gene silencing.

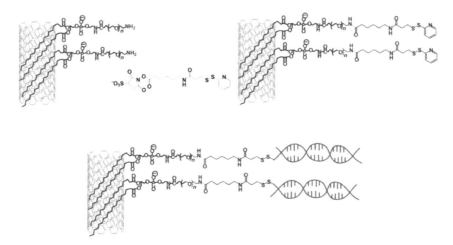

Figure 9.12 A scheme showing siRNA conjugation to SWCNTs through a disulphide bond. PL-PEG2000-amine-functionalised SWCNTs are activated by the sulpho-LC-SPDP bifunctional linker. The pyridyl disulphide group can then be coupled to thiolated siRNA to create a disulphide linkage through a thiol exchange reaction. Reproduced from Liu *et al.*[70] with permission. See also Colour Insert.

9.2.2.2 CNTs for bioimaging and biosensing

The same research group has also developed an interesting microarray for sensitive protein detection, based on functionalised, macromolecular SWCNTs as multicolour Raman labels.[71] SWCNTs have intrinsic electronic and spectroscopic properties that render them ideal for surface-enhanced Raman scattering (SERC). The main advantage of this technique is that SWCNTs provide multiple-colour SWCNT Raman labels, while requiring only a single excitation source. Moreover, the SWCNT Raman excitation and scattering photons are in the NIR region, which is the most transparent optical window for biological species both *in vitro* and *in vivo*. More precisely, SWCNTs were functionalised with PL-PEG chains, to which a secondary antibody, goat anti-mouse immunoglobulin G (GaM-IgG), was conjugated (Fig. 9.13). Proteins, such as polyclonal mouse IgG or human serum albumin (HSA), were detected by Raman scattering upon the binding of GaM-IgG-conjugated SWCNTs (GaM-IgG–SWCNTs). Surprisingly, the detection limit of the SWCNT Raman assay was three orders of magnitude superior than standard fluorescence assays. Moreover, the authors suggested that, although this work focused on antibody–antigen interactions, it could be applied for probing protein–protein interactions and nucleic acid hybridisation as well. That is why in a further experiment,[72] the researchers labelled cancer cells bearing specific receptors with three differently "coloured" SWCNTs conjugated with various targeting ligands; they included Herceptin (anti-Her2), Erbitux (anti-Her1) and RGD peptide. This allowed multicolour Raman imaging of cells in a multiplexed manner (Fig. 9.14).

Alternatively, semiconducting SWCNTs were proposed as NIR fluorescent tags for selective probing of cell surface receptors and cell imaging; this was demonstrated by the conjugation of SWCNT-PEG with antibodies such as Rituxan, which was able to selectively recognise CD20 receptors present on the cell surface of B cells, with minimal non-specific binding.[73]

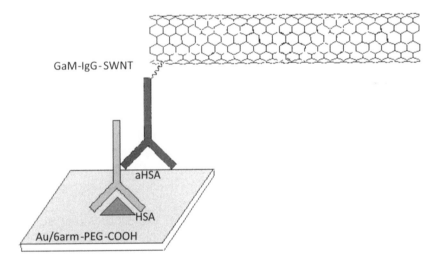

Figure 9.13 Sandwich assay scheme. Immobilised proteins in a surface spot were used to capture an analyte (antibody) from a serum sample. Detection of the analyte by Raman scattering measurement was carried out after incubation of SWCNTs conjugated to goat anti-mouse antibody (GaM-IgG–SWCNTs), specific to the captured analyte. Figure redrawn from Chen *et al.*[71] See also Colour Insert.

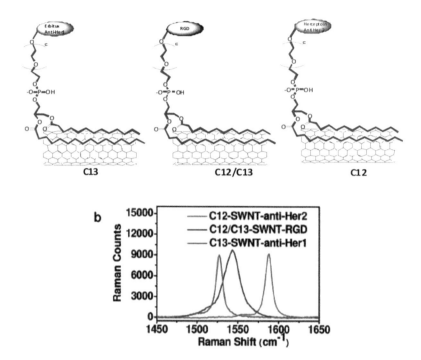

Figure 9.14 SWCNTs with different Raman colours. (a) Schematic of SWCNTs conjugated with different targeting ligands. (b) Solution-phase Raman spectra of the three SWCNT conjugates under 785 nm laser excitation. Figure partially modified from Liu *et al.*[72] with permission. See also Colour Insert.

The Raman technique was also used to evaluate tumour targeting and localisation of SWCNTs in living mice, after intravenous injection.[74] Since Raman imaging was able to detect increased accumulation of targeting SWCNT–RGD in tumour as opposed to plain SWCNTs (with remarkable statistical significance, $p < 0.05$), these results encourage the development of a new preclinical tool based on the Raman imager.

Another recent development in the field of diagnostics is represented by the label-free biosensor obtained from SWCNTs coated with a specific rheumatoid arthritis (RA)-peptide and deposited on a quartz crystal microbalance (QCM) sensing crystal.[75] Results deriving from the detection of RA antibodies in serum by QCM sensing indicated a higher sensitivity (in the femtomolar range) of the nanotube-based sensor, in comparison with both the native peptide and other conventional approaches such as ELISA and microarray; moreover, the CNT-based device demonstrated a better identification of eventual false negatives (i.e., patients not having the disease), thus suggesting a disease-specific detection with higher accuracy.

Besides the above-mentioned techniques, photoacoustic imaging of living subjects offers a very good spatial resolution and allows deeper tissues to be imaged compared with most optical imaging techniques.[76] To that purpose, many contrast agents for photoacoustic imaging have been explored,[77] but most were not able to target a diseased site in living subjects. Conversely, Dai's group demonstrated that SWCNTs conjugated with the RGD peptide (Fig. 9.15) can be effectively used as a contrast agent after intravenous administration to mice bearing tumors.[67] Results from photacoustic imaging were confirmed by two-dimensional Raman images of the excised tumours; on the whole, they showed eight times greater photoacoustic signal in the tumour than mice injected with plain nanotubes. It was observed a temporary photoacoustic signal for plain single-walled carbon nanotubes, attributable to circulating nanotubes that were eventually cleared from the bloodstream. Differently, SWCNT–RGD, bound to the tumour vasculature, originated a consistent photoacoustic signal from the tumour, thus providing sensitive and non-invasive cancer imaging and envisaging future monitoring of nanotherapeutics in living subjects.

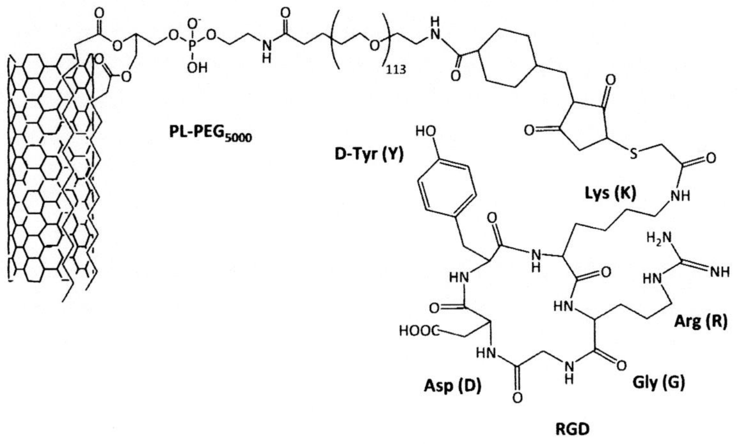

Figure 9.15 Structure of SWCNT-RGD. Figure redrawn from De La Zerda *et al.*[67] See also Colour Insert.

9.2.2.3 Electronics and optical properties of CNTs

A considerable part of the research conducted in Dai's group is devoted to the development of CNT-FETs. One limiting aspect associated with this topic is the difficulty to isolate single-chirality components from a mixture of tubes.

Yet, the researchers have reported a method to isolate, with 99% of purity, SWCNTs with specific chirality (10, 5).[78] Such result was achieved through ion exchange chromatography (IEC), by conjugating HiPco tubes of 1.03 nm with the single-stranded DNA (TTTA)$_3$T sequence, which can recognise only (10, 5) tubes. This also represents the first time that SWCNT-FETs with single-chirality SWCNTs have been achieved.

With the aim of obtaining SWCNTs for potentially high performance electronics, the authors have carried out separation of DNA-functionalised HiPco SWCNTs by (i) length, using size-exclusion chromatography (SEC), and (ii) diameter, metallicity and chirality, utilising IEC.[79] For SEC, longer nanotubes were eluted earlier than shorter nanotubes, while IEC method was effective in separating semiconducting SWCNTs by chirality and diameter. SEC and IEC were essentially independent of each other for length and metallicity separations, respectively. Photoluminescence excitation/emission (PLE) spectroscopy was used for the first time to characterise the (n, m) chiralities of SWCNTs following the combination of SEC/IEC separations. The subsequent fabrication of FETs with these separated tubes displayed high I_{on}/I_{off} ratios, owing to semiconducting SWCNTs enriched in only a few (n, m) chiralities, thus confirming successful separation and paving the way for high-performance nanoelectronics.

Besides showing interesting properties for the fabrication of advanced FETs, SWCNTs possess the intrinsic characteristic of emitting light, by assuming a wide number of electronic states (known as van Hove singularities) within narrow ranges of energy. In particular, in semiconducting tubes interband transitions and excitons could be observed by photo- and electroluminescence, while they are not detected in case of metallic SWCNTs due to their non-radiative relaxation. However, it is also possible to measure the emission of light from a suspended quasi-metallic SWCNT (QM-SWCNT) under low bias voltages, owing to Joule heating.[80] As a consequence of this phenomenon, strong peaks could be identified in the visible and IR regions, corresponding to interband transitions. This is a result of thermal light emission in a 1D system, which could not be observed in case of large bundles of SWCNTs or multi-walled nanotubes.

Additionally, it was possible to convert metallic tubes into semiconducting SWCNTs by co-doping the tubes with boron and nitrogen (BCN-SWCNTs).[81] The obtained FETs showed pure semiconducting behaviour, indicating a great advance in comparison with the merely 70% achieved by undoped SWCNTs.

Normally, raw CNTs are characterised by a huge aspect ratio, due to their narrow diameter (of a few nanometres) in comparison with their length (typically 1–5 μm). In a new investigation,[82] the researchers have instead considered ultra-short SWCNTs about 7.5 nm in length, obtained with relatively low density gradient ultracentrifugation (DGU) technique and long centrifugation times. UV-vis-NIR absorption and photoluminescence (PL) peaks of these quasi zero-dimensional SWCNTs showed a blue shift up to ~30 meV compared with long nanotubes, owing to quantum confinement effects along the length of ultra-short SWCNTs. On the whole, these carbon "capsules" corresponded to SWCNT quantum dots.

9.2.3 USA

Mark HERSAM
@ Northwestern University-Illinois
Website: http://www.hersam-group.northwestern.edu/index.html
Expertise: CNT "sorting" for electronics and single-molecule manipulation

The Hersam Research Group at Northwestern University joins materials science and engineering, with the purpose of applying organic molecules to inorganic substrates for a better functionality of the resulting hybrid system. This unique combination is enabled by a successful integration of advanced techniques and sophisticated instruments ranging from ultra-high-vacuum (UHV) scanning tunneling microscopy (STM) to atomic force microscopy (AFM). Research projects involve mainly fundamental studies (e.g., single-molecule manipulation with UHV STM) and applied technology (e.g., optimisation of CNT materials for electronic and optical devices). On the whole, this highly interdisciplinary research group has been promoting worldwide impact in the fields of information technology, bionanotechnology and alternative energy.

9.2.3.1 CNT sorting

In order to achieve optimal performance in all conceivable applications, SWCNTs need to be monodisperse in their diameter, electronic behaviour and chiral handedness. This is because SWCNTs with extremely similar (but not identical) diameters can show diverse chiral vectors and thus different electronic characteristics. On the other hand, although sharing the same chiral vectors, CNTs might differ for chiral handedness, which means that they will interact with circularly polarised light differently. Hence, there is the need for an effective SWCNTs sorting strategy, which should incorporate a few crucial aspects: (i) scalability (to respond to an anticipated increase in demand), (ii) compatibility of dimensions as raw material, (iii) durability, (iv) reproducibility, (v) affordability.[83]

These goals can be achieved, provided SWCNTs undergo strategic functionalisations, and they are processed through several techniques, including ultracentrifugation, electrophoretic separation and/or chromatographic methods. In particular, covalent sidewall chemistries have been shown to afford the best selectivity: for example, the functionalisation with *p*-hydroxybenzene diazonium salt induces a negative charge on metallic SWCNTs through deprotonation in alkaline solutions, thus enabling tube separation by electrophoresis.[84] Alternatively, diazonium compounds with a long alkyl tail render the metallic SWCNTs selectively soluble in tetrahydrofuran (THF).[85] Otherwise, dichlorocarbene converts metallic tubes into semiconducting SWCNTs, while nitronium ions selectively attack small-diameter metallic SWCNTs, reducing them to amorphous carbon.[86] In addition, separation of semiconducting from metallic SWCNTs has been achieved by cycloaddition of azomethine ylides.[87] However, since covalent functionalisation affects nanotube integrity remarkably, non-covalent strategies (e.g., using surfactants,[88] wrapped polymers,[89] solvents[90] and porphyrins[91]) have been pursued, though the vast majority of work on selective non-covalent functionalisation has been focused on relatively small-diameter SWCNTs (diameter < 1.2 nm).

Among the techniques of CNT separation, conventional electrophoresis sorts SWCNTs in response to a direct current (DC) electric field. The tubes with the smallest molecular weight travel most quickly, thus leading to sorting on the basis of SWCNTs' length and diameter. Moreover, capillary electrophoresis can also provide separation of individual SWCNTs from bundles.[92] It has been demonstrated that alternating current (AC) dielectrophoresis can sort SWCNTs on the basis of their dielectric constants, which are considerably different between metallic and semiconducting SWCNTs. In fact, the more polarisable species (i.e., metallic tubes) are selectively deposited onto the substrate between the microelectrodes, while the electric field also induces SWCNT alignment. A big disadvantage of this technique is the limited throughput, for which as low as 100 pg of metallic SWCNTs could be isolated.[92] Therefore, recent work has attempted to scale up dielectrophoresis using, for example, dielectrophoretic field-flow fractionation (FFF).[93]

Chromatographic techniques (e.g., IEC) have also provided successful separations of SWCNTs, especially when the tubes are pre-encapsulated with single-stranded DNA in aqueous solution. Separation by electronic type and diameter can be followed through optical absorbance measurements, which demonstrated that the sorting quality depends on DNA sequence, with a sequence of (GT)*n* (where *n* is between 10 and 45) leading to the best results.[94] Even better separations have been achieved with the association of SEC,[95] which, in combination with IEC, has enabled selective sorting of SWCNTs with narrow diameter, length and chiral angle distributions. At the moment, however, the success of chromatographic sorting is limited to small-diameter (<1.2 nm) tubes.

Moreover, Hersam *et al.* have proposed another post-synthetic sorting approach through ultracentrifugation. The efficacy of this technique suffers from convolution among multiple structural parameters (e.g., diameter and length), because it sorts SWCNTs according to their sedimentation coefficient, which is influenced by several parameters. Therefore, in order to be successful, it has to be associated, for example, with surfactant encapsulation or chemical derivatisation of SWCNTs. This has allowed the separation between metallic and semiconducting SWCNTs, with 99% purity.[96]

In particular, density gradient ultracentrifugation (DGU), combined with surfactant-encapsulated SWCNTs, has demonstrated several advantages for large-scale production due to its compatibility with several raw materials and non-covalent and reversible functionalisations. In fact, co-surfactant mixtures of sodium cholate (SC) and sodium dodecyl sulphate (SDS) have led to diameter and electronic-type sorting by DGU. The mechanism at the basis of this successful separation could be attributed to different binding of the two surfactants as a function of the SWCNTs' polarisability, which is directly correlated with the tubes' electronic properties.

In addition to their separating properties, sodium cholate encapsulated SWCNTs have been investigated to determine both the linear density of surfactant molecules along the length of the tubes and the molar volume of sodium cholate on SWCNTs' surfaces.[97] The obtained results, of 3.6 ± 0.8 nm^{-1} and 270 ± 20 cm$^3 \cdot$mol^{-1}, respectively, represent a valuable contribution in understanding the hydrodynamic properties of SWCNTs and the interactions between SWCNTs and surfactants in aqueous solution.

In a very recent publication,[98] DGU showed the ability to separate double-walled CNTs (DWCNTs), which was very useful to study the inter-wall interactions in CNTs. Usually, it is very difficult to synthesise or separate this type of tubes through conventional techniques, since mixtures with single- and multi-walled tubes are generally produced. Typically, a DWCNT with the same outer diameter of a SWCNT possesses similar SC encapsulation properties to the SWCNT. However, the DWCNT is expected to present a much higher density, given by the contribution of the additional DWCNT's inner wall. Taking advantage of this difference, DGU was able to selectively separate DWCNTs from a mixture of tubes. Four sets of bands located at different densities were clearly observed in the centrifuge tube. The upper two bands corresponded to small- and large-diameter SWCNTs, the third band represented the DWCNTs and the final thick black region was populated by CNT bundles, multi-walled tubes (MWCNTs) and carbonaceous impurities (as confirmed by optical absorbance measurements). DWCNTs were found to be, on average, ~44% longer than SWCNTs, thus providing two to four times higher conductivity in transparent conductors than SWCNTs and unsorted nanotubes.

With regard to this last aspect, the authors have described that the performance of transparent conductors, produced using monodisperse metallic SWCNTs, can be increased by 5.6 and 10 times in the visible and IR regions, respectively.[99] Moreover, provided monodisperse metallic SWCNTs with Angstrom-level resolution in diameter are used, semitransparent conductive coatings with tunable optical transmittance can be produced.

Finally, the use of special catalysts has resulted in remarkable progress towards selective growth of monodisperse samples.[100] As concerns this aspect, the authors have described how DGU process could easily separate SWCNTs produced by carbon monoxide (CO) disproportionation on bimetallic Co-Mo catalysts (CoMoCAT growth strategy).[101]

Interestingly, the accurate comparison between NIR spectra and AFM images revealed large variations among growth methods and effective sorting by DGU.[102] To reiterate, the researchers have elaborated a counting-based method, aimed at quantifying the semiconducting tubes present in several fractions of as-grown or processed SWCNTs, providing useful information about sample compositions in view of future techniques that should control the sample's characteristics precisely.

A few ongoing projects are also related with optical and electronic properties of CNTs.

More precisely, the optical properties of semiconducting SWCNTs are governed by excitons. The excitons created in semiconductors with an ultra-short laser pulse initially possess a definite phase relationship between themselves. However, scattering among the tubes and with charged carriers, phonons, impurities and defects can lead to dephasing and, eventually, population relaxation. An experimental study of ultrafast exciton dephasing in semiconducting SWCNTs utilising a femtosecond four-wave mixing (FWM) technique has been recently reported.[103] Throughout this investigation, the Hersam group discovered that both exciton–exciton and exciton–phonon scattering have profound effects on the dephasing process.

Another study on near-field photoluminescence (PL) of semiconducting single SWCNTs revealed large DNA-wrapping-induced red shifts of the exciton energy, which were two times higher than the value indicated by confocal microscopy.[104] As a consequence, two distinct PL bands (attributable to DNA-wrapped and unwrapped nanotube segments) were clearly identified, with a transition between these two energy levels smaller than the spatial resolution of about 15 nm. This was confirmed by another experiment on the exciton energy transfer in pairs of semiconducting nanotubes, using high-resolution optical microscopy.[105] PL bands were observed, with intensities correlated with inter-tube distance. As expected, the efficient energy transfer was limited to a few nanometres because of competing fast non-radiative relaxation responsible for low photoluminescence quantum yield. On the other hand, low-energy PL bands were produced in the same type of tubes by pulsed excitation.[106] They corresponded to dark excitons that became bright due to defects on the tubes, while displaying longer lifetimes and lack of thermal equilibrium with the bright state.

9.2.4 USA

> **The Smalley Institute**
> @Rice University for Nanoscale Science and Technology
> Website: http://smalley.rice.edu/content.aspx?id=842
> Expertise: Synthesis, functionalisation and optical properties of CNTs

As reported on the website, research activity at the Smalley Institute is aimed at actively supporting and promoting researchers using nanotechnology to explore five grand challenges – (i) energy, (ii) water, (iii) environment, (iv) disease and (v) education – by providing experienced and knowledgeable leadership, a solid administrative framework, world-class scientific infrastructure, and productive community, industry and government relations. In order to guarantee the achievement of many ambitious goals, the Smalley Institute comprises several centres (e.g., Center for Biological and Environmental Nanotechnology [CBEN], International Council on Nanotechnology [ICON]) and components and is affiliated with various organisations (e.g., Advanced Energy Consortium [AEC] or Consortium for Nanoma Terials for Aerospace Commerce and Technology [CONTACT]) with unique objectives.

The high quality of the research developed in this institute is provided by world-class researchers in physics, chemistry, biology and engineering who collaborate on a variety of projects dealing with the field of nanotechnology. A huge number of articles have been published by this research group in the years 2007–2009, and their complete evaluation would require a whole book rather than a chapter. For that reason, we have selected a few (not exhaustive) highlighting publications, chosen on the basis of the topics covered by this book, i.e., synthesis, functionalisation and optical properties of CNTs. Many more articles, including polarisers, lithium batteries, hydrogen storage, etc., have been omitted in this section.

9.2.4.1 Synthesis of CNTs

Researchers at the Smalley Institute are involved in the growth of CNTs through different procedures and techniques. For example, mechanical abrasion of stainless steel (SS) surfaces has shown efficient deposition of catalyst to support growth of CNTs through water-assisted catalytic chemical vapour deposition (Fig. 9.16).[107] In all cases of Fe-containing materials abraded on Al_2O_3 substrates, CNT growth was observed, demonstrating a method that is (i) simple, (ii) accurate in catalyst patterning and also (iii) proficient for large area catalyst deposition. An additional advantage is represented by the high quality of the tubes, as demonstrated by Raman spectroscopy. Interestingly, the same procedure of chemical vapour deposition can be exploited to enable the reuse of the catalyst for multiple regrowths of vertically aligned CNT arrays (carpets).[108] The as-produced CNTs could be transferred from the growth substrate, while the catalyst and the substrate are reactivated by annealing in air, which controls length and diameter distribution of nanotubes accurately due to an additional size-dependent process of iron carbide particle reoxidation. However, a limitation of this study was the inability to achieve indefinite regrowth due to catalyst dynamics taking place during the growth and regrowth processes. Conversely, the high density and continued growth of CNTs were achieved by cutting the SWCNT fibres with a focused ion beam technique followed by etching for removing amorphous carbon and opening the ends of the SWCNTs.[109] Through this procedure, nanoscopically flat open-ended SWCNT substrates were efficiently prepared.

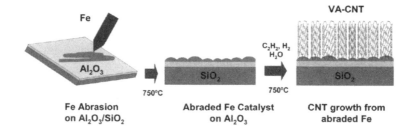

Figure 9.16. Scheme that describes catalyst deposition through abrasion. Any object with Fe metal content abrades a catalyst in the form of particles that can be used as nucleation sites for CNT growth. Reproduced from Alvarez *et al.*[107] with permission.

Another example of CNT "flying carpets" has been recently described by the research group at the Smalley Institute, where chemical vapour deposition (CVD) with roll-to-roll e-beam deposition was employed to produce the flakes on which dense and aligned tubes were able to grow.[110] A theoretical study regarding nanotubes' growth has been developed at Rice University by Yakobson and collaborators,[111] who proved that a CNT's growth rate is proportional to the chiral angle of the tube, as shown by *ab initio* energy calculations. The predicted values were in full agreement with experimental measurements, thus strengthening the solid consistency of the study.

In order to compare a range of catalyst compositions, CNTs have been grown using Fe, Co, Ni and Co/Fe spin-on-catalyst (SOC) systems, involving the metal salt dispersed with a spin-on-glass precursor.[112] During initial growth runs (CH_4/H_2/900°C), the CNT yield followed the order Co-SOC > Fe-SOC >> Ni-SOC. The Fe catalysts produced the longest nanotubes at the expense of a larger average CNT diameter. A series of binary Co/Fe-SOCs was also prepared. After initial CNT growth, the original samples were subjected to additional growth runs. Four individual reactions were observed in the Fe-SOC and binary Co/Fe-SOC (Fig. 9.17): regrowth (amplification), double growth (a second CNT growing from a previously active catalyst), CNT etching, and nucleation from initially inactive catalysts (new growth). CNT etching was observed for the mixed catalyst systems (Co/Fe-SOC) but not for either Fe-SOC or Co-SOC. During the regrowth experiments, some CNTs were observed in positions where they were not present after the initial growth run, thus suggesting that the catalysts that were initially inactive towards nucleation of CNTs in the original growth run became activated when placed back into the furnace and submitted to regrowth under identical conditions.

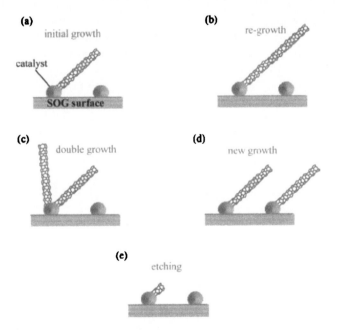

Figure 9.17 Schematic representations of (a) initial growth, (b) regrowth (amplification), (c) double growth, (d) new growth (nucleation from initially inactive catalysts) and (e) CNT etching. The red CNT represents material grown during the first growth run, whereas the green CNT represents new material formed during subsequent growth runs. Reproduced from Crouse *et al.*[112] with permission. See also Colour Insert.

9.2.4.2 Functionalisation of CNTs

The functionalisation of CNTs, either via covalent or non-covalent procedures, represents a crucial step to achieve the physicochemical properties required in almost every application. Many different methodologies have been investigated, and the groups at the Smalley Institute have contributed to the development of some of them. An example is given by the production of single-walled nanotube salts and their subsequent interaction with different molecules to yield transient radical anions that dissociate into nitrogen-centred free radicals and halides.[113] More precisely, SWCNT salts could be prepared by the reduction of pristine tubes (**1**) (Fig. 9.18) using lithium in liquid ammonia. They subsequently react with *N*-bromosuccinimide (NBS) to yield nanotubes functionalised by nitrogen-centred free radicals. Addition of NBS to a suspension of the SWCNT salts led to NBS radical anions that eventually produce succinimidyl radicals. When succinimidyl functionalised SWCNTs (**3**) were treated with hydrazine, aminated SWCNTs (**4**) were formed. Through a similar procedure, hydroxyl functionalised SWCNTs could be prepared using SWCNT salts and then performing hydrolysis in fuming H_2SO_4. As a result, the aminated and hydroxylated SWCNTs could be further functionalised with other groups, amino acids and DNA for chemical and biological applications.

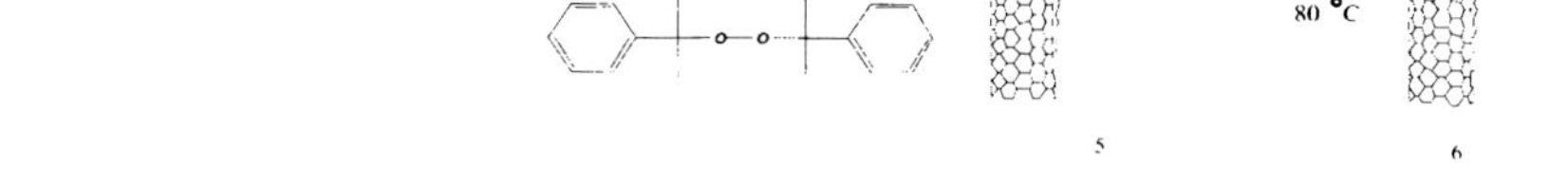

Figure 9.18 Scheme of functionalisation of SWCNT salts.

Another functionalisation consists in the recently reported photohydroxylation reactions by degassing the dispersion of SWCNTs/surfactant (e.g., sodium dodecyl sulphate, SDS).[114] The degassing method was based on the desorption of O_2, CO_2 and endoperoxide from the solution and the surface of SWCNTs. In this way, solutions remained fluorescent in acidic conditions (pH 3) demonstrating that the vast majority of the O_2

was removed from the solution. On the other hand, degassing the SWCNT/ surfactant solution increased its fluorescence intensity, especially when the temperature was raised to 80°C, after which it started to decay. The collected spectroscopic data suggested that the photoreaction was driven by absorption of UV light by SWCNTs. Although the mechanism of this photoreaction is not yet understood, it is thought that, similarly to the previous example, radicals are generated during the irradiation process. This assumption was confirmed by the suppression of the reaction when irradiations were performed in air rather than degassed solutions: O_2 in fact is a well-known radical scavenger, so it quenched the radical reactive intermediates.

Overall, these results indicated that the reaction was initiated by the SWCNT photon absorption, leading to the generation of an excited state, which thereafter activated small molecules, likely water, in their close surroundings. Semiconducting SWCNTs seem good candidates as their excited-state lifetimes, reported in the literature, are in the picosecond range, while metallic tubes have shorter lifetimes (femtosecond range).[115] Such difference provided semiconducting tubes enough time to react with the molecules in their surroundings and give selective reactions, while metallic SWCNTs remained essentially unreactive toward photohydroxylation. This work represented the first selective reaction in the liquid phase that takes advantage of an intrinsic property of the tubes to develop both separation processes and selective functionalisation chemistries.

In a different experiment, it was observed that the addition of diazonium SWCNTs/surfactant suspensions quenched the intrinsic NIR fluorescence of semiconducting SWCNTs through sidewall reactions.[116] Fluorescence spectroscopy was employed to quantify the structure-dependent relative reactivities of these reactions. In previous studies, it had been suggested that the mechanism behind this reaction involved injection of electrons from the metallic SWCNTs into the diazonium salt (Fig. 9.19a), implying that the semiconducting SWCNT species with smaller band gaps would react more readily. However, the observed results showed the opposite trend. Alternatively, the electron-deficient diazonium salt becomes an electron-rich diazotate that reacts with another molecule of diazonium salt to yield the diazoanhydride (Fig. 9.19b). This electron-rich anhydride could preferentially form charge transfer complexes with the larger-band-gap, electron-poor SWCNTs, implying that greater reaction selectivity should occur for diazonium salts in which the R group more strongly promotes diazoanhydride formation. Since this hypothesis was not confirmed, the authors concluded that it seems likely that different reaction mechanisms are dominant in the reactions of diazonium salts with metallic versus semiconducting SWCNTs.

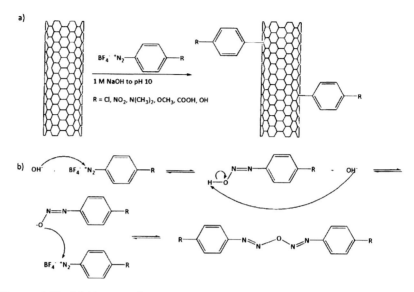

Figure 9.19. (a) Functionalisation of surfactant-wrapped SWCNTs by addition of 4-substituted benzenediazonium salts at basic pH. (b) Diazoanhydride formation. Figure modified from Doyle *et al.*[116] with permission.

Another method has made use of aryl diazonium salts to introduce repetitive functionalisations for multifunctional SWCNTs.[117] The obtained products displayed enhanced water solubility and provided a protective environment towards those molecules that, otherwise, would not be very stable. The bulky assembles were characterised using Raman spectroscopy, thermogravimetric analysis (TGA) and X-ray photoelectron spectroscopy (XPS).

Covalent functionalisation of the nanotube sidewall has been achieved through the partial substitution of fluorine in fluoro-CNTs (F-SWCNTs) by other moieties, including urea, guanidine and thiourea (Fig. 9.20).[118] The successful functionalisation has been confirmed by several techniques, including Raman spectroscopy, Fourier transform IR (FTIR), TGA, scanning electron microscopy (SEM), XPS, transmission electron microscopy (TEM) and AFM. F-SWCNTs have been also characterised by magic angle spinning ^{13}C NMR spectroscopy,[119] from which an interesting conclusion could be derived: where a comparison of samples with a high degree of functionalisation is required, NMR provides a much better quantification than Raman. However, where a comparison between samples with low levels of functionalisation or large differences in degree of functionalisation is required, Raman provides a much better quantification than NMR.

By comparison among the three synthesised samples, it emerged that the urea-functionalised SWCNTs (U-F-SWCNTs) showed the highest stability of the tubes' dispersion in DMF, water and aqueous urea solutions; though urea is extremely soluble in water, it can be still drained to the van der Waals forces within the F-SWCNT bundles, intercalate, wrap and thus separate them, thereby creating new opportunities for biomedical applications with nanotubes.

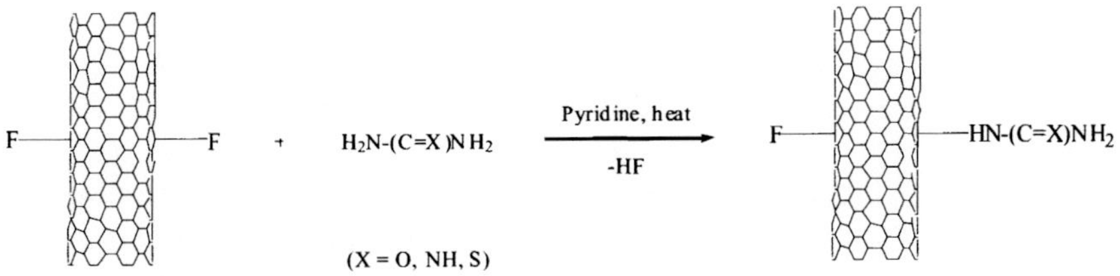

Figure 9.20 Functionalisation of F-SWCNTs with urea (X = O), guanidine (X = NH) and thiourea (X = S).

The same F-SWCNTs were also treated with either branched (MW between 600 and 25,000 Da) or linear (MW = 25,000 Da) polyethyleneimine (PEI),[120] providing the covalent attachment of the polymer to the sidewalls of the nanotubes in the form of PEI-SWCNTs. As expected, the number of polymeric units per SWCNT was larger for low molecular weight PEI. However, above 1,800 Da, such value did not vary remarkably, as confirmed by Raman spectroscopy. Solid-state ^{13}C NMR indicated the presence of carboxylate substituents as a consequence of the reversible CO_2 adsorption to the primary amine substituents of the PEI. The authors also showed how to accomplish the desorption of CO_2 by heating under argon at 75°C, opening the way for further investigations.

9.2.4.3 Optical properties of nanomaterials

Carbon nanotubes represent a remarkable section towards advances in analytical nanotechnology, especially on the basis of their optical properties. SWCNTs have been recently shown to be exceptional NIR fluorophores that display high optical anisotropy[121] and an absence of fluorescence intermittency,[122] which are characteristics not present in quantum dots or fluorescent dyes and make them extremely promising for applications as fluorescent markers and biosensors. Weisman and collaborators have been focusing on the photoluminescence properties of these carbon-based nanomaterials, demonstrating that CNTs' structure can be modified to detect individual molecules. It is important to note that, as the excitation intensity increases, photoemission from SWCNTs reaches a saturation value, suggesting an upper limit on the exciton density for each nanotube species. Kono and co-workers not only demonstrated this phenomenon, but they also developed a model based on diffusion-limited exciton–exciton annihilation, which allowed estimating exciton densities in the saturation regime.[123] It is also possible, by applying a magnetic field (**B**) along the tube axis, to enhance the brightness of the dark state of SWCNTs, leading to a new bright peak with increasing intensity till a dominant role at **B** > 3T.[124] This behaviour could be considered universal, since it was observed for more than 50 different tubes, showing values of 1–4 meV for tube diameters of 1.0–1.3 nm. This emphasised how the surrounding environment could influence the excitonic fine structure of the tubes.

In addition to these observations associated with SWCNTs of specific chirality, the inner shells of DWCNTs have also shown some NIR emission,[125] although such fluorescence is weaker than SWCNTs' by a factor of at least 10,000. Better profiles have been obtained with self-assembling peptides, which show the ability to suspend the tubes in solution while preserving strong NIR luminescence.[126] Until now, a limiting aspect of CNTs has been their highly hydrophobic character and their tendency to aggregate into nonemissive bundles. Therefore, CNTs need to be functionalised in order to become dispersible and biocompatible. Covalent functionalisation, although providing a stable coating against displacement or disruption, tends to quench the NIR nanotube emission. Previous attempts to suspend SWCNTs included pluronics (synthetic nonionic surfactants),[127] single-stranded DNA[128] and bovine serum albumin (BSA).[129] In addition, the ionic surfactant sodium dodecylbenzene sulphonate (SDBS, which is not biocompatible) allowed by far the strongest NIR emission. In fact, it was used in another experiment to suspend SWCNTs in aqueous solution and enabled the calculation of the product of absorption cross section and fluorescence quantum yield for 12 (n, m) SWCNTs,[130] thus deriving the empirical calibration factors needed to deduce quantitative (n, m) distributions from bulk fluorimetric intensities. In comparison with this and other suspending agents, peptides show the advantage of a simple and versatile chemical assembly of biocompatible amino acids into a variety of engineered structures with tailored functionalities. These peptides are organised with an A-B-A block motif, in which the peripheral A domain contains electrostatically charged amino acids while the interior B domain consists of alternating hydrophilic and hydrophobic amino acids, which cause the peptide to assume a β-sheet conformation. Interestingly, these peptides are not acting as simple amphiphilic surfactants. Instead, they interact with the nanotubes in a more specific fashion, in which their tendency to self-assemble is reinforced by the presence of the SWCNTs. As a proof of that, it was observed that the peptides that allowed the brightest SWCNT emission were not those that displayed the strongest tendency to self-assemble, but rather those that used SWCNT surface as a template to form a stable coating.

As an alternative, Schmidt and Pasquali used a new strategy to obtain suspensions of highly luminescent SWCNTs by using a combination of surfactant (SDBS) and a biocompatible polymer (polyvinylpyrrolidone, PVP) or by polymerising *in situ* its monomer vinyl pyrrolidone (ISPVP).[131] PVP or VP adsorbed strongly to the external surface of the SDBS micelle due to charge transfer between the sulphate group of SDBS and the nitrogen of PVP and/or VP (Fig. 9.21). At neutral pH values, the PVP slightly disrupted the SDBS micelle and affected the tubes' luminescence, while the monomer did not. In acidic conditions, conformational changes of PVP and VP polymerisation determined efficient wrapping of the SDBS-SWCNTs, providing additional protection against the surrounding environment. Therefore, stable suspensions in a wide range of pH were obtained, while avoiding photoluminescence changes or flocculation of the samples. The results showed incredible stability of the ISPVP-SDBS-SWCNT complex even after extended dialysis and after lyophilisation and resuspension in deionised water. Moreover, using NIR photoluminescence microscopy, it was demonstrated that ISPVP-SDBS-SWCNT suspensions were stable and strongly luminescent in living cell cultures, where they interacted efficiently on cell membranes and did not display any sign of aggregation.

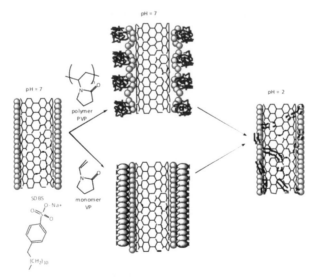

Figure 9.21 Scheme of (Left) SDBS-SWCNT at pH 7. (Centre-up and right) PVP-SDBS-SWCNT at pH 7 and 2. (Centre-below and right) VP-SDBS-SWCNT at pH 7 and 2 for which cationic polymerisation occurred. Figure modified from Duque *et al.*[131] with permission. See also Colour Insert.

On the other hand, it has been proved that CNT photoluminescence is quenched by external electric fields, as demonstrated by the drastic reduction in emission intensity when fields $\leq 10^7$ V/m were applied.[132] In more detail, photoluminescence intensity was found to follow a reciprocal hyperbolic cosine dependence on the applied electric field. Therefore, the use of NIR luminescence provides useful information that substantiated the application of CNTs as fluorescent markers. At the same time, it also offers the opportunity to explore biological systems through biosensing and accurate bio-imaging. With regards to this aspect, the research goup has performed some biocompatibility studies in *Drosophila melanogaster* (fruit flies).[133] In the experiments, *Drosophila* larvae were nourished on food containing approximately 10 ppm of disaggregated SWCNTs, showing normal growth following nanotube ingestion. NIR nanotube fluorescence was imaged from intact living larvae, and it proved extremely useful at identifying and counting isolated SWCNTs to gain an estimation of their biodistribution.

However, while the NIR luminescence of CNTs has been deeply investigated, the exploration on far-field near-IR of SWCNTs has been limited by the diffraction limit to resolution. Once again, the researchers were able to overcome the problem through a new analytic method that allowed subwavelength ($<\lambda/10$) mapping of single-molecule and the precise localisation of excitonic luminescence regions along the nanotube axis.[134] This new method seems very useful at revealing lengths, curvatures and defects of luminescent SWCNTs with unprecedented details.

9.2.5 USA

> **Michael S. STRANO**
> @ Massachussets Institute of Technology (MIT)
> Website (research activities): http://web.mit.edu/stranogroup/research.html
> Expertise: Optical modulation of nanostructures for sensor applications, chemistry of single-walled carbon nanotubes, and molecular interactions with nanoelectronic sensor arrays

9.2.5.1 CNT-based sensors

Strano's research group is mainly focused on the development of carbon-based sensors for biological detection. This is because SWCNTs possess many advantages when used as optical sensors, including photostable intrinsic NIR emission for prolonged detection through biological media and single-molecule sensitivity. However, a big challenge in the fabrication of such devices is the ability to impart selective analyte binding without disrupting optical and electronic properties. A recent example towards that goal is the synthesis of a 3,4-diaminophenyl-functionalised dextran (DAP-dex) wrapping around SWCNTs in such a way as to provide a rapid and selective fluorescence detection of nitric oxide (NO), which is a vital biological messenger.[135] The advantage of this functionalisation relies on the immediate bleaching of the NIR fluorescence of SWCNT-DAP-dex complex by NO, but without affecting other reactive nitrogen or oxygen species. This indicates that the resulting optical sensor is able to detect the NO molecules deriving from NO synthase in macrophages in real time and it could be eventually extended for *in vivo* detection in animal models.

During some investigations on SWCNTs sensors, it was noticed that, once photogenerated excitons in SWCNTs undergo one-dimensional quantum confinement, the signals deriving even from an individual molecule could be amplified tremendously, also in case of living cells and tissues. In view of this observation, the authors have prepared a collagen film, containing SWCNTs able to sense single-molecule adsorption of quenching species. More precisely, H_2O_2, H^+ and $Fe(CN)_6^{3-}$ were investigated as quenching molecules.[136] H_2O_2 displayed the highest quenching equilibrium constant of 1.59 at 20 µM. Interestingly, reverse (unquenching) rate constants were concentration-independent while the forward (quenching) rate varied along with concentration and the redox potential of the quencher.

Taking advantage of SWCNTs presenting a 1D electronic structure sensitive to molecular adsorption, the research group has identified at least four modes in SWCNTs that can be modulated to uniquely fingerprint agents by the degree to which they alter either the emission band intensity or wavelength.[137] This is particularly useful for those species (e.g., reactive oxygen species [ROS]) that present half-lives between a nanosecond and a millisecond and form adducts with DNA, since it acts as a label-free sensor able to convert chemical information into an NIR signal. Therefore, exposing the SWCNT-DNA complex to several analytes simultaneously has been demonstrated to achieve signal multiplexing.

As reported in Fig. 9.22, in the first reaction the $d(GT)_{15}$ oligonucleotide-bound nanotube (SWCNT-DNA)[128] is exposed to a chemotherapeutic alkylating agent, which reacts with the guanine nucleobase and will lead to eventual strand breakage, resulting in apoptosis of cancer cells. The interaction of such agent with the SWCNT-oligonucleotide complex results in a uniform red shift in the photoluminescence bands of both (6, 5) and (7, 5) nanotubes (Fig. 9.22b). The second reaction involves direct adsorption of H_2O_2 by the nanotube, which results in attenuation of both the nanotubes' emission and a slight concomitant energy shift (Fig. 9.22c). Singlet oxygen, generated in the third reaction by exposing the nanotube complexes to Cu^{2+} and H_2O_2, causes a pronounced red shift of the (6, 5) nanotube emission, but no corresponding shift in the (7, 5) band (Fig. 9.22d). Finally, hydroxyl radicals, produced in the presence of SWCNT by Fe^{2+} and H_2O_2, damage the DNA backbone and attenuate emission from both nanotubes but preferentially affect the (7, 5) emission, without energy shifts (Fig. 9.22e).

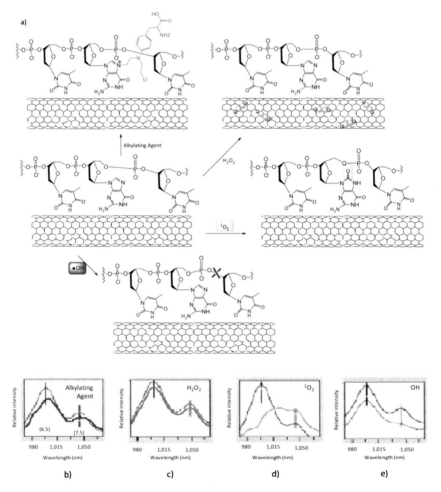

Figure 9.22 Multimodal detection of four reaction pathways. (a) Scheme of interactions on the SWCNT-DNA complex – an alkylating agent reaction with guanine, H_2O_2 adsorption on the nanotube sidewall, a singlet oxygen ($^1O_2$) reaction with DNA, and hydroxyl radical (•OH) damage to DNA. (b–e) SWCNT-DNA photoluminescence spectra before (grey) and after introducing the alkylating agent (red) (b), H_2O_2 (blue) (c), singlet oxygen (yellow) (d) and hydroxyl radicals (green) (e). Figure redrawn from Heller *et al.*[137] See also Colour Insert.

The identification method was validated *in vitro* by demonstrating the detection of six genotoxic analytes. Results showed that chemotherapeutic alkylating agents were detected immediately upon exposure to the SWCNT-DNA complex. Emission red shifts of up to 6 meV and concomitant attenuation were observed using NIR spectrofluorimetry. Hydrogen peroxide was detected by the SWCNT-DNA complex through attenuation of both (6, 5) and (7, 5) fluorescence bands to similar extents with slight shifting of peak wavelengths (Fig. 9.22c). Singlet oxygen induced a pronounced redshift in (6, 5) nanotube emission with virtually no (7, 5) nanotube shift (Fig. 9.22d), as confirmed by a three-dimensional photoluminescence profile. In addition, exposure to the singlet oxygen scavenger sodium azide significantly reduces the magnitude of the (6, 5) band shift. Finally, it was possible to detect these analytes in real time within live 3T3 cells, demonstrating multiplexed optical detection from a nanoscale biosensor and the first label-free tool able to optically discriminate among several short-living agents.

In another study, the authors evaluated the *in vivo* pharmacological effects on the immune response deriving from the individual, simultaneous or sequential delivery of dexamethasone (DX) and vascular endothelial growth factor (VEGF).[138] For the first time, imaging of a SWCNT fluorescence device implanted subcutaneously in a rat was monitored, together with the localisation of the vascular network formed around the hydrogel-coated, microcapillary implant of SWCNTs. For tissue response studies, the chick embryo chorioallantoic membrane (CAM) was used as a tissue model for an 8-day implantation period. The average vascular density of the tissue surrounding the sensor with simultaneous, sequential or no delivery of DX and VEGF was $1.24 \pm 0.35 \times 10^{-3}$ vessels/μm^2, $1.15 \pm 0.30 \times 10^{-3}$ vessels/μm^2 and $0.71 \pm 0.20 \times 10^{-3}$ vessels/μm^2, respectively. Calculations of the therapeutic index (i.e., vasculature/inflammation ratio), which reflects promotion of angiogenesis versus the host immune response, demonstrated that sequential DX/VEGF delivery was 60.3% and 139.3% higher than that of VEGF and DX release alone, respectively, and was also 32.1% higher when compared with simultaneous administration, proving to be a more effective strategy in optimising the delivery of pharmacologically active molecules.

The adsorption of several species on the surface of CNTs is often associated with an irreversible interaction, which is at the basis of electronic sensors or thin-film transistors although without a clear mechanism. In order to have a better understanding of the process involved, the research group investigated the reversible adsorption of thionyl chloride ($SOCl_2$) via non-covalent functionalisation with molecules bearing free amino-groups.[139] The evaluation of 11 of such molecules demonstrated that the analyte ($SOCl_2$) adsorption is deeply affected by the basicity (expressed as pKb of the base or as pKa of its conjugated acid) of the groups at the surface. More precisely, only the amine-containing molecules with pKa <

8.8 were effective at inducing irreversible to reversible transition. In other cases, the molecules irreversibly adsorbed upon the entire tube surface so that no bare SWCNT surface was exposed for $SOCl_2$ adsorption. It also suggested that thionyl chloride is expected to adsorb on the amine layer rather than on SWCNT surface (Fig. 9.23). Interestingly, the irreversible-to-reversible transition upon surface functionalisation is not limited only to thionyl chloride but is a general phenomenon. Therefore, a similar strategy could demonstrate that the reversibility of SWCNT gas sensor can be tuned by variation of the surface chemistry. In fact, the authors translated these results into new types of nanoelectronic devices for analyte detection using microelectromechanical system (MEMS)-based micro-gas chromatography (µGC).[140,141] They demonstrated the unprecedented reversible detection of as few as 10^9 molecules of dimethyl methylphosphonate (DMMP), a nerve agent simulant, at the end of a µGC column. An SWCNT network was formed across the electrodes through AC dielectrophoresis. Polypyrrole (PPy), an amine of pKb = 5.4, was selected as a functionalisation material for DMMP binding.[142] The sensor showed high sensitivity and complete reversibility (self-regenerating) at the ppb level.

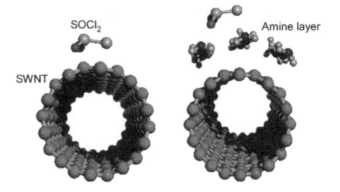

Figure 9.23 Thionyl chloride adsorbed on a bare SWCNT (*left*) and on a functionalised SWCNT (*right*). Reproduced from Lee *et al.*[139] with permission.

The combination of SWCNTs and magnetic iron oxide nanoparticles gives origin to heterostructured complexes that can be used as multimodal bioimaging agents.[143] These complexes are composed of catalyst (Fe) nanoparticles attached to SWCNTs, and they were stably dispersed in aqueous solution by wrapping oligonucleotides with the sequence $d(GT)_{15}$, followed by magnetic separation. The resulting nanotube complexes show distinct NIR fluorescence, Raman scattering and visible/NIR absorbance features, corresponding to the various nanotube species. Therefore, they were incubated with macrophages and identified using magnetic resonance imaging (MRI) and NIR fluorescence.

The results provided clear images and suggested that, by incorporating further functional groups and/or species, the complexes could act as both targeting agent and biosensing devices. In addition, through NIR irradiation upon application of an external magnetic field, the complexes could induce phototherapy and hyperthermia in cells.

9.2.5.2 Single-particle tracking

Single-particle tracking by individual SWCNTs using their NIR band gap fluorescence represents an interesting strategy to understand how this carbon-based material interacts with several molecules, including living systems. This methodology was initially developed by the R. B. Weisman laboratory at Rice University, and it still represents one of the few available tools to study individual SWCNT molecules in solution over a prolonged time, since SWCNTs have no apparent irreversible photobleaching effect and no intrinsic blinking mechanism.[144] Recent advances have enabled real-time measurement of tubes' physical characteristics (e.g., length) and mechanical properties, while combined translational and rotational measurements revealed the influence of local environment on nanotube mobility.[145] Moreover, single-particle tracking of SWCNTs intrinsic photoluminescence was used to map the trajectories of the tubes, as they were internalised inside and expelled from NIH-3T3 cells in real time.[146] The authors found that cellular uptake of SWCNTs was nearly 1,000 times faster than the rate of Au nanoparticles reported in literature; conversely, nanotubes exocytosis showed an expulsion rate that matched the endocytotic process. Moreover, the experiments were associated with a detailed evaluation of the trajectories of CNTs in their internalisation pathway:[147] over 10,000 individual trajectories of non-photobleaching SWCNTs were tracked as they were incorporated into and expelled from NIH-3T3 cells in real time on a perfusion microscope stage. The subsequent analysis of the mean square displacement allowed the complete construction of the mechanistic steps involved. Data collected from this study allowed to justify the observed accumulation of SWCNTs inside the cells via endocytotic and exocytotic pathways. In addition, the study provided useful information for the design of future systems with lower toxicity and improved efficacy.

9.2.6 Mexico

TERRONES (Humberto & Mauricio)
@ Advanced Materials Department, IPICYT, Mexico
Website: http://www.uc3m.es/portal/page/portal/home/chairs_
excellence/chairs_excellence_09/mauricio_terrones
Expertise: Functionalisation and electrical properties of carbon
nanotubes

Terrones' group is involved in theoretical and experimental research of new materials, with the aim to understand and predict the mechanical, electronic and magnetic properties of nanostructures, and to apply them to novel technologies; the final purpose is to obtain more compact tools and better performance.

In more detail, the main research areas include computational materials science, low dimensional systems, magnetism in new materials, nanoscience and nanotechnology, superconductivity, electron microscopy and optical properties.

9.2.6.1. Doping of CNTs

Terrones' group is specialised on CNTs' "doping", which consists in the incorporation of electron acceptors or electron donor species such as nitrogen (N), boron (B) and phosphorus (P) (Fig. 9.24). This process enables the modulation of the tubes' physical, chemical and electronic properties for a better characterisation of the samples and for potential applications in chemical sensors. In particular, it seems that chemical attributes of BN-doped tubes are very important for the production of reinforced composites with insulating characteristics.[148] From the theoretical point of view, it has emerged that (BN)-C-hetero-nanotubes could have important implications for advanced nanoelectronics. Recent articles reporting data obtained from energy-dispersive X-ray analysis (EDX) and electron energy loss spectroscopy (EELS) have shown that P and N can be homogeneously incorporated into the lattice of MWCNTs (with diameters usually above 20 nm).[149] In that case, MWCNTs doped with phosphorus (P) and nitrogen (N) could be synthesised using a solution of ferrocene ($Fe(C_5H_5)_2$), triphenylphosphine ($P(C_6H_5)_3$) and benzylamine ($C_6H_5CH_2NH_2$) in conjunction with spray pyrolysis. Iron phosphide (Fe_3P) nanoparticles acted as catalysts during nanotube growth, leading to the formation of novel heterodoped PN-MWCNTs. Differently from P, the doping of CNTs with N has been already highly documented and subjected to intense debates regarding its role in the formation of unique structural features, including shortened tubes, smaller diameters and even bamboo-like multilayered nanotubules.[150] Both theoretical and experimental techniques have concluded that N acts as a surfactant during growth, favouring the formation of tubes with reduced diameters. Additionally, it could also induce tube closure, which implies a relatively large amount of N atoms into the tube lattice, leading to bamboo-like structures.

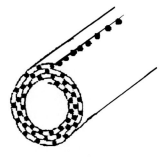

Figure 9.24 Schematic representation of doped carbon nanotubes.

Despite the successful incorporation of P in MWCNTs, doping SWCNTs with large contents of P is difficult because P atoms are larger than C ones, thus increasing the disorder within the hexagonal carbon structure. Moreover, P by itself decreases the catalytic efficacy of iron during nanotube growth. In addition, smaller amounts of doping elements have to be inserted into the growth environment of SWCNTs in order to obtain samples of acceptable quality, but they are harder to detect. Anyhow, the authors have elaborated a successful methodology for a uniform doping of SWCNTs with P.[151]

It emerged that in comparison with N-doped nanotubes, both P and N doping created new states due to the presence of the doping species, which appeared close to the Fermi energy. However, while for N this state lied beneath the conduction bands, causing all N-doped nanotubes to be metallic, P created a highly localised state that did not affect the conduction bands, and therefore they did not influence the semiconductive or metallic character of the nanotubes. This was in agreement with another study, in which density functional theory (DFT) calculations on phosphorus (P) and phosphorus–nitrogen (PN)-doped SWCNTs were reported.[152] The results of this study demonstrated that substitutional P and PN doping created localised electronic states that acted as scattering centres. It was also predicted that the energy for P-doped tube formation is lower as the curvature increases, thus making narrow-diameter nanotubes preferable when compared with large-diameter nanotubes. Nevertheless, for low doping concentrations (1 doping site per ~200 atoms), the PN- and P-doped nanotubes displayed promising properties (e.g., 50% reduction in the elongation upon fracture, constant Young's modulus, stable conductance) for components in sensors operating at the molecular level.

Although less discussed, doping of CNTs could be responsible for changes in the magnetic properties of CNTs. In order to address this point, a theoretical study was carried out for >80 sp^2-like species (in the range of 600–2,300 atoms) to estimate the magnetic response when additional electrons are involved, i.e., in case of the so-called charge transfer doping.[153] Results showed that a large paramagnetism was found if a small percentage of electrons (0.02–4%) was introduced to the C nanosystem, thus confirming that doping of CNTs can affect their physicochemical properties.

The authors have also described an *in vitro* phagocytosis assay with *Entamoeba histolytica* to assess the biocompatibility of undoped MWCNTs and N-doped (CN$_x$) MWCNTs.[154] *E. histolytica* is a protozoan parasite whose life cycle oscillates between the infective (cyst) and the invasive (trophozoite) stages. The latter is a mobile form. *E. histolytica* trophozoites are one of the most actively phagocytic and proteolytic cells in nature, able to phagocyte cells of the same size or even bigger, as well as to destroy almost every human tissue, including bones. In most cases, they damage the intestinal epithelial surface, causing intestinal amoebiasis or dysentery.

The nanotubes used in these experiments were pure MWCNTs and CN$_x$ MWCNTs, with nitrogen content of 2–4 wt%. The samples were subjected to strong acidic conditions (H$_2$SO$_4$: HNO$_3$ in a 3:1 ratio), which are known to cut the tubes. Results after 24 hours of incubation showed that the viability of the trophozoites was not affected by the CN$_x$ MWCNTs in the culture medium. Moreover, the mobility and shape of amoebas were not altered by the presence of CNx MWCNTs. In contrast, undoped (pure carbon) MWCNTs severely affected the viability of the trophozoites, with a significant decrease (about 50%) after 8 hours of incubation. This seems to be due to stronger van der Waals interactions in pure MWCNTs, which promoted a larger agglomeration of the nanotubes. In every case, the length of both types of nanotube found inside the cells is very similar (<2 mm). These findings encourage further study using CNx MWCNTs as drug delivery systems and cell transporters.

9.2.6.2 Electrical properties of CNTs

Spectroscopy, especially Raman scattering, has been largely used to study defects in sp^2 carbon materials, and also doping phenomena. This is because the so-called G band is a well-known feature that is observed in the Raman spectra of all sp^2 carbon materials, including nanotubes, nanohorns and nanoribbons. After doping, a new G$_D$ (where D stands for "defect") peak is observed at lower/higher frequency for N/P doping.[155] As a defect site is a structure localised at the atomic level, it cannot be detected with accuracy; therefore, usually near-field Raman and near-field photoluminescence imaging and spectroscopy are employed to spatially resolve local defect sites along an individual SWCNT. In fact, bright photoluminescence emission originates from a defect site along the SWCNT. Once the defect site is located, it is possible to monitor the G band and show that a clear change in the G line shape is measured near the defect site, as expected.

Interestingly, pristine, N-doped and P-doped SWCNTs all exhibit a lower frequency G_D Raman peak with similar frequencies, thus indicating that the defect site is caused by the presence of one extra electron. On the whole, the behaviour of the G_D peak is an effective probe to accurately measure the local effect of a single defect (including the one determined by dopants) on the electronic and vibrational properties of sp^2-hybridised carbon material.

In addition, taking into account the effects determined by localised defects, the authors have suggested introducing such defects "artificially" at specific sites, where they behave as bouncing centres, thus modulating the electric current along precise paths.[156] This type of defect can be incorporated while preserving the connectivity of each carbon atom embedded within the graphitic lattice. At the same time, it stimulates the design of defect-based nanostructures for nanoelectronic devices.

Although not mentioned previously, doping of CNTs with different moieties could occur internally or to the external wall of the tubes. In the second case, we refer to exohedral doping. In an interesting paper,[157] the authors combined density functional theory (DFT) and resonance Raman spectroscopy to disclose the effects of exohedral doping on M@S and S@M DWCNTs), where a metallic (M) tube is either inside or outside a semiconducting (S) one. More precisely, there exist four possible configurations (M@M, M@S, S@S and S@M), which should present different electronic properties. The authors have recently reported a detailed Raman spectroscopy analysis carried out on individual DWCNTs, where both concentric tubes were measured under resonance conditions, to understand the dependence of their electronic and optical properties according to their configuration.[158] Interestingly, they found that metallic nanotubes were extremely sensitive to doping even when they were inner tubes, in sharp contrast to semiconducting nanotubes, which were not affected by doping when the outer shell was a metallic nanotube (screening effects).

The researchers have also elaborated a strategy based on DWCNTs fluorination, which enabled the specific suppression of the radial breathing mode (RBM) and absorption peaks from the outer tubes of DWCNTs.[159] Conversely, Raman signals from the inner tube showed no difference from the pristine DWCNTs, thus suggesting fluorination as a useful method to preserve the inner tube from having unwanted contact with other substances that may distract from the inner tube's own characteristics.

In a parallel experiment, DWCNTs were proposed as the preferable reinforcing filler in polymers.[160] In fact chemical moieties, which were selectively introduced on the outer tubes (i.e., exohedral incorporation), imparted effective anchoring sites for strong bonding with polymers, while the optical properties of the inner tubes of the DWCNTs remained unchanged. On the whole, these findings promote further exohedral investigations towards novel nanoscale electronics.

9.2.6.3 Junctions between CNTs or between metals and CNTs

Through different procedures, it is possible to obtain heterojunctions between CNTs and different metal nanocrystals m (Fe, Co, Ni and FeCo) (Fig. 9.25).[161] The heterojunctions are formed from metal-filled MWCNTs via intense electron-beam irradiation at temperatures in the range of 450–700°C, while real-time *in situ* imaging could be carried out. The time required to obtain the final heterojunction was 11 minutes. DFT calculations predicted that these structures are mechanically strong, the bonding at the interface is covalent (directional) and the electronic states at and around the Fermi level are delocalised across the entire system. This confirms the metallic nature of the heterostructure, thus promoting the generation of these MWCNT-*m*-MWCNT junctions (where *m* is the metal particle) with promising transport properties for robust nanotube–metal composite materials.

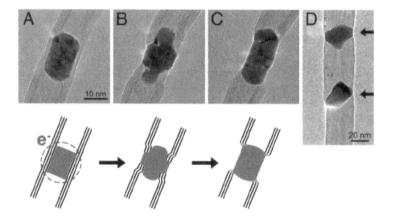

Figure 9.25 Formation of a MWCNT-*m*-MWCNT junction from a magnetic-particle-filled MWCNT subjected to electron irradiation (200 keV) at 700°C. (A) Beginning of the experiment. (B) The focused electron beam damages the tube, and the *m* nanowire is expelled to the surface, experiencing shape changes after 6 minutes of irradiation. (C) Finally, the *m* (Co) particle acts as a link between the two MWCNT segments. The sketch at the bottom shows the mechanism of the process. The circle indicates the zone subjected to electron irradiation. Reproduced from Rodriguez-Manzo *et al.*[161] with permission.

A further development of this study was the elaboration of Y junctions through analogous electron-beam irradiation of SWCNTs at high temperatures.[162] To achieve such shape, a merging process of asymmetric into symmetric rings using CNTs was applied and confirmed by TEM. On the basis of such results, many diversified structures of interferometers with defined ring symmetries could be envisaged.

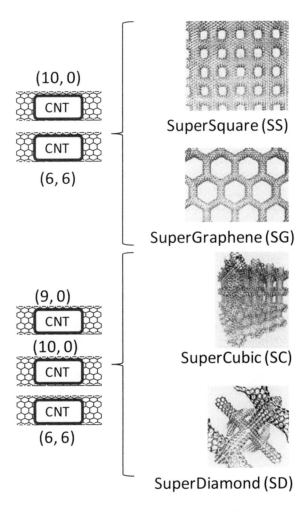

Figure 9.26 Ordered networks based on CNTs (1D nanostructures). (a, b) Super-square and super-graphene correspond to 2D networks, whereas (c, d) super-cubic and super-diamond represent 3D network examples. The four families are constructed from either armchair or zigzag CNTs in order to study the chirality effects. The red rings point out the non-hexagonal carbon rings in each node. Figure modified from Romo-Herrera *et al.*[164] with permission. See also Colour Insert.

Y and T morphologies were also displayed by the coaxial structures obtained from the combination of MWCNTs doped with N (MWCNTs-CNx) as core element, and concentric carbon-based shells formed externally by pyrolysis of toluene over Fe-coated MWCNTs-CNx.[163] The advantage of this methodology is that it is possible to control both growth and dimensions of these nanotube networks, with potential applications in nanoelectronics.

To a wider extent, 2D and 3D ordered networks could be generated from 1D units that were connected covalently. The research group has indeed created multi-terminal junctions containing 1D carbon blocks in the form of ordered networks based on carbon nanotubes (ON-CNTs).[164] These branched structures have demonstrated specific super-architectures (hexagonal, cubic, square and diamond type; Fig. 9.26), which influence both mechanical and electronic characteristics. Hence, charges follow specific paths through the nodes of these multi-terminal systems, which could result in complex integrated nanoelectronic circuits. The 3D architectures revealed their ability to support extremely high unidirectional stress when their mechanical properties were tested. These networks performed better than standard carbon aerogels because of their low mass densities, continuous porosities and high surface areas.

Despite several patterning methods available for the fabrication of nano-ensembles on various substrates, researchers working in Terrones' laboratories have decided to explore low-cost and scalable routes that avoid the use of complicated lithographic approaches. To that purpose, they reported the growth of aligned MWCNTs-CNx presenting fascinating structures such as cactus, flower, volcano, and cake.[165] These architectures were obtained by pyrolyzing ferrocene/toluene/benzylamine solutions at 850°C over electrochemically HF-etched Si substrates. By changing the current densities, it was possible to control the formation and morphology of the circular patterns exhibiting different SiO_x distributions.

9.2.6.4 Incorporation of CNTs with different species

Researchers at Terrones' laboratory demonstrated that molybdenum (Mo) has the ability to self-assemble into nanowires once restrained in a limited space like the interior of CNTs.[166] In such a situation, it tends to form nanowires that consist of a subunit of a Mo body-centred cubic crystal rather than assuming a linear chain conformation.[167] Quantum mechanical calculations have also demonstrated that tubes need to have a diameter ≥0.7 nm to accommodate the smallest of such wires and thus to make this phenomenon happen.

On the other hand, when lithium and ammonia are intercalated inside MWCNTs, the tubes could be opened longitudinally by exfoliation, especially when acid treatment and abrupt heating are applied.[168] In this way, the authors were able to obtain several structures, consisting of (i) multilayered flat graphitic structures (nanoribbons), (ii) partially open MWCNTs and (iii) graphene flakes. On the whole, the completely unwrapped nanotubes were called "ex-MWCNTs", the utility of which should be deeply investigated.

The authors also reported the synthesis and characterisation of selenium-carbon (Se-C) nanocables,[169] which consisted of a SWCNT as core structure, surrounded by a trigonal selenium shell made of selenium oxide. This external part covered preferably semiconducting SWCNTs, suggesting its role in the fabrication of photonic devices as an interface between electronic and photonic materials.

Among other atoms, sulphur has been identified as a triggering agent in the formation of branched nanotube networks with stacked-cone morphologies made up of heptagons (negative curvature) and pentagons (positive curvature).[170] The molecular dynamics (MD) simulations confirmed that sulphur atoms promote the formation of heptagonal rings, and they are also more abundant in curved regions (either heptagons or pentagons).

Finally, it is worth mentioning the investigation devoted to the evaluation of "nucleation", the first and most important step of nanotube growth; the mechanism behind such process has never been accessible to observation and has remained unaddressed. Therefore, Terrones' research group has created CNTs by the injection of carbon atoms into the catalytic metal nanoparticles and subsequent electron irradiation of CNTs partly filled with transition metals such as Co, Fe, FeCo and Ni.[171] When MWCNTs containing metal particles were irradiated with an electron beam, carbon from graphitic shells surrounding the metal particles was ingested into the body of the particle and subsequently emerged as SWCNTs or MWCNTs inside the host nanotubes. This process was monitored for the first time with spatial resolution in real-time studies. Images under TEM (Fig. 9.27) showed a CNT growing from the metal nanowire in the inner core of the tube (growth speeds were of only 1 nm/min). The growth of SWCNTs was also observed inside MWCNTs filled with Fe crystals. Here the growth of the tube was accompanied by the appearance and migration of an oval domain inside the metallic core.

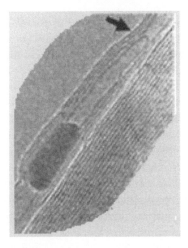

Figure 9.27 SWCNT growth from a metal crystal. Image after SWCNT has reached a length of 11 nm. Figure modified from Rodriguez-Manzo *et al.*[171]

9.3 EUROPE

Maurizio PRATO
@Department of Pharmaceutical Sciences, University of Trieste, Italy
Website: http://www2.units.it/~pratoweb/Maurizio_Prato/Home.
html
Expertise: Functionalisation of carbon-based materials for biomedical
applications

Prato's research group includes about 20 members focused on organic functionalisation of carbon based-materials (mainly fullerenes, carbon nanotubes and graphite) for potential applications in materials sciences and medicinal chemistry. Representative examples are energy storage devices, based on electron-acceptor materials such as fullerene–porphyrin and fullerene–ferrocene hybrids, or fulleropyrrolidines possessing a DNA minor groove binder and an oligonucleotide chain or even unnatural fullero aminoacids as building blocks for the construction of nanotube- or fullerene-based peptides. The group, working in strong collaboration with Dr Alberto Bianco (CNRS, Strasbourg, France) and Prof. Kostas Kostarelos (School of Pharmacy, London), has gained attention worldwide by developing purification processes and synthetic procedures (e.g., 1,3-dipolar cycloaddition of azomethine ylides, which has been extensively used) of diversified carbon allotropes, in view of promising therapeutic applications.

9.3.1. Drug Delivery and Other Biomedical Applications

Among the latest projects developed by this group, it is worth mentioning the delivery of siRNA from CNT-based dendron (more details on this article have been provided in chapter 5).[172] Branched structures, made up of polyamidoamine (PAMAM) dendrons, were covalently linked to the surface of MWCNTs via the 1,3-dipolar cycloaddition of azomethine ylides. The following incorporation of ethylenediamine and methyl acrylate in a divergent approach originated more dendritic-type structures. The presence of PAMAM dendrons guaranteed the solubility of the tubes in aqueous media, while the fluorescent siRNA was subsequently conjugated with the MWCNT dendrons at a 1:16 mass ratio on the basis of previous optimisation.[173] It was observed that the cellular uptake of the siRNA was nanotube-dependent, i.e., the higher the extent of the branching, the more efficient the delivery of siRNA. Notably, almost no uptake of siRNA alone was observed under the same experimental conditions, thus suggesting that an efficient release of siRNA inside target cells might trigger the desired effect of suppressing a gene implicated in a particular disease. More precisely, on the basis of previous evidence that CNTs promote highly efficient cytoplasmic transport of siRNA, the researchers have comparatively studied the *in vivo* cytotoxic effects of complexes including a toxic siRNA sequence (siTOX) with either functionalised MWCNTs or cationic liposomes (DOTAP:Chol).[173] Specificity for all human tumour cell lines was evidenced by the dramatic enhancement of cytotoxicity compared with that obtained for the panel of murine cells. In addition, Calu 6 (human lung carcinoma) cells were chosen to prepare human tumour xenografts and comparatively study the *in vivo* cytotoxic activity of siTOX delivered by either liposomes or MWCNT-NH$_3^+$. Results showed that the liposome:siTOX complexes had maximum activity at 72 hours post-transfection, while the MWCNT-NH$_3^+$:siRNA complexes reached their optimum after 24 hours, suggesting a faster translocation through the plasma membrane in comparison with liposomes. A MWCNT-NH$_3^+$:siRNA at 8:1 mass ratio was used, demonstrating that mice prolonged survival only for the MWCNT-NH$_3^+$:siTOX 50 days after Calu 6 xenograft implantation, with a much higher potency compared with cationic liposomes. Overall, siTOX delivery via cationic MWCNT-NH$_3^+$ was successful at triggering an apoptotic cascade with extensive necrosis of the human tumour cells and a tumour collapse from the inside of the cancerous mass.

In terms of drug delivery, the authors recently described a previously unreported supramolecular complex formed by the non-covalent functionalisation of MWCNTs and the drug doxorubicin for cancer therapy.[174] The advantage of using doxorubicin is that this anthracycline is a fluorescent molecule in view of its chromophore composed of three aromatic hydroxyanthraquinone rings, and hence it can be used as a drug as well as a fluorescent tag to monitor its supramolecular interaction with MWCNTs. It was observed that CNTs were able to quench doxorubicin to an extent proportional to their concentration, most probably due to the π–π stacking interaction between the aromatic chromophore groups and the CNT backbone. Importantly, mitochondrial reductase activity (MTT) assays of these supramolecular complexes showed enhancement in cytotoxic capability of the drug, but tubes alone did not affect cell viability. These data also suggested that MWCNTs can mediate the delivery of doxorubicin and hence improve the cellular uptake of the drug.

The group has also reported a compelling mechanism by which CNTs seem to be excreted through the normal renal pathway.[175] In previous studies, it was seen that radio-labelled CNTs were found in the urine samples after intravenous (tail vein) administration in mice.[176] In addition, intravenously administered MWCNTs, functionalised with the alkylating agent diethylenetriaminepentaacetic acid (DTPA) and radio-labelled with indium [^{111}In], were dynamically tracked *in vivo* using a micro–single photon emission tomography scanner.[177] Imaging showed that nanotubes entered the systemic blood circulation and within 5 minutes began to permeate through the renal glomerular filtration system into the bladder. In their recent publication, based on histological TEM evaluation in absence of any probe molecules, the researchers indicated that factors including size, surface charge, shape of materials and extent of aggregation play a crucial role in determining the fate of their elimination and/or accumulation. The tubes used in these experiments were MWCNTs with diameter of 20–30 nm and length of 0.5–2 μm, and they were subjected to 1,3-dipolar cycloaddition reaction. Hence, tubes presented lengths significantly larger than the dimensions of the glomerular capillary wall (minimum diameter of fenestra is 30 nm, the thickness of the glomerular basement membrane in rodents and humans is 200–400 nm, and the width of the epithelial podocyte filtration slits is 40 nm).[178] Therefore, the only way they could penetrate through such tiny system was by adopting a spatial conformation in which the longitudinal CNT dimension is perpendicular to the endothelial fenestrations, while the traverse dimension of CNT (cross section of 20–30 nm) is small enough to allow permeation. Therefore, it was shown that MWCNTs are capable of reorientation while in blood circulation (Schematic representation in Fig. 9.28).

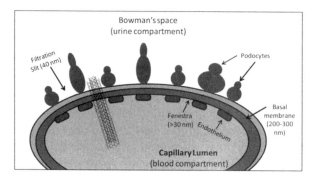

Figure 9.28 Schematic representation of renal clearance of MWCNTs. MWCNT (in black, with diameter ≤40 nm) a few minutes after intravenous (tail vein) injection. The tubes can crossthe renal filtration membrane after orienting perpendicularly. See also Colour Insert.

Despite these results, the impact of administered CNTs on the physiological functions of different organs and tissue is still to be addressed. That is why the group has performed a short-term (24 hours) investigation on various types of MWCNTs tail-vein injected in healthy mice.[179] Results indicated that organ accumulation was proportional to the degree of ammonium (NH_3^+) functionalisation at the tubes' surface: the higher the degree of functionalisation, the less their accumulation in tissues, without any hazardous alteration even at the highest doses ever injected so far *in vivo* (20 mg/kg). Conversely, non-functionalised tubes accumulated almost entirely in the lung and liver in form of large dark clusters. In a parallel study,[180] the authors disclosed that the degree of tube functionalisation, rather than the nature of the functional groups attached to the CNTs, was the critical factor leading to less tissue accumulation and normal tissue physiology within the first 24 hours post-administration (Fig. 9.29 and Table 9.1).

It is important to note that the chemical functionalisation onto CNTs might provide, by itself, useful signals for tracking the presence of functionalised tubes inside the cells: for example, the functionalisation of CNTs (single- and multi-walled) via the 1,3-cycloaddition reaction has led to the appearance of a luminescence signal with characteristic broad 395/485 nm (excitation/emission) peaks that could be detected by confocal laser scanning microscopy (CLSM).[181] In other words, the intrinsic SWCNT-NH$_3^+$ luminescence enables the intracellular visualisation and tracking of the SWCNT-NH$_3^+$ signal by fluorescence spectrophotometry. As expected, the detected green signal increased in proportion to the concentration of SWCNT-NH$_3^+$ allowed for interaction with the cells, following a dose-response trend.

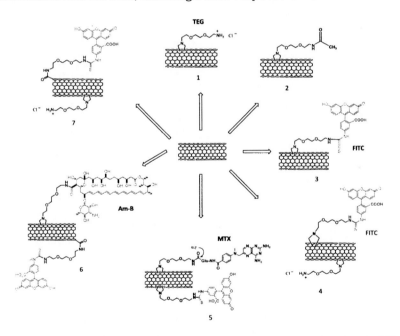

Figure 9.29 Molecular structures of CNTs covalently functionalised with different types of small molecules. (1) Ammonium-functionalised CNTs, (2) acetamido-functionalised CNTs, (3) CNTs functionalised with fluorescein isothiocyanate (FITC), (4) CNTs bifunctionalised with ammonium groups and FITC, (5) CNTs bifunctionalised with methotrexate (MTX) and FITC, (6) shortened CNTs bifunctionalised with amphotericin B (AmB) and FITC, and (7) shortened CNTs bifunctionalised with ammonium groups (via 1,3-dipolar cycloaddition) and FITC (through an amide linkage). Figure redrawn from Kostarelos *et al*.[180] See also Colour Insert.

Table 9.1 Cellular uptake and internalisation of various functionalised CNTs. Functionalised (f)-CNTs interacted with adherent mammalian cell monolayers (A549, HeLa, MOD-K), mammalian cell suspensions (Jurkat), fungal cells (*Cryptococcus neoformans*), yeast (*Saccharomyces cerevisiae*) and bacteria (*Escherichia coli*) between 1 and 4 hours at 37°C

	Type of CNT	Loading characteristics (mmol/g) (amino groups by quantitative Kaiser test)	Cell type	Observations
1	f-SWNT f-MWNT	0.45–0.55 0.85–0.95	A549 (20 µg/mL) Fibroblasts HeLa CHO HEK293	f-CNTs were able to be internalised in all cell types. The nature of the functional group on the CNTs did not influence their internalisation. Up-take of f-CNTs by fungi and yeast cells (which contain polysaccharides unable of active energy-dependent mechanism [i.e., endocytosis]) suggested that the mechanism involved in the CNTs' internalisation is different from endocytosis.
2	f-SWNT f-MWNT	0.45–0.55 0.85–0.95	A549 (20 µg/mL) Fibroblasts HeLa CHO HEK293	
3	f-SWNT f-MWNT	0.45–0.55 0.85–0.95	HeLa (2 µg/mL) Fibroblasts CHO HEK293 Keratinocytes A549	
4	f-MWNT	0.95 (0.65)	MOD-K (5 µg/mL) Jurkat	
5	f-MWNT	0.71 (0.25)	Jurkat (5 µg/mL)	
6	f-MWNT		Jurkat (20 µg/mL)	
7	f-MWNT		*C. neoformans* (10 µg/mL) *E. coli* *S. cerevisiae*	

9.3.2 Neuronal Applications

Neural cells are known to be electroactive, and the electronic properties of nanostructures can be tailored to match the charge transport requirements of electrical cellular interfacing. Nanostructures offer the advantage of translocating across the blood–brain barrier (BBB) and eventually migrating along axons and dendrites and even crossing synapses.[182] Aside from establishing an electronic communication link with neural tissue, they seem to be particularly suitable for developing neuronal electrodes (NEs) to be applied in case of devastating diseases (e.g., Parkinson's and Alzheimer's diseases, sclerosis and paralysis). In this context, CNTs are excellent candidates for interfacing with neural systems for the development of biocompatible, durable and robust neuroprosthetic devices. This is because they present extraordinary strength, toughness, electrical conductivity and high surface area. The influence of CNTs in neuronal living networks is developed in chapter 7. Quite recently (in 2000) CNTs were promoted as substrates for neuronal growth,[183] since the neuronal bodies were able to adhere to CNTs' surface with neurites extending through the bed of CNTs and developing into many branches. The neurons remained alive on CNTs for at least 11 days, stimulating other scientists to investigate the role of CNTs in the central nervous system (CNS). In fact, Prato's group demonstrated the spontaneous post-synaptic currents (PSCs) from a single neuron associated with CNTs, thus indicating functional synapse formation.[184] The sample used consisted of CNTs initially functionalised and subsequently subjected to thermal treatment in order to recover unfunctionalised and purified tubes. They provided an efficient substrate for hippocampal cell growth with dimensions comparable with the controls (CNT-free neurones) and promoted a significant increase in brain network operation. An additional interesting aspect is that, despite being highly electrically conductive, pristine CNTs can be chemically modified with different biomolecules while maintaining their intrinsic properties. To that purpose, Prato's group has explored the functionalisation of multi-walled carbon nanotubes (*f*-MWCNTs) with cell adhesion peptides, to be potentially delivered at the site of nerve injury to promote local tissue repair.[185] In these experiments, MWCNTs were preferred because of their wider external surface. They underwent both oxidation (to cut them into smaller tubes) and 1,3-dipolar cycloaddition of azomethine ylides (Fig. 9.30), after which the free amino group was coupled with the maleimido function and then the selected peptides. The first sequence corresponded to Gly-Arg-Gly-Asp-Ser-Pro (GRGDSP, Pep 1), which is a fibronectin-derived peptide capable of increasing integrin-mediated cell adhesion and spreading via the cell-binding domain RGD residues. Coating surfaces with RGD-based sequences have shown to promote not only cell adhesion but also neurite outgrowth.[186] The second selected sequence corresponded to Ile-Lys-Val-Ala-Val (IKVAV, Pep 2) contained in laminin.

With the aim to design peptide–MWCNT conjugates for interacting with neuronal cells, hippocampal neurons were treated with soluble peptide–MWCNTs 5 and 6, with the control peptides alone, and with MWCNTs 3 without peptides. Results showed that cortical neurons, in the presence of MWCNTs functionalised with peptides, did not alter neuronal spontaneous activity. These data substantiated previous investigations aimed to test whether electric stimulation delivered via SWCNTs could produce neuronal signaling.[187] In that case, experiments were performed on hippocampal cells grown on pure SWCNT substrates, showing that SWCNTs could directly stimulate brain circuit activity. Indeed, in the latest study, *f*-MWCNTs do not appear to alter neuronal morphology, viability and basic functions, so they represent a promising candidate for the exploitation of novel drug-delivery systems or for designing new generations of self-assembling nerve "bridges".

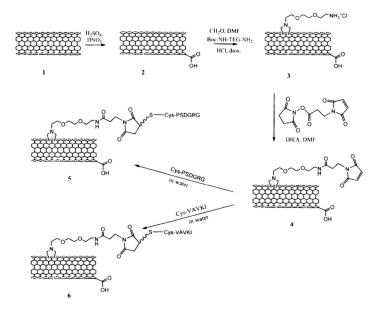

Figure 9.30 Preparation of peptide–MWCNTs.

9.3.3 Photovoltaic Applications

Although representing a small portion of the research performed in Prato's group, the characterisation and manipulation of carbon-based materials in order to change their electronic properties has led to excellent discoveries, originating, for example, from the strategic combination of SWCNTs with electron donors or acceptors.[188] One major limitation to the widespread use of CNTs is represented by the polydispersive nature of the tubes, which exist as bundles and aggregates of different sizes. In order to overcome this problem, the authors have complexed the tubes with a perylene dye, which offers several advantages towards the development of electronics and photovoltaics: first of all, it presents tremendous electron-accepting and electron-conducting features with a five-fused ring π-system. Additionally, it favours the dispersion of SWCNTs by π–π interactions, thus determining individual SWCNTs in solution. Therefore, it guarantees uniform samples and ideal conductivity towards advanced (opto)electronic devices.

π–π interactions were also employed to anchor an electron donor (based on π-extended tetrathiafulvalene [exTTF]) to the surface of SWCNTs.[189] A full spectroscopic characterisation showed that new conduction band electrons, injected from photoexcited exTTF, shifted the transitions normally associated with the van Hove singularities to lower energies.

Alternative photoactive electrodes have been produced by SWCNTs bearing phthalocyanine chromophores[190] or by grafting polymers onto SWCNTs either via free-radical polymerisation of (vinylbenzyl)trimethylam monium chloride (VBTA) into (PVBTA^{n+}) or by simply wrapping them around the tubes in the form of aqueous suspensions of positively charged SWCNT/PVBTA^{n+}.[191] The supramolecular complexes consisted of versatile donor–acceptor nanohybrids obtained by electrostatic/van der Waals interactions between covalent SWCNT-PVBTA^{n+} and/or non-covalent SWCNT/PVBTA^{n+} and porphyrins (H$_2$P^{8-} and/or ZnP^{8-}). Photoexcitation of these nanohybrids afforded radical ion pairs, similarly to those produced by the above-mentioned perylene dye.

More complex nanometre scale structures, such as SWCNT-ZnPc conjugates[192] have been produced through click chemistry of SWCNTs with 4-(2-trimethylsilyl)ethynylaniline and the subsequent attachment of a zinc-phthalocyanine (ZnPc) derivative using the Huisgen 1,3-dipolar cycloaddition ("click chemistry"). The complex showed remarkable properties, as demonstrated by monochromatic internal photoconversion efficiencies of 17.3% when the SWCNT-ZnPc hybrid material was tested as photoactive material in an ITO photoanode.

9.3.4 Functionalisation of CNTs

In Prato's laboratory, scientists are also developing new procedures for the strategic functionalisation of CNTs. As mentioned before, they could be considered the pioneers of the 1,3-dipolar cycloaddition of azomethine ylides applied onto fullerenes first and nanotubes later. Besides that, many other methodologies have been explored and applied on the surface and/or tips of CNTs, among which a recent publication reported a microwave (MW) process to functionalised nanostructures in ionic liquids.[193] The application of MW irradiation and ionic liquids (IL) on the cycloaddition of azomethine ylides to [60]fullerene has provided a conversion of up to 98 % in 2–10 minutes, by using a 1:3 mixture of the IL 1-methyl-3-*n*-octyl imidazolium tetrafluoroborate ([omim]BF$_4$) and *o*-dichlorobenzene, and an applied power as low as 12 W. The mono- versus poly-addition, specific for [60]fullerene, could be tuned as a function of fullerene concentration. A similar procedure could be applied to SWCNTs, yielding group coverages of up to one functional group per 60 carbon atoms of the SWCNT network, enabling the introduction of fluorous-tagged (FT) pyrrolidine moieties onto the tubes' surface in good yields (1/108 functional coverage).

In another article, the authors have described the formation of supramolecular hybrids through the functionalisation with thymine units[194] or adenine,[195] the last representing the first nucleobases covalently attached to the exosurface of SWCNTs. The samples were extensively characterised, indicating the presence of one base unit for all 26–37 carbon atoms. Also, a pattern of silver nanoparticles was found localised over the surface of the CNT network, suggesting CNTs' ability to coordinate with metal ions in view of interesting applications in the development of novel electronic devices and as new supports for different catalytic transformations.

9.4 ASIA

9.4.1 Japan

> **Sumio IIJIMA**
> @ Meijo University, Japan
> Expertise: Encapsulation and reactions inside carbon nanotubes

Very little information is available on the websites about the actual research group of Prof. Iijima, the first scientist who officially revealed the identity of this new allotrope of carbon in 1991. More precisely, the sample that was the origin of this discovery had been produced by Prof. Ando, one of the professors of the Department of Materials Science and Engineering in Japan. Since then, Meijo University has been called "the birthplace of nanotubes".

9.4.1.1 Encapsulation and reactions inside CNTs

An interesting aspect associated with SWCNTs is that the space within the nanotubes has a size comparable to that of organic molecules, and therefore it suggests a favourable interaction between guest species and host SWCNTs. To that purpose, Raman spectroscopy is one of the most suitable and efficient techniques, in which the characteristic radial breathing mode (RBM) phonon of SWCNTs is inversely correlated with the tube diameter and it is sensitive to molecular encapsulation. However, since RBM frequencies of SWCNTs are also deeply influenced by tube chirality, excitation wavelength and mixture of tubes with different diameters, it is necessary to measure the Raman spectra by using several laser excitation lines. In particular, the encapsulation of fullerene C_{60} molecules inside SWCNTs (the complex formed is called nanopeapod) under laser excitations tuned from 1.19 to 1.65 eV was recently investigated.[196]

Figure 9.31 depicts 2D RBM intensity maps of SWCNTs both as control tubes in solution (9.31a) and as nanopeapods (9.31b). Typical $2n+m$ family patterns were observed between 175 and 185 cm^{-1} under the excitation of 1.23–1.27 eV and between 181 and 199 cm^{-1} under excitation wavelengths of 1.31–1.36 eV. On the basis of the photoluminescence excitation (PLE) and Raman results reported in a previous study,[197] the observed signals were attributed to the $2n+m$ = 28, 29, 31, 32 and 34 families. In the case of the C_{60} nanopeapods, the RBM frequencies between 170 and 183 cm^{-1} showed lower shifts of approximately 1–7 cm^{-1}. Moreover, two new peaks were observed at 193 and 198 cm^{-1} under the excitation energies of ~1.28 and ~1.24 eV, respectively. It was important to note that the RBM phonon frequency on encapsulation was observed only in the case of SWCNTs with diameters greater than 1.25 nm, which corresponded to the smallest limit for C_{60} encapsulation. In other words, only tubes with a diameter bigger than 1.25 nm were able to encapsulate the fullerene, with a diameter dependence trend that was totally comparable to the RBM phonon frequencies: such signals were upshifted in small tubes after C_{60} encapsulation because of the steric hindrance provided by the C_{60} molecules. In contrast, larger-diameter tubes showed downshifts in RBM phonon frequency, which could be attributed to the hybridisation of the electronic states of the SWCNTs and those of C_{60} on the fullerene encapsulation.

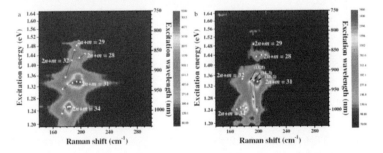

Figure 9.31 2D Raman intensity maps of RBM phonon regions of (a) SWCNT control samples and (b) C_{60} nanopeapods in SDBS-D$_2$O solution. Reproduced with permission from Joung *et al.*[196] Copyright (2009) by the American Physical Society. See also Colour Insert.

Besides fullerenes, other molecules have been encapsulated inside CNTs: an interesting example is provided by the self-assembly of SiO_2-based nanomaterial inside single- and double-walled CNTs.[198] Once again, tubes' diameter played a crucial role, by influencing the final structure of the original cubic octameric $H_8Si_8O_{12}$ molecules formed inside the CNTs: in case of CNTs having inner diameters of 1.2–1.4 nm, a new ordered self-assembled structure composed of $H_8Si_{4n}O_{8n-4}$ molecules was formed, while for tubes larger than 1.7 nm, a disordered structure originated inside the tubes. Such result was confirmed by high-resolution transmission electron microscopy (HR-TEM), FT-IR spectroscopy and Raman spectroscopy, suggesting strong interactions occurring between CNTs and $H_8Si_{4n}O_{8n-4}$ molecules.

In another study, the research group investigated the migration of individual cesium ions (Cs^+) both inside and outside C_{60} nanopeapods.[199] Cs^+ ions inside the nanopeapods were detected as dark dots inside tubes of about 1.5 nm, but not in the inner wall of SWCNTs of 1.38 ± 0.03 nm. It was suggested that these tubes were too small to accommodate the migration of Cs^+ ions, while smaller K^+ ions were already demonstrated to be internalised inside tubes of 1.4 nm.[200] TEM images showed an ordered Cs_2C_{60} structure with planar Cs^+ configuration in which each C_{60} cage was adjacent to four Cs^+ ions (Fig. 9.32).

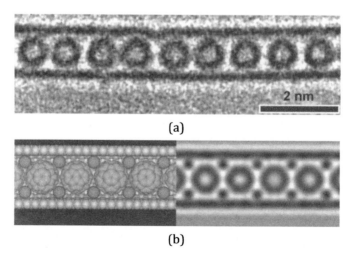

(a)

(b)

Figure 9.32 (a) TEM images of C_{60} nanopeapods after doping with cesium INSIDE. (b) Schematic illustration of the ordered Cs_2C_{60} structure inside a SWCNT (*right*) and its simulated TEM image (*left*). Reproduced from Sato *et al.*[199] with permission. See also Colour Insert.

Sequential TEM images of these samples indicated that Cs^+ ions were influenced by the deformation of the tubes, which fluctuated between 1.44 and 1.56 ± 0.03 nm due to elastic deformation during the TEM. On the other hand, Cs^+ ions migrate randomly on the outer nanotube wall, without retaining their position for more than 2 seconds. Conversely, they were trapped at defective parts of the nanotubes (Fig. 9.33).

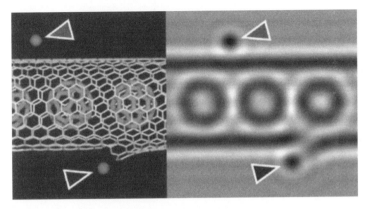

Figure 9.33 Images of C_{60} nanopeapods after doping with cesium OUTSIDE. Cs^+ ions trapped at the defects in the nanotubes are indicated by blue triangles, and that of the defect-free part of the nanotube wall is indicated by a red triangle. Reproduced with permission from Sato *et al.*[199] See also Colour Insert.

Ask for copyright permission to Sato, Yuta; Suenaga, Kazu; Bandow, Shunji; Iijima, Sumio. Site-dependent migration behavior of individual cesium ions inside and outside C_{60} fullerene nanopeapods. Small, 2008, 4(8), 1080-1083.

It has been repeatedly demonstrated that C_{60} fullerenes can be encapsulated in SWCNTs to form a 1D hybrid nanostructure, the so-called carbon nanopeapods.[201] However, the coalescence of C_{60} molecules assisted by intratubular doped iodine atom(s) inside SWCNTs has been much less investigated. The research group has adopted HR-TEM and Raman spectroscopy to demonstrate that the doped iodine atoms (even one atom!) can induce the structure transformation of C_{60} fullerenes inside SWCNTs.[202] The C_{60} peapods were immersed in molten iodine in an evacuated glass tube at 150°C over 24 hours for doping, after which some dark spots appeared, presumably attributable to iodine (Fig. 9.34). It seemed that SWCNTs with diameter larger than 1.6 nm could accommodate the iodine molecules, although in an inhomogeneous way. In order to verify that C_{60}'s transformation into tubular shape was induced by iodine, both intact C_{60} peapods and their iodine-doped counterparts were incubated for 24 hours at 550°C, which represents a value below the reported minimum coalescence temperature (800°C) of the intact C_{60}@SWCNTs. As expected, in the case of the iodine-doped samples, a prominent structure transformation was revealed by the Raman spectra and HR-TEM observation (Fig. 9.35).

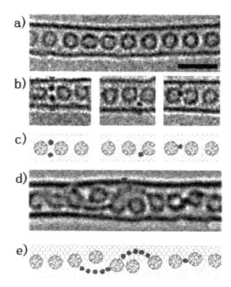

Figure 9.34. (a) An HR-TEM image of the intact C_{60} peapods. (b) HR-TEM images of the iodine-doped C_{60} peapods with low doping ratio; the black spots (indicated by arrows) are ascribed to the doped iodine atoms. (c) The best fit schematic model (purple, iodine; gray, carbon). (d) An HR-TEM image of the iodine-doped C_{60} peapods with higher doping ratio; the bent chains (indicated by arrows) are ascribed to the doped iodine species. (e) The schematic model. Scale bar = 2 nm. Reproduced from Guan *et al.*[202] with permission.

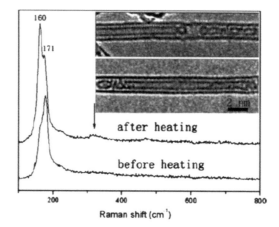

Figure 9.35. Raman spectra of the iodine-doped C_{60}@SWCNTs before and after heat treatment. The arrow indicates the newly appearing RMB mode as the result of heat treatment. Inserted are typical HR-TEM images of the iodine-doped C_{60}@SWCNTs after heating. Reproduced from Guan *et al.*[202] with permission.

Imaging via TEM could be extended to monitor the time-dependent orientational changes of encapsulated molecules, like those reported for D_{5d}-C_{80} and I_h-Er$_3$N@C$_{80}$ fullerenes and demonstrated by aberration-corrected TEM at real atomic-level resolution.[203] In that particular case, careful analysis under TEM enabled the unambiguous identification of the ellipsoidal D_{5d}-C_{80} fullerene inside CNTs.

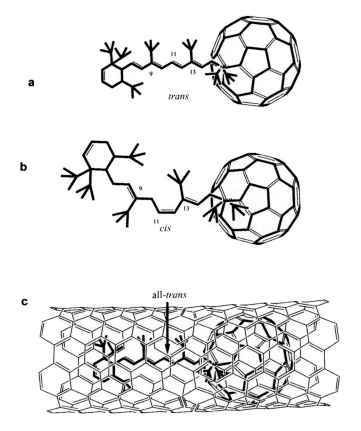

Figure 9.36 (a, b) Two isomeric forms of retinal attached onto C$_{60}$ molecules: *trans* (a) and *cis* (b). (c) Best-fit model of a retinal-C$_{60}$ molecule inside an SWCNT, suggesting that the image in C is of the all-*trans* isomer. Figure redrawn from Liu *et al.*[204] See also Colour Insert.

Another interesting example has been provided by the molecule retinal, which is associated with conformational changes once it absorbs light.[204] This study was performed in collaboration with Prof. Kataura (reported in the following section), thus demonstrating the synergistic effect derived from outstanding scientists. In their investigation, the retinal chromophore was attached to C_{60} molecules (Fig. 9.36a,b) and subsequently incorporated inside SWCNTs (Fig. 9.36c–e); this second step was monitored by HR-TEM subjected to sub-second time frame recording, which revealed the dynamic behaviour of the molecule during its cis–trans isomerisation. The use of SWCNTs provided the double advantage of a low background noise in the images and a restricted molecular motion inside their inner space. At HR-TEM, the C_{60} molecules appeared as a circular shape, while the retinyl group appeared in either "strong" or "weak" contrast, which suggested its existence in one of two states when inserted inside the SWCNTs.

A careful evaluation of the images correlated well with an all-*trans* isomer. The *cis/trans* structural transformation, which normally requires photoactivation, was shown to occur under these conditions; this unprecedent result has important implications, since it provides a preliminary insight into how retinal undergoes different conformational changes when it is placed in a confined space (i.e., inside SWCNTs).

Analogously, the information deriving from HR-TEM and EELS has proved advantageous in the characterisation of atomic wires produced by encapsulation of lanthanum atoms attached to fullerene C_{80} inside CNTs.[205] These techniques allowed for the identification of the internalised complex $(La_2@C_{80})$ in the form of dimers and the preservation of their original valence state.

An additional interesting study with iodine demonstrated that single, double and triple helixes of iodine chains inside CNTs are influenced by SWCNTs' diameter remarkably, as shown by their crystallisation in the presence of tubes larger than 1.45 nm.[206] Therefore, a critical size cut-off of the hollow space inside SWCNTs was identified, on the basis of which polymorphic and crystal structures of iodine can be generated in a controlled manner.

9.4.1.2 Synthesis of CNTs

Part of the ongoing projects at Iijima's laboratory deals with the development of several synthetic procedures for the fabrication and characterisation of CNTs.

A commonly used technique for the preparation of CNTs consists of CVD from hydrocarbons. In this regard, the group has selected nine aromatic hydrocarbons as carbon sources.[207] These hydrocarbons mainly differed in terms of hybridisation of the atoms attached to benzene rings. In particular, hydrocarbons with sp^3 moieties included toluene, *p*-xylene, ethylbenzene and *n*-propylbenzene, while species with sp^2 moieties were styrene, allylbenzene and 1,4-divinylbenzene. In order to compare their SWCNT production efficiencies, their thermal decomposition pathway was considered, which indicated that C_2H_3/C_2H_4 were the major products formed, and thus the major precursors for SWCNT growth. The authors also showed that the diameter of SWCNTs could be tuned by changing the C_2H_4 concentration.

Interestingly, the researchers are responsible for the first identification of the smallest CNTs and their characterisation through aberration-corrected HR-TEM.[208] This tube species showed a chiral index of (3, 3), (4, 3) or (5, 1). The tubes were grown inside SWCNTs with the diameter of 1.0–1.2 nm (Fig. 9.37), although the (3, 3) nanotube was rather unstable under the electron beam and, therefore, could not survive alone without the protection of outer nanotube. The closing cap of such tube was demonstrated to correspond to a half-dome of C_{20} fullerene, which consists of six pentagons only.

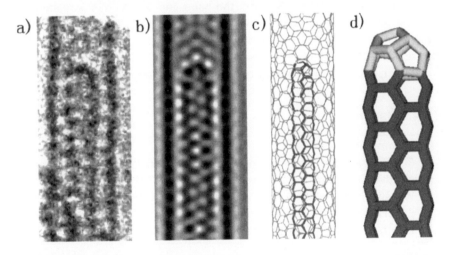

Figure 9.37 (a) HR-TEM image of the cap region of (3, 3)@(10, 6) DWCNT, (b) simulated images, and (c) schematic model. (d) The inner cap structure can be modelled as half a C_{20} fullerene consists of six adjacent pentagons (shown in yellow). Reproduced from Guan *et al.*[208] with permission.

As just mentioned, the authors have reported efficient methods for the growth of SWCNTs, but they have also disclosed the mechanism behind this process through *ex situ* microscopic and spectroscopic analyses. More precisely, the research group has elucidated the effect of water on the catalysts employed in CVD.[209] They proved that catalyst deactivation is a consequence of carbon coating. Water molecules are said to assist this process by removing such coating and reviving catalysts' activity during CVD.

Another in-depth investigation has been recently performed to disclose the mechanism associated with the growth of CNTs in general and of their extremities in particular.[210] Surprisingly, it was found that the tips evolve inhomogeneously, with a few precise sites developing faster than others. As a result, the caps remain closed specifically when catalysts are not employed, while they maintain a round shape during the shrinking process.

The authors also investigated the filling capability of different CNTs with diameters >1.4 nm under photoluminescence spectroscopy.[211] Previously, Okada and co-workers predicted that C_{60} molecules can enter the interior space of SWCNTs without friction at diameter ≥ 1.3 nm[212-214] without changing upon fullerene insertion. In this diameter region, efficient coupling (hybridisation) between the π states of C_{60} and the nearly free electron (NFE) states of SWCNTs occurs. The results of this study revealed the characteristic band gap shifts of SWCNTs upon C_{60} encapsulation. For nanopeapods with smaller diameters ($d_t \leq 1.32$ nm), the band gap of SWCNTs was modified by the mechanical strain because of encapsulated C_{60}. For larger-diameter regimes ($d_t \geq 1.32$ nm), hybridisation between the π state of C_{60} and the NFE state of SWCNTs induced the same effects as the small reduction of the tube diameter, resulting in the family type-dependent band gap shifts.

9.4.2 Japan

Hiromichi KATAURA

@ *Self-assembled Nano-electronics group* @ National Research Institute
@ Advanced Industrial Science and Technology (AIST)
Website: http://staff.aist.go.jp/h-kataura/kataurae.html
Expertise: Encapsulation of and reactions inside carbon nanotubes and their electronic properties

---

Prof. Kataura is the leader of the Self-assembled Nano-electronics group at the University of Tsukuba. This is one of the 15 groups that constitute the National Research Institute (NRI). The NRI emphasises the foresight role played by theory and computational science in nanotechnology, develops novel nano-processing and characterisation technologies and promotes substantial R&D aimed at the development of novel nanomaterials, nanodevices and nanobiotechnology. NRI is an essential component of Advanced Industrial Science and Technology (AIST), which is a non-government institution including over 50 autonomous research units (e.g., NRI) in various innovative research fields; the units are located at nine research bases and several sites (smaller than research bases) of AIST all over Japan. About 2,500 research scientists and over 3,000 visiting scientists, post-doctoral fellows and students work at AIST.

9.4.2.1 Investigations on molecules@CNT conjugates

Similarly to Iijima, Kataura's group is mainly focused on the in-depth investigation of molecules encapsulated inside CNTs; the group's studies explore the electronic properties of the newly formed nanocomposites and their characterisation down to the atomic level. With regard to this topic, SWCNT functionalisation via encapsulation is beneficial for tuning the physical properties of SWCNTs without losing their unique structure, as demonstrated by several techniques. Among the available tools, EELS is a very useful technique to examine the chemical composition and bonding states of matter. Its combination with TEM enables even the identification of the atomic components of single molecules. However, in order to realise single-atom spectroscopy, the application of high-energy incident electrons is essential, with consequent massive structural damage of the specimens. Decreasing the accelerating voltage below the typical displacement energy threshold for carbon atoms (~80–120 kV) has been suggested as a possible solution,[215] but with drastic loss of resolution. It was proposed the doping of SWCNTs with various metallofullerene molecules so that single atoms could be isolated in each fullerene cage.[216] As a proof of that, of the authors were able to visualize erbium (Er) atoms incorporated in fullerenes encapsulated inside SWCNTs (Er@C_{82}@SWCNTs), without displaying massive irradiation damage.

However, despite the good resolution for the lanthanoids, it was much more difficult to detect other molecules (e.g., potassium K or calcium Ca) which are more common in biological systems. The reason for that could be attributable to the fact that the atom might escape from the electron probe (especially if such probe is too narrow); therefore, the time of exposure under the electron beam should be increased to achieve better resolution. Alternatively, a bigger probe could result beneficial, since it could cover the overall space occupied by the fullerene and it does not allow the encapsulated atoms to escape during the time scale of the spectrum acquisition. In every case, EELS was demonstrated to be a useful technique to discriminate between single atoms of different elements. Both lanthanum (La) (atomic number $Z = 57$) and Er ($Z = 68$) atoms were co-doped within fullerenes inside SWCNTs (La and Er@C_{82}@SWCNTs). Although it was not possible to visualise individual atoms, the EELS chemical maps differentiated them on the basis of different absorption energies (99 and 168 eV for La and Er, respectively), thus demonstrating an efficient "atom-by-atom" labelling. The situation became much more complicated with La and Cerium (Ce) ($Z = 58$), in which atoms have a smaller atomic number difference ($\Delta Z = 1$) and close absorption energies (99 and 109 eV, respectively). However, chemical identification was still possible through ADF images, thus indicating a successful discrimination with this technique.

Besides experimental microscopic detection, a theoretical simulation could provide useful insights on the structural and electronic properties of different material. For example, molecular dynamics (MD) calculations have been performed on ice polygons inside CNTs.[217] The so-formed "ice-nanotubes" (ice-NTs) consisted of polygonal (from pentagonal to heptagonal) water rings placed along the SWCNT long axis. In the presence of an external electric field and in a temperature range between 100 and 350 K, these ice-NTs represented the smallest ferroelectronic with spontaneous polarisation of around 1 $\mu C/cm^2$, thus opening the way to the encapsulation of other dielectric materials in view of compact electronic devices.

In another study, the authors have incorporated water molecules inside SWCNTs in spite of their hydrophobic nature, showing that encapsulated water is interestingly transformed from a liquid-like state to a tube-like structure (in agreement with the already mentioned ice-NTs). Having a shape similar to hollow cylinders with diameters comparable to those of gas molecules, they have stimulated a further evaluation of the competing behaviour between water and gas molecules inside CNTs.[218] More precisely, SWCNTs bearing an average diameter of 13.5 and 14.4 Å were purified and subsequently heated in air for 20–40 minutes between 350 and 450°C, in order to open their extremities. After mixing methane (CH_4) and water molecules within the same SWCNT, the results strongly indicated a so-called filling–ejecting type transition of water. In other words, below a critical temperature (T_c), the water was ejected from the inside of the SWCNTs by the entering gas (e.g., CH_4) molecules. The ejected water could be located around the open ends of the SWCNTs, where it could reversibly re-enter the SWCNTs. Such process should be possible when the gas-molecule–SWCNT interactions have comparable strength to that of the water–SWCNT interactions, so that water and gases are competing with each other to fill the SWCNTs.

In contrast, for a weaker gas (Neon, Ne) such replacement was not observed. The observed on–off-like behaviour makes water–SWCNTs act as "valves" for gas molecules, suggesting potential applications as gas selective molecular nanovalves.

Throughout their investigations on molecules encapsulated inside CNTs, the researchers have also carefully evaluated the behaviour of fullerene C_{60} inside carbon nanopeapods (NPPs).[219] Surprisingly, by using inelastic neutron scattering to probe the dynamics of nanopeapods (NPPs), they observed high orientational mobility of the C_{60} molecules when they are confined inside a nanotube. In other words, these molecules undergo "free" rotations down to very low temperatures, implying that the transition is continuously different in nature. On the other hand, the nanotube field restricts the centre of mass position of the fullerenes on a 1D chain along the nanotube axis, which means that the molecular orientation of these molecules is influenced by neighboring C_{60}–C_{60} interactions, so that these molecules can statistically adopt a large number of orientations. At low temperature, however, the potential barrier that a C_{60} has to overcome to change from one orientation to another becomes non-negligible.

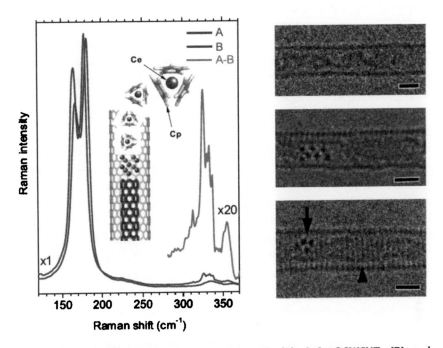

Figure 9.38 *Left*: Raman spectra for Ce@DWCNTs (A), CeCp$_3$@SWCNTs (B), and their difference (A–B) at the radial breathing mode region. *Right*: TEM micrographs of CeCp$_3$@SWCNT (*top*) and Ce@DWCNT (*middle* and *bottom*). Scale bar = 1 nm. Reprinted with permission from Shiozawa et al.[220] Copyright (2009) by the American Physical Society. See also Colour Insert.

Because of their excellent structural and chemical integrity, CNTs also provide an ideal 1D environment to explore chemistry at the nanoscale. To that purpose, the researchers have doped SWCNTs with tris(cyclopentadie nyl)cerium (CeCp$_3$) through van Hove singularity (VHS), which has shown the ability to modulate the conductivity of SWCNTs.[220] In more details, the left-hand panel of Fig. 9.38 shows the Raman spectra of a CeCp$_3$@SWCNT sample before and after annealing at 1000°C in vacuum. Both spectra contain the expected radial breathing modes of pristine tubes at frequencies ranging from 130 to 210 cm^{-1}. After annealing, sharp lines appeared in the range of 300–370 cm^{-1}, owing to the formation of inner tubes and conversion into DWCNTs. HR-TEM confirmed such results by showing a typical CeCp$_3$@ SWCNT on the top-right panel, while the middle and bottom panels showed the sample transformed into Ce@DWCNT.

The production of DWCNTs can also be induced by peapods[221] or by encapsulation or ferrocene molecules as precursors.[222] This catalytic process offers new opportunities for controlled materials design. Moreover, from the electronic point of view, this structural change from SWCNTs to DWCNTs is accompanied by the conversion of semiconducting tubes into metallic ones, as a consequence of the doping-induced transition into a metallic phase. This is confirmed by a massive enhancement in conductivity of doped SWCNTs, thus providing new insights on 1D electronic systems.

By changing the nanotube diameter it is possible to modulate the tubes' properties. Along that direction, the research group has attempted to clarify whether dye molecules encapsulated in SWCNTs can exhibit a photosensitising function.[223] For photosensitisation to occur, it is necessary that the absorption band of the encapsulated dye is located in the spectral range where the semiconducting SWCNTs have a low extinction coefficient. In the experiments, squarylium (SQ) dye was chosen as favourite dye because of its small dimensions (thus suitable for encapsulation inside SWCNTs of 1.4 nm) and its absorption bands not overlapping with SWCNTs. Photoluminescence measurements clearly revealed that the energy of the excited state of the encapsulated SQ dye (SQ@SWCNTs) was transferred to the SWCNTs, thus exhibiting the photosensitising function inside the SWCNTs.

X-ray diffraction, instead, indicated that the dye inside the tubes was placed off-centre in comparison with chloroform molecules of pristine tubes. The difference could be attributed to the π-conjugated structure, which could determine π–π interactions with the tube wall,[224] thus shifting SQ molecules off the SWCNT centre.

A small part of Kataura's research is devoted to the separation of CNTs with different techniques. One example is the recent gel-based separation of metallic and semiconducting SWCNTs on the basis of the selective adsorption of semiconducting SWCNTs on agarose gel.[225] A gel containing a mixture of SWCNTs and sodium dodecyl sulphate was frozen, thawed and squeezed, affording a solution containing 70% of pure metallic SWCNTs and a gel with 95% of pure semiconducting SWCNTs. Interestingly, field-effect transistors, produced on the so-separated semiconducting SWCNTs, demonstrated that their electronic properties were preserved.

Another separating procedure consisted in the density gradient ultracentrifugation, in which the obtained fractions were characterised by aberration-corrected TEM;[226] results showed that armchair (n, n) and chiral $(n, n - 3)$ SWCNTs with large chiral angles (>20°) were the dominant metallic nanotubes, while no particular preference was detected for semimetallic tubes. This work was performed in collaboration with Prof. Iijima, thus substantiating his undiscussed leadership in this field of research.

References

1. Ajayan, P. M. (1999) Nanotubes from carbon, *Chem. Rev.*, **99**, 1787–1800.

2. Jacoby, M. (2002) Nanoscale electronics, *Chem. Eng. News*, **80**, 38–43.

3. Mitchell, D. T., Lee, S. B., Trofin, L., Li, N., Nevanen, T. K., Soderlund, H., and Martin, C. R. (2002) Smart nanotubes for bioseparations and biocatalysis, *J. Am. Chem. Soc.*, **124**, 11864–11865.

4. Lee, S. B., Mitchell, D. T., Trofin, L., Nevanen, T. K., Soderlund, H., and Martin, C. R. (2002) Antibody-based bio/nanotube membranes for enantiomeric drug separation, *Science*, **296**, 2198–2200.

5. Karousis, N., Tagmatarchis, N., Tasis, D. (2010) Current Progress on the Chemical Modification of Carbon Nanotubes. *Chem. Rev.* **110**, 5366–5397.

6. Holzinger, M., Vostrowsky, O., Hirsch, A., Hennrich, F., Kappes, M., Weiss, R., and Jellen, F. (2001) Sidewall functionalization of carbon nanotubes, *Angew. Chem. Int. Ed.*, **40**, 4002–4005.

7. Holzinger, M., Abraham, J., Whelan, P., Graupner, R., Ley, L., Hennrich, F., Kappes, M., and Hirsch, A. (2003) Functionalization of single-walled carbon nanotubes with (R-)oxycarbonyl nitrenes, *J. Am. Chem. Soc.*, **125**, 8566–8580.

8. Holzinger, M., Steinmetz, J., Samaille, D., Glerup, M., Paillet, M., Bernier, P., Ley, L., and Graupner, R. (2004) [2+1] cycloaddition for cross-linking SWNTs, *Carbon*, **42**, 941–947.

9. Dyke, C. A., and Tour, J. M. (2003) Solvent-free functionalization of carbon nanotubes, *J. Am. Chem. Soc.*, **125**, 1156–1157.

10. Avouris, P. (2007) Electronics with carbon nanotubes, *Phys. World*, **20**, 40–45.

11. Avouris, P., Chen, Z., and Perebeinos, V. (2007) Carbon-based electronics, *Nat. Nanotechnol.*, **2**(10), 605–615.

12. Landauer, R. (1988) Spatial variation of currents and fields due to localized scatterers in metallic conduction, *IBM J. Res. Dev.*, **32**, 306.

13. Buttiker, M. (1988) Symmetry of electrical conduction, *IBM J. Res. Dev.*, **32**, 317.

14. Ilani, S., Donev, L. A. K., Kindermann, M., and McEuen, P. L. (2006) Measurement of the quantum capacitance of interacting electrons in carbon nanotubes, *Nat. Phys.*, **2**, 687–691.

15. Burke, P. (2003) An RF circuit model for carbon nanotubes, *IEEE Trans. Nanotechnol.*, **2**, 55–58.

16. Tans, S. J., Verscheuren, A. R. M., and Dekker, C. (1998) Room-temperature transistor based on a single carbon nanotube, *Nature*, **393**, 49–52.

17. Martel, R., Schmidt, T., Shea, H. R., Hertel, T., and Avouris, P. (1998) Single- and multi-wall carbon nanotube field-effect transistors, *Appl. Phys. Lett.*, **73**, 2447–2449.

18. Sze, S. M. (1981) *Physics of Semiconductor Devices*, Wiley, New York.

19. Appenzeller, J., Knoch, J., Radosavljević, M., and Avouris, P. (2004) Multimode transport in Schottky-barrier carbon-nanotube field-effect transistors, *Phys. Rev. Lett.*, **92**, 226802.

20. Lee, K., Wu, Z., Chen, Z., Ren, F., Pearton, S. J., and Rinzler, A. G. (2004) Single wall carbon nanotubes for p-type ohmic contacts to gan light-emitting diodes, *Nano Lett.*, **4**, 911–914.

21. Li, J., Hu, L., Wang, L., Zhou, Y, Grüner, G., and Marks, T. J. (2006) Organic light-emitting diodes having carbon nanotube anodes, *Nano Lett.*, **6**, 2472–2477.

22. Chen, J., Perebeinos, V., Freitag, M., Tsang, J., Fu, Q, Liu, J., and Avouris, P. (2005) Bright infrared emission from electrically induced excitons in carbon nanotubes, *Science*, **310**, 1171–1174.

23. Freitag, M., Chen, J., Tersoff, J., Tsang, J. C., Fu, Q., Liu, J., and Avouris, P. (2004) Mobile ambipolar domain in carbon-nanotube infrared emitters, *Phys. Rev. Lett.*, **93**, 076803.

24. Marty, L., Adam, E., Albert, L., Doyon, R., Ménard, D., and Martel, R. (2006) Exciton formation and annihilation during 1D impact excitation of carbon nanotubes, *Phys. Rev. Lett.* **96**, 136803.

25. Freitag, M., Tsang, J. C., Kirtley, J., Carlsen, A., Chen, J., Troeman, A., Hilgenkamp, H., and Avouris, P. (2006) Electrically excited, localized infrared emission from single carbon nanotubes, *Nano Lett.*, **6**, 1425–1433.

26. Pop, E., Mann, D., Cao, J., Wang, Q., Goodson, K., and Dai, H. (2005) Negative differential conductance and hot phonons in suspended nanotube molecular wires, *Phys. Rev. Lett.*, **95**, 155505.

27. Lazzeri, M., Pisanec, S., Mauri, F., Ferrari, A. C., and Robertson, J. (2005) Electron transport and hot phonons in carbon nanotubes, *Phys. Rev. Lett.*, **95**, 236802.

28. Steiner, M., Freitag, M., Perebeinos, V., Tsang, J. C, Small, J. P., Kinoshita, M., Yuan, D., Liu, J., and Avouris, P. (2009) Phonon populations and electrical power dissipation in carbon nanotube transistors, *Nat. Nanotechnol.*, **4**(5), 320–324.

29. Bonini, N., Lazzeri, M., Marzari, N., and Mauri, F. (2007) Phonon anharmonicities in graphite and graphene, *Phys. Rev. Lett.*, **99**, 176802.

30. Avouris, P., Freitag, M., and Perebeinos, V. (2008) Carbon-nanotube photonics and optoelectronics, *Nat. Photonics*, **2**(6), 341–350.

31. Hertel, T., Perebeinos, V., Crochet, J., Arnold, K., Kappes, M., and Avouris, P. (2008) Intersubband decay of 1-d exciton resonances in carbon nanotubes, *Nano Lett.*, **8**, 87–91.

32. Perebeinos, V., and Avouris, P. (2008) Phonon and electronic nonradiative decay mechanisms of excitons in carbon nanotubes, *Phys. Rev. Lett.*, **101**(5), 057401/1-057401/4.

33. Ostojic, G. N., Zaric, S., Kono, J. , Moore, V. C. , Hauge, R. H. and Smalley, R. E. (2005). Stability of High-Density One-Dimensional Excitons in Carbon Nanotubes under High Laser Excitation, *Phys. Rev.Lett.*, **94**, 097401-097404.

34. Perebeinos, V., and Avouris, P. (2007) Exciton ionization, Franz-Keldysh, and Stark effects in carbon nanotubes, *Nano Lett.*, **7**(3), 609-613.

35. Freitag, M., Steiner, M., Naumov, A., Small, J. P., Bol, A. A., Perebeinos, V., and Avouris, P. (2009) Carbon nanotube photo- and electroluminescence in longitudinal electric fields, *ACS Nano*, **3**(11), 3744-3748.

36. Xia, F., Steiner, M., Lin, Y.-M., and Avouris, P. (2008) A microcavity-controlled, current-driven, on-chip nanotube emitter at infrared wavelengths, *Nat. Nanotechnol.*, **3**(10), 609-613.

37. Brorson, S. D., Yokoyama, H., and Ippen, E. (1990) Spontaneous emission rate alteration in optical waveguide structures, *IEEE J. Quantum Electron.*, **26**, 1492-1499.

38. Yamamoto, Y., Machida, S., and Bjork, G. (1992) Micro-cavity semiconductor lasers with controlled spontaneous emission, *Opt. Quantum Electron.* **24**, S215-S243.

39. Björk, G. (1994) On the spontaneous lifetime change in an ideal planar microcavity-transition from a mode continuum to quantized modes, *IEEE J. Quantum Electron.*, **30**, 2314-2318.

40. Abram, I., Robert, I., and Kuszelewicz, R. (1998) Spontaneous emission control in semiconductor microcavities with metallic or Bragg mirrors, *IEEE J. Quantum Electron.*, **34**, 71-76.

41. Engel, M., Small, J. P., Steiner, M., Freitag, M., Green, A. A., Hersam, M. C., and Avouris, P. (2008) Thin film nanotube transistors based on self-assembled, aligned, semiconducting carbon nanotube arrays, *ACS Nano*, **2**(12), 2445-2452.

42. Tulevski, G. S., Hannon, J., Afzali, A., Chen, Z., Avouris, P., and Kagan, C. R. (2007) Chemically assisted directed assembly of carbon nanotubes for the fabrication of large-scale device arrays, *J. Am. Chem. Soc.*, **129**(39), 11964-11968.

43. Liu, Z., Winters, M., Holodniy, M., and Dai, H.J. (2007) siRNA delivery into human T cells and primary cells with carbon-nanotube transporters, *Angew. Chem. Int. Ed. Engl.*, **46**, 2023-2027.

44. Liu, Z., Davis, C., Cai, W., He, L., Chen, X., and Dai, H. (2008) Circulation and long-term fate of functionalized, biocompatible single-walled carbon nanotubes in mice probed by Raman spectroscopy, *Proc. Natl. Acad. Sci. USA*, 105, 1410-1415.

45. Liu, Z., Cai, W., He, L., Nakayama, N., Chen, K., Sun, X., Chen, X., and Dai, H. (2007) In vivo biodistribution and highly efficient tumour targeting of carbon nanotubes in mice, *Nat. Nanotechnol.*, **2**, 47-52.

46. Bianco, A., Kostarelos, K., Partidos, C.D., and Prato, M. (2005) Biomedical applications of functionalised carbon nanotubes, *Chem. Commun.*, 571–577.

47. Liu, Y., Wu, D. C., Zhang, W. D., Jiang, X., He, C. B., Chung, T. S., Goh, S. H., and Leong, K. W. (2005) Polyethylenimine-grafted multiwalled carbon nanotubes for secure noncovalent immobilization and efficient delivery of DNA, *Angew. Chem. Int. Ed. Engl.*, **44**, 4782–4785.

48. Liu, Z., Tabakman, S., Welsher, K., and Dai, H. (2009) Carbon nanotubes in biology and medicine: in vitro and in vivo detection, imaging and drug delivery, *Nano Res.*, **2**, 85–120.

49. Liu, Z., Tabakman, S. M., Chen, Z., and Dai, H. (2009) Preparation of carbon nanotube bioconjugates for biomedical applications, *Nat. Protoc.*, **4**(9), 1372–1381.

50. Liu, Z., Sun, X., Nakayama-Ratchford, N., and Dai, H. (2007) Supramolecular chemistry on water-soluble carbon nanotubes for drug loading and delivery, *ACS Nano*, **1**(1), 50–56.

51. Morelli, D., Menard, S., Colnaghi, M. I., and Balsari, A. (1996) Oral administration of anti-doxorubicin monoclonal antibody prevents chemotherapy-induced gastrointestinal toxicity in mice. *Cancer Res.*, **56**, 2082–2085.

52. Von Hoff, D. D., Layard, M.W., Basa, P., Davis, H. L., Von Hoff, A. L., Jr., Rozencweig, M., and Muggia, F. M. (1979) Risk factors for doxorubicin-induced congestive heart failure, *Ann. Intern. Med.*, **91**, 710–717.

53. Liu, Z., Fan, A. C., Rakhra, K., Sherlock, S., Goodwin, A. P., Chen, X., Yang, Q., Felsher, D. W., and Dai, H. (2009) Supramolecular stacking of doxorubicin on carbon nanotubes for in vivo cancer therapy, *Angew. Chem., Int. Ed. Engl.*, **48**(41), 7668–7672.

54. Nakayama-Ratchford, N., Bangsaruntip, S., Sun, X., Welsher, K., and Dai, H. (2007) Noncovalent functionalization of carbon nanotubes by fluorescein-polyethylene glycol: supramolecular conjugates with pH-dependent absorbance and fluorescence, *J. Am. Chem. Soc.*, **129**(9), 2448–2449.

55. Prencipe, G., Tabakman, S. M., Welsher, K., Liu, Z., Goodwin, A. P., Zhang, L., Henry, J., and Dai, H. (2009) PEG branched polymer for functionalization of nanomaterials with ultralong blood circulation, *J. Am. Chem. Soc.*, **131**(13), 4783–4787.

56. Liu, Z., Davis, C., Cai, W. B., He, L., Chen, X. Y., and Dai, H. J. (2008) Circulation and long-term fate of functionalized, biocompatible single-walled carbon nanotubes in mice probed by Raman spectroscopy, *Proc. Natl. Acad. Sci. USA*, **105**(5), 1410–1415.

57. Welsher, K., Liu, Z., Sherlock, S. P., Robinson, J. T., Chen, Z., Daranciang, D., Dai, H. (2009) A route to brightly fluorescent carbon nanotubes for near-infrared imaging in mice, *Nat. Nanotechnol.*, **4**(11), 773–780.

58. Goodwin, A. P., Tabakman, S. M., Welsher, K., Sherlock, S. P., Prencipe, G., and Dai, H. (2009) Phospholipid-dextran with a single coupling point: a useful amphiphile for functionalization of nanomaterials, *J. Am. Chem. Soc.*, **131**(1), 289–296.

59. Yalpani, M., and Brooks, D. E. (1985) Selective chemical modifications of dextran, *J. Polym. Sci., Part A: Polym. Chem.* **23**, 1395–1405.

60. Liu, Z., Chen, K., Davis, C., Sherlock, S., Cao, Q., Chen, X., and Dai, H. (2008) Drug delivery with carbon nanotubes for in vivo cancer treatment, *Cancer Res.*, **68**(16), 6652–6660.

61. Le Garrec, D., Gori, S., Luo, L., Lessard, D., Smith, D. C., Yessine, M. A., Ranger, M., and Leroux, J. C. (2004) Poly(N-vinylpyrrolidone)-block-poly(D,L-lactide) as a new polymeric solubilizer for hydrophobic anticancer drugs: in vitro and in vivo evaluation, *J. Controlled Release*, **99**, 83–101.

62. Sugahara, S., Kajiki, M., Kuriyama, H., Kobayashi, T. R. (2002) Paclitaxel delivery systems: the use of amino acid linkers in the conjugation of paclitaxel with carboxymethyldextran to create prodrugs, *Biol. Pharm. Bull.*, **25**, 632–41.

63. Sharma, A., and Straubinger, R. M. (1994) Novel taxol formulations: preparation and characterization of taxol-containing liposomes, *Pharm. Res.*, **11**, 889–896.

64. Schipper, M. L., Nakayama-Ratchford, N., Davis, C. R., Kam, N. W. S., Chu, P., Liu, Z., Sun, X., Dai, H., and Gambhir, S. S. (2008), A pilot toxicology study of single-walled carbon nanotubes in a small sample of mice, *Nat. Nanotechnol.*, **3**(4), 216–221.

65. Liu, Z., Davis, C., Cau, W., He, L., Chen, X., and Dai, H. (2008) Circulation and long-term fate of functionalized, biocompatible single-walled carbon nanotubes in mice probed by Raman spectroscopy, *Proc. Natl. Acad. Sci. USA*, **105**(5), 1410–1415.

66. Liu, Z., Cai, W., He, L., Nakayama, N., Chen, K., Sun, X., Chen, X., and Dai, H. (2007) In vivo biodistribution and highly efficient tumour targeting of carbon nanotubes in mice, *Nat. Nanotechnol.*, **2**(1), 47–52.

67. De La Zerda, A., Zavaleta, C., Keren, S., Vaithilingam, S., Bodapati, S., Liu, Z., Levi, J., Smith, B. R., Ma, T.-J., Oralkan, O., Cheng, Z., Chen, X., Dai, H., Khuri-Yakub, B. T., and Gambhir, S. S. (2008) Carbon nanotubes as photoacoustic molecular imaging agents in living mice, *Nat. Nanotechnol.*, **3**(9), 557–562.

68. Dhar, S., Liu, Z., Thomale, J., Dai, H., and Lippard, S. J. (2008) Targeted single-wall carbon nanotube-mediated Pt(IV) prodrug delivery using folate as a homing device, *J. Am. Chem. Soc.*, **130**(34), 11467–11476.

69. Feazell, R. P., Nakayama-Ratchford, N., Dai, H., and Lippard, S. J. (2007) Soluble single-walled carbon nanotubes as longboat delivery systems for platinum(IV) anticancer drug design, *J. Am. Chem. Soc.*, **129**(27), 8438–8439.

70. Liu, Z., Winters, M., Holodniy, M., and Dai, H. (2007) siRNA delivery into human T cells and primary cells with carbon-nanotube transporters, *Angew. Chem., Int. Ed. Engl.* **46**(12), 2023–2027.

71. Chen, Z., Tabakman, S. M., Goodwin, A. P., Kattah, M. G., Daranciang, D., Wang, X., Zhang, G., Li, X., Liu, Z., Utz, P. J., Jiang, K., Fan, S., and Dai, H. (2008) Protein microarrays with carbon nanotubes as multicolor Raman labels, *Nat. Biotechnol.*, **26**(11), 1285–1292.

72. Liu, Z., Li, X., Tabakman, S. M., Jiang, K., Fan, S., and Dai, H. (2008) Multiplexed multicolor Raman imaging of live cells with isotopically modified single walled carbon nanotubes, *J. Am. Chem. Soc.*, **130**(41), 13540–13541.

73. Welsher, K., Liu, Z., Daranciang, D., and Dai, H. (2008) Selective probing and imaging of cells with single walled carbon nanotubes as near-infrared fluorescent molecules, *Nano Lett.*, **8**(2), 586–590.

74. Zavaleta, C., de la Zerda, A., Liu, Z., Keren, S., Cheng, Z., Schipper, M., Chen, X., Dai, H., and Gambhir, S. S. (2008) Noninvasive Raman spectroscopy in living mice for evaluation of tumor targeting with carbon nanotubes, *Nano Lett.*, **8**(9), 2800–2805.

75. Drouvalakis, K. A., Bangsaruntip, S., Hueber, W., Kozar, L. G., Utz, P. J., and Dai, H. (2008) Peptide-coated nanotube-based biosensor for the detection of disease-specific autoantibodies in human serum, *Biosens. Bioelectron.*, **23**(10), 1413–1421.

76. Xu, M. H., and Wang, L. H. V. (2006) Photoacoustic imaging in biomedicine, *Rev. Sci. Instrum.*, **77**, 041101.

77. Eghtedari, M., Oraevsky, A., Copland, J. A., Kotov, N. A., Conjusteau, A., and Motamedi, M. (2007) High sensitivity of in vivo detection of gold nanorods using a laser optoacoustic imaging system, *Nano Lett.*, **7**, 1914–1918.

78. Zhang, L., Tu, X., Welsher, K., Wang, X., Zheng, M., and Dai, H. (2009) Optical characterizations and electronic devices of nearly pure (10.5) single-walled carbon nanotubes, *J. Am. Chem. Soc.*, **131**(7), 2454–2455.

79. Zhang, L., Zaric, S., Tu, X., Wang, X., Zhao, W., and Dai, H. (2008) Assessment of chemically separated carbon nanotubes for nanoelectronics, *J. Am. Chem. Soc.*, **130**(8), 2686–2691.

80. Mann, D., Kato, Y. K., Kinkhabwala, A., Pop, E., Cao, J., Wang, X., Zhang, L., Wang, Q., Guo, J., and Dai, H. (2007) Electrically driven thermal light emission from individual single-walled carbon nanotubes, *Nat. Nanotechnol.*, **2**(1), 33–38.

81. Xu, Z., Lu, W., Wang, W., Gu, C., Liu, K., Bai, X., Wang, E., and Dai, H. (2008) Converting metallic single-walled carbon nanotubes into semiconductors by boron/nitrogen co-doping, *Adv. Mater.*, **20**(19), 3615–3619.

82. Sun, X., Zaric, S., Daranciang, D., Welsher, K., Lu, Y., Li, X., and Dai, H. (2008) Optical properties of ultrashort semiconducting single-walled carbon nanotube capsules down to sub-10 nm, *J. Am. Chem. Soc.*, **130**(20), 6551–6555.

83. Hersam, M. C. (2008) Progress towards monodisperse single-walled carbon nanotubes, *Nat. Nanotechnol.*, **3**(7), 387–394.

84. Kim, W. J., Usrey, M. L., and Strano, M. S. (2007) Selective functionalization and free solution electrophoresis of single-walled carbon nanotubes: separate enrichment of metallic and semiconducting SWNT, *Chem. Mater.*, **19**, 1571–1576.

85. Toyoda, S. Yamaguchi, Y., Hiwatashi, M., Tomonari, Y., Murakami, H., and Nakashima, N. (2007) Separation of semiconducting single-walled carbon nanotubes by using a long alkyl-chain benzenediazonium compound, *Chem. Asian J.*, **2**, 145–149.

86. An, K. H., Yang, C.-M., Lee, J. Y., Lim, S. C., Kang, C., Son, J.-H., Jeong, M. S., and Lee, Y. H. (2006) A diameter-selective chiral separation of single-wall carbon nanotubes using nitronium ions, *J. Electron. Mater.*, **35**, 235–242.

87. Ménard-Moyon, C., Izard, N., Doris, E., and Mioskowski, C. (2006) Separation of semiconducting from metallic carbon nanotubes by selective functionalization with azomethine ylides, *J. Am. Chem. Soc.*, **128**, 6552–6553.

88. McDonald, T. J., Engtrakul, C., Jones, M., Rumbles, G., and Heben, M. J. (2006) Kinetics of PL quenching during single-walled carbon nanotube rebundling and diameter-dependent surfactant interactions, *J. Phys. Chem. B*, **110**, 25339–25346.

89. Chen, F. M., Wang, B., Chen, Y., and Li, L. J. (2007) Toward the extraction of single species of single-walled carbon nanotubes using fluorene-based polymers, *Nano Lett.*, **7**, 3013–3017.

90. Hwang, J. Y., Nish, A., Doig, J., Douven, S., Chen, C.-W., Chen, L.-C., and Nicholas, R. J. (2008) Polymer structure and solvent effects on the selective dispersion of single-walled carbon nanotubes, *J. Am. Chem. Soc.*, **130**, 3543–3553.

91. Li, H. P., Zhou, B., Lin, Y., Gu, L., Wang, W., Fernando, K. A. S., Kumar, S. K., Allard, L. F., and Sun, Y.-P. (2004) Selective interactions of porphyrins with semiconducting single-walled carbon nanotubes, *J. Am. Chem. Soc.*, **126**, 1014–1015.

92. Krupke, R., Hennrich, F., von Lohneysen, H., and Kappes, M. M. (2003) Separation of metallic from semiconducting single-walled carbon nanotubes, *Science*, **301**, 344–347.

93. Peng, H. Q., Alvarez, N. T., Kittrell, C., Hauge, R. H., and Schmidt, H. K. (2006) Dielectrophoresis field flow fractionation of single-walled carbon nanotubes, *J. Am. Chem. Soc.*, **128**, 8396–8397.

94. Zheng, M., Jagota, A., Strano, M. S., Santos, A. P., Barone, P., Chou, S. G., Diner, A., Dresselhaus, M. S., McLean, R. S., Onoa, G. B., Samsonidze, G. G., Semke, E. D., Usrey, M., and Walls, D. J. (2003) Structure-based carbon nanotube sorting by sequence-dependent DNA assembly, *Science*, **302**, 1545–1548.

95. Huang, X. Y., McLean, R. S., and Zheng, M. (2005) High-resolution length sorting and purification of DNA-wrapped carbon nanotubes by size-exclusion chromatography, *Anal. Chem.*, **77**, 6225–6228.

96. Arnold, M. S., Green, A. A., Hulvat, J. F., Stupp, S. I., and Hersam, M. C. (2006) Sorting carbon nanotubes by electronic structure using density differentiation, *Nat. Nanotechnol.*, **1**, 60–65.

97. Arnold, M. S., Suntivich, J., Stupp, S. I., and Hersam, M. C. (2008) Hydrodynamic characterization of surfactant encapsulated carbon nanotubes using an analytical ultracentrifuge, *ACS Nano,* **2**(11), 2291–2300.

98. Green, A. A., and Hersam, M. C. (2009) Processing and properties of highly enriched double-wall carbon nanotubes. *Nat. Nanotechnol.*, **4**(1), 64–70.

99. Green, A. A., and Hersam, M. C. (2008) Colored semitransparent conductive coatings consisting of monodisperse metallic single-walled carbon nanotubes, *Nano Lett.*, **8**(5), 1417–1422.

100. Kitiyanan, B., Alvarez, W. E., Harwell, J. H., and Resasco, D. E. (2000) Controlled production of single-wall carbon nanotubes by catalytic decomposition of CO on bimetallic Co-Mo catalysts, *Chem. Phys. Lett.*, **317**, 497–503.

101. Green, A. A., and Hersam, M. C. (2007) Ultracentrifugation of single-walled nanotubes, *Mater. Today*, **10**(12), 59–60.

102. Naumov, A.V., Kuznetsov, O. A., Harutyunyan, A. R., Green, A. A., Hersam, M. C., Resasco, D. E., Nikolaev, P. N., and Weisman, R. B. (2009) Quantifying the semiconducting fraction in single-walled carbon nanotube samples through comparative atomic force and photoluminescence microscopies, *Nano Lett.*, **9**(9), 3203–3208.

103. Ma, Y.-Z., Graham, M. W., Fleming, G. R., Green, A. A., and Hersam, M. C. (2008) Ultrafast exciton dephasing in semiconducting single-walled carbon nanotubes, *Phys. Rev. Lett.*, **101**(21), 217402/1–217402/4.

104. Qian, H., Araujo, P. T., Georgi, C., Gokus, T., Hartmann, N., Green, A. A., Jorio, A., Hersam, M. C., Novotny, L., and Hartschuh, A. (2008) Visualizing the local optical response of semiconducting carbon nanotubes to DNA-wrapping, *Nano Lett.*, **8**(9), 2706–2711.

105. Qian, H., Georgi, C., Anderson, N., Green, A. A., Hersam, M. C., Novotny, L., and Hartschuh, A. (2008) Exciton energy transfer in pairs of single-walled carbon nanotubes, *Nano Lett.*, **8**(5), 1363–1367.

106. Harutyunyan, H., Gokus, T., Green, A. A., Hersam, M. C., Allegrini, M., and Hartschuh, A. (2009) Defect-induced photoluminescence from dark excitonic states in individual single-walled carbon nanotubes, *Nano Lett.*, **9**(5), 2010–2014.

107. Alvarez, N. T., Pint, C. L., Hauge, R. H., and Tour, J. M. (2009) Abrasion as a catalyst deposition technique for carbon nanotube growth, *J. Am. Chem. Soc.*, **131**(41), 15041–15048.

108. Pint, C. L., Nicholas, N., Duque, J. G., Parra-Vasquez, A. N. G., Pasquali, M., and Hauge, R. (2009) Recycling ultrathin catalyst layers for multiple single-walled carbon nanotube array regrowth cycles and selectivity in catalyst activation, *Chem. Mater.*, **21**(8), 1550–1556.

109. Kim, M. J., Haroz, E., Wang, Y., Shan, H., Nicholas, N., Kittrell, C., Moore, V. C., Jung, Y., Luzzi, D., Wheeler, R., BensonTolle, T., Fan, H., Da, S., Hwang, W.-F., Wainerdi, T. J., Schmidt, H., Hauge, R. H., and Smalley, R. E. (2007) Nanoscopically flat open-ended single-walled carbon nanotube substrates for continued growth, *Nano Lett.*, **7**(1), 15–21.

110. Pint, C. L., Pheasant, S. T., Pasquali, M., Coulter, K. E., Schmidt, H. K., and Hauge, R. H. (2008) Synthesis of high aspect-ratio carbon nanotube "flying carpets" from nanostructured flake substrates, *Nano Lett.*, **8**(7), 1879–1883.

111. Ding, F., Harutyunyan, A. R., and Yakobson, B. I. (2009) Dislocation theory of chirality-controlled nanotube growth. *Proc. Natl. Acad. Sci. USA*, **106**(8), 2506–2509.

112. Crouse, C. A., Maruyama, B., Colorado, R., Jr., Back, T., and Barron, A. R. (2008) Growth, new growth, and amplification of carbon nanotubes as a function of catalyst composition, *J. Am. Chem. Soc.*, **130**(25), 7946–7954.

113. Mukherjee, A., Combs, R., Chattopadhyay, J., Abmayr, D. W., Engel, P. S., and Billups, W. E. (2008) Attachment of nitrogen and oxygen centered radicals to single-walled carbon nanotube salts, *Chem. Mater.*, **20**(23), 7339–7343.

114. Alvarez, N. T., Kittrell, C., Schmidt, H. K., Hauge, R. H., Engel, P. S., and Tour, J. M. (2008) Selective photochemical functionalization of surfactant-dispersed single wall carbon nanotubes in water, *J. Am. Chem. Soc.*, **130**(43), 14227–14233.

115. Kono, J., Ostojic, G. N., Zaric, S., Strano, M. S., Moore, V. C., Shaver, J., Hauge, R. H., and Smalley, R. E. (2004) Ultra-fast optical spectroscopy of micelle-suspended single-walled carbon nanotubes, *Appl. Phys. A: Mater. Sci. Process.* **78**, 1093–1098.

116. Doyle, C. D., Rocha, J.-D. R., Weisman, R. B., and Tour, J. M. (2008) Structure-dependent reactivity of semiconducting single-walled carbon nanotubes with benzenediazonium salts, *J. Am. Chem. Soc.*, **130**(21), 6795–6800.

117. Stephenson, J. J., Hudson, J. L., Leonard, A. D., Price, B. K., and Tour, J. M. (2007) Repetitive functionalization of water-soluble single-walled carbon nanotubes. Addition of acid-sensitive addends, *Chem. Mater.* **19**(14), 3491–3498.

118. Pulikkathara, M. X., Kuznetsov, O. V., and Khabashesku, V. N. (2008) Sidewall covalent functionalization of single wall carbon nanotubes through reactions of fluoronanotubes with urea, guanidine, and thiourea, *Chem. Mater.* **20**(8), 2685–2695.

119. Alemany, L. B., Zhang, L., Zeng, L., Edwards, C. L., and Barron, A. R. (2007) Solid-state NMR analysis of fluorinated single-walled carbon nanotubes: assessing the extent of fluorination, *Chem. Mater.* **19**(4), 735–744.

120. Dillon, E. P., Crouse, C. A., and Barron, A. R. (2008) Synthesis, characterization, and carbon dioxide adsorption of covalently attached polyethyleneimine-functionalized single-wall carbon nanotubes, *ACS Nano*, **2**(1), 156–164.

121. Lefebvre, J., Fraser, J. M., Finnie, P., and Homma, Y. (2004) Photoluminescence from an individual single-walled carbon nanotube, *Phys. Rev. B*, **69**, 075403-1–075403-5.

122. Cognet, L., Tsyboulski, D., Rocha, J.-D. R., Doyle, C. D., Tour, J. M., and Weisman, R. B. (2004) Stepwise quenching of exciton fluorescence in carbon nanotubes by single-molecule reactions, *Science*, **316**, 1465–1468.

123. Murakami, Y., and Kono, J. (2009) Nonlinear photoluminescence excitation spectroscopy of carbon nanotubes: exploring the upper density limit of one-dimensional excitons, *Phys. Rev. Lett.*, **102**(3), 037401/1–037401/4.

124. Srivastava, A., Htoon, H., Klimov, V. I., and Kono, J. (2008) Direct observation of dark excitons in individual carbon nanotubes: inhomogeneity in the exchange splitting, *Phys. Rev. Lett.*, **101**(8), 087402/1–087402/4.

125. Tsyboulski, D. A., Hou, Y., Fakhri, N., Ghosh, S., Zhang, R., Bachilo, S. M., Pasquali, M., Chen, L., Liu, J., Weisman, R. B. (2009) Do inner shells of double-walled carbon nanotubes fluoresce? *Nano Lett.*, **9**(9), 3282–3289.

126. Tsyboulski, D. A., Bakota, E. L., Witus, L. S., Rocha, J.-D. R., Hartgerink, J. D., and Weisman, R. B. (2008) Self-assembling peptide coatings designed for highly luminescent suspension of single-walled carbon nanotubes, *J. Am. Chem. Soc.*, **130**(50), 17134–17140.

127. Cherukuri, P., Bachilo, S. M., Litovsky, S. H., and Weisman, R. B. (2004) Near-infrared fluorescence microscopy of single-walled carbon nanotubes in phagocytic cells, *J. Am. Chem. Soc.*, **126**, 15638–15639.

128. Zheng, M., Jagota, A., Semke, E. D., Diner, B. A., McClean, R. S., Lustig, S. R., Richardson, R. E., and Tassi, N. G. (2003) DNA-assisted dispersion and separation of carbon nanotubes, *Nat. Mater.*, **2**, 338–342.

129. Matsuura, K., Saito, T., Okazaki, T., Ohshima, S., Yumura, M., and Iijima, S. (2006) Selectivity of water-soluble proteins in single-walled carbon nanotube dispersions, *Chem. Phys. Lett.*, **429**, 497–502.

130. Tsyboulski, D. A., Rocha, J.-D. R., Bachilo, S. M., Cognet, L., and Weisman, R. B. (2007) Structure-dependent fluorescence efficiencies of individual single-walled carbon nanotubes, *Nano Lett.*, **7**(10), 3080–3085.

131. Duque, J. G., Cognet, L., Parra-Vasquez, A. N. G., Nicholas, N., Schmidt, H. K., and Pasquali, M. (2008) Stable luminescence from individual carbon nanotubes in acidic, basic, and biological environments, *J. Am. Chem. Soc.*, **130**(8), 2626–2633.

132. Naumov, A. V., Bachilo, S. M., Tsyboulski, D. A., and Weisman, R. B. (2008) Electric field quenching of carbon nanotube photoluminescence, *Nano Lett.*, **8**(5), 1527–1531.

133. Leeuw, T. K., Reith, R. M., Simonette, R. A., Harden, M. E., Cherukuri, P., Tsyboulski, D. A., Beckingham, K. M., and Weisman, R. B. (2007) Single-walled carbon nanotubes in the intact organism: near-IR imaging and biocompatibility studies in drosophila, *Nano Lett.*, **7**(9), 2650–2654.

134. Cognet, L., Tsyboulski, D. A., and Weisman, R. B. (2008) Subdiffraction far-field imaging of luminescent single-walled carbon nanotubes, *Nano Lett.*, **8**(2), 749–753.

135. Kim, J.-H., Heller, D. A., Jin, H., Barone, P. W., Song, C., Zhang, J., Trudel, L. J., Wogan, G. N., Tannenbaum, S. R., and Strano, M. S. (2009) The rational design of nitric oxide selectivity in single-walled carbon nanotube near-infrared fluorescence sensors for biological detection, *Nat. Chem.*, **1**(6), 473–481.

136. Jin, H., Heller, D. A., Kim, J.-H., and Strano, M. S. (2008) Stochastic analysis of stepwise fluorescence quenching reactions on single-walled carbon nanotubes: single molecule sensors, *Nano Lett.*, **8**(12), 4299–4304.

137. Heller, D. A., Jin, H., Martinez, B. M., Patel, D., Miller, B. M., Yeung, T.-K., Jena, P. V., Hoebartner, C., Ha, T., Silverman, S. K., and Strano, M. S. (2009) Multimodal optical sensing and analyte specificity using single-walled carbon nanotubes, *Nat. Nanotechnol.*, **4**(2), 114–120.

138. Sung, J., Barone, P. W., Kong, H., and Strano, M. S. (2009) Sequential delivery of dexamethasone and VEGF to control local tissue response for carbon nanotube fluorescence based micro-capillary implantable sensors, *Biomaterials*, **30**(4), 622–631.

139. Lee, C. Y., and Strano, M. S. (2008) Amine basicity (pK_b) controls the analyte binding energy on single walled carbon nanotube electronic sensor arrays, *J. Am. Chem. Soc.*, **130**(5), 1766–1773.

140. Lee, C. Y., Scharma, R., Radadia, A. D., Masel, R. I., and Strano, M. S. (2008) On-chip micro gas chromatograph enabled by a noncovalently functionalized single-walled carbon nanotube sensor array, *Angew. Chem. Int. Ed.* **47**(27), 5018–5021.

141. de Marcos, S., and Wolfbeis, O. S. (1996) Optical sensing of pH based on polypyrrole films, *Anal. Chim. Acta.*, **334**, 149–153.

142. Collins, G. E., and Buckley, L. J. (1996) Conductive polymer-coated fabrics for chemical sensing, *Synth. Met.*, **78**, 93–101.

143. Choi, J. H., Nguyen, F. T., Barone, P. W., Heller, D. A., Moll, A. E., Patel, D., Boppart, S. A., and Strano, M. S. (2007) Multimodal biomedical imaging with asymmetric single-walled carbon nanotube/iron oxide nanoparticle complexes, *Nano Lett.*, **7**(4), 861–867.

144. Strano, M. S., and Jin, H. (2008) Where is it heading? Single-particle tracking of single-walled carbon nanotubes, *ACS Nano*, **2**(9), 1749–1752.

145. Tsyboulski, D. A., Bachilo, S. M., Kolomeisky, A. B., and Weisman, R. B. (2008) Translational and rotational dynamics of individual single-walled carbon nanotubes in aqueous suspension, *ACS Nano*, **2**(9), 1770–1776.

146. Jin, H., Heller, D. A., Sharma, R., and Strano, M. S. (2009) Size-dependent cellular uptake and expulsion of single-walled carbon nanotubes: single particle tracking and a generic uptake model for nanoparticles. *ACS Nano*, **3**(1), 149–158.

147. Jin, H., Heller, D. A., and Strano, M. S. (2008) Single-particle tracking of endocytosis and exocytosis of single-walled carbon nanotubes in NIH-3T3 cells, *Nano Lett.*, **8**(6), 1577–1585.

148. Terrones, M., Romo-Herrera, J. M., Cruz-Silva, E., Lopez-Urias, F., Munoz-Sandoval, E., Velazquez-Salazar, J. J., Terrones, H., Bando, Y., and Golberg, D. (2007) Pure and doped boron nitride nanotubes, *Mater. Today*, **10**(5), 30–38.

149. Cruz-Silva, E., Cullen, D. A., Gu, L., Romo-Herrera, J. M., Munoz-Sandoval, E., Lopez-Urias, F., Sumpter, B. G., Meunier, V., Charlier, J.-C., Smith, D. J., Terrones, H., and Terrones, M. (2008) Heterodoped nanotubes: theory, synthesis, and characterization of phosphorus-nitrogen doped multiwalled carbon nanotubes, *ACS Nano*, **2**(3), 441–448.

150. Sumpter, B. G., Meunier, V., Romo-Herrera, J. M., Cruz-Silva, E., Cullen, D. A., Terrones, H., Smith, D., J., and Terrones, M. (2007) Nitrogen-mediated carbon nanotube growth: diameter reduction, metallicity, bundle dispersability, and bamboo-like structure formation, *ACS Nano*, **1**(4), 369–375.

151. Maciel, I. O., Campos-Delgado, J., Cruz-Silva, E., Pimenta, M. A., Sumpter, B. G., Meunier, V., Lopez-Urias, F., Munoz-Sandoval, E., Terrones, H., Terrones, M., and Jorio, A. (2009) Synthesis, electronic structure, and Raman scattering of phosphorus-doped single-wall carbon nanotubes, *Nano Lett.*, **9**(6), 2267–2272.

152. Cruz-Silva, E., Lopez-Urias, F., Munoz-Sandoval, E., Sumpter, B. G., Terrones, H., Charlier, J.-C., Meunier, V., and Terrones, M. (2009) Electronic transport and mechanical properties of phosphorus- and phosphorus-nitrogen-doped carbon nanotubes, *ACS Nano*, **3**(7), 1913–1921.

153. Rodriguez-Manzo, J. A., Lopez-Urias, F., Terrones, M., and Terrones, H. (2007) Anomalous paramagnetism in doped carbon nanostructures, *Small*, **3**(1), 120–125.

154. Elias, A. L., Carrero-Sanchez, J. C., Terrones, H., Endo, M., Laclette, J. P., and Terrones, M. (2007) Viability studies of pure carbon- and nitrogen-doped nanotubes with Entamoeba histolytica: from amoebicidal to biocompatible structures, *Small*, **3**(10), 1723–1729.

155. Maciel, I. O., Anderson, N., Pimenta, M. A., Hartschuh, A., Qian, H., Terrones, M., Terrones, H., Campos-Delgado, J., Rao, A. M., Novotny, L., and Jorio, A. (2008) Electron and phonon renormalization near charged defects in carbon nanotubes, *Nat. Mater.*, **7**(11), 878–883.

156. Romo-Herrera, J. M., Terrones, M., Terrones, H., and Meunier, V. (2008) Guiding electrical current in nanotube circuits using structural defects: a step forward in nanoelectronics, *ACS Nano*, **2**(12), 2585–2591.

157. Souza Filho, A. G., Meunier, V., Terrones, M., Sumpter, B. G., Barros, E. B., Villalpando-Paez, F., Mendes Filho, J., Kim, Y. A., Muramatsu, H., Hayashi, T., Endo, M., and Dresselhaus, M. S. (2007) Selective tuning of the electronic properties of coaxial nanocables through exohedral doping, *Nano Lett.*, **7**(8), 2383–2388.

158. Villalpando-Paez, F., Son, H., Nezich, D., Hsieh, Y. P., Kong, J., Kim, Y. A., Shimamoto, D., Muramatsu, H., Hayashi, T., Endo, M., Terrones, M., and Dresselhaus, M. S. (2008) Raman spectroscopy study of isolated double-walled carbon nanotubes with different metallic and semiconducting configurations, *Nano Lett.*, **8**(11), 3879–3886.

159. Hayashi, T., Shimamoto, D., Kim, Y. A., Muramatsu, H., Okino, F., Touhara, H., Shimada, T., Miyauchi, Y., Maruyama, S., Terrones, M., Dresselhaus, M. S., and Endo, M. (2008) Selective optical property modification of double-walled carbon nanotubes by fluorination, *ACS Nano*, **2**(3), 485–488.

160. Jung, Y. C., Shimamoto, D., Muramatsu, H., Kim, Y. A., Hayashi, T., Terrones, M., and Endo, M. (2008) Robust, conducting, and transparent polymer composites using surface-modified and individualized double-walled carbon nanotubes, *Adv. Mater.*, **20**(23), 4509–4512.

161. Rodriguez-Manzo, J. A., Banhart, F., Terrones, M., Terrones, H., Grobert, N., Ajayan, P. M., Sumpter, B. G., Meunier, V., Wang, M., Bando, Y., and Golberg, D. (2009) Heterojunctions between metals and carbon nanotubes as ultimate nanocontacts, *Proc. Natl. Acad. Sci. USA*, **106**(12), 4591–4595.

162. Grimm, D., Venezuela, P., Banhart, F., Grobert, N., Terrones, H., Ajayan, P. M., Terrones, M., and Latge, A. (2007) Synthesis of SWCNT rings made by two Y junctions and possible applications in electron interferometry, *Small*, **3**(11), 1900–1905.

163. Lepro, X., Vega-Cantu, Y., Rodriguez-Macias, F. J., Bando, Y., Golberg, D., and Terrones, M. (2007) Production and characterization of coaxial nanotube junctions and networks of CNx/CNT, *Nano Lett.*, **7**(8), 2220–2226.

164. Romo-Herrera, J. M., Terrones, M., Terrones, H., Dag, S., and Meunier, V. (2007) Covalent 2D and 3D networks from 1D nanostructures: designing new materials, *Nano Lett.*, **7**(3), 570–576.

165. Munoz-Sandoval, E., Agarwal, V., Escorcia-Garcia, J., Ramirez-Gonzalez, D., Martinez-Mondragon, M. M., Cruz-Silva, E., Meneses-Rodriguez, D., Rodriguez-Manzo, J. A., Terrones, H., and Terrones, M. (2007) Architectures from aligned nanotubes using controlled micropatterning of silicon substrates and electrochemical methods, *Small*, **3**(7), 1157–1163.

166. Meunier, V., Muramatsu, H., Hayashi, T., Kim, Y. A., Shimamoto, D., Terrones, H., Dresselhaus, M. S., Terrones, M., Endo, M., and Sumpter, B. G. (2009) Properties of one-dimensional molybdenum nanowires in a confined environment, *Nano Lett.*, **9**(4), 1487–1492.

167. Muramatsu, H., Hayashi, T., Kim, Y. A., Shimamoto, D., Endo, M., Terrones, M., and Dresselhaus, M. S. (2008) Synthesis and isolation of molybdenum atomic wires, *Nano Lett.*, **8**(1), 237–240.

168. Cano-Marquez, A. G., Rodriguez-Macias, F. J., Campos-Delgado, J., Espinosa-Gonzalez, C. G., Tristan-Lopez, F., Ramirez-Gonzalez, D., Cullen, D. A., Smith, D. J., Terrones, M., and Vega-Cantu, Y. I. (2009) Ex-MWNTs: graphene sheets and ribbons produced by lithium intercalation and exfoliation of carbon nanotubes, *Nano Lett.*, **9**(4), 1527–1533.

169. Rodrigues, O. E. D., Saraiva, G. D., Nascimento, R. O., Barros, E. B., Mendes Filho, J., Kim, Y. A., Muramatsu, H., Endo, M., Terrones, M., Dresselhaus, M. S., and Souza Filho, A. G. (2008) Synthesis and characterization of selenium-carbon nanocables, *Nano Lett.*, **8**(11), 3651–3655.

170. Romo-Herrera, J. M., Sumpter, B. G., Cullen, D. A., Terrones, H., Cruz-Silva, E., Smith, D. J., Meunier, V., and Terrones, M. (2008) An atomistic branching mechanism for carbon nanotubes: sulfur as the triggering agent, *Angew. Chem., Int. Ed. Engl.* **47**(16), 2948–2953.

171. Rodriguez-Manzo, J., Terrones, M., Terrones, H., Kroto, H. W., Sun, L., and Banhart, F. (2007) In situ nucleation of carbon nanotubes by the injection of carbon atoms into metal particles, *Nat. Nanotechnol.*, **2**(5), 307–311.

172. Herrero, M. A., Toma, F. M., Al-Jamal, K. T., Kostarelos, K., Bianco, A., Da Ros, T., Bano, F., Casalis, L., Scoles, G., and Prato, M. (2009) Synthesis and characterization of a carbon nanotube-dendron series for efficient siRNA delivery, *J. Am. Chem. Soc.*, **131**(28), 9843–9848.

173. Podesta, J. E., Al-Jamal, K. T., Herrero, M. A., Tian, B., Ali-Boucetta, H., Hegde, V., Bianco, A., Prato, M., and Kostarelos, K. (2009) Antitumor activity and

prolonged survival by carbon nanotube-mediated therapeutic siRNA silencing in a human lung xenograft model, *Small*, **5**, 1176–1185.

174. Ali-Boucetta, H., Al-Jamal, K. T., McCarthy, D., Prato, M., Bianco, A., and Kostarelos, K. (2008) Multiwalled carbon nanotube-doxorubicin supramolecular complexes for cancer therapeutics, *Chem. Commun.*, **4**, 459–461.

175. Lacerda, L., Herrero, M. A., Venner, K., Bianco, A., Prato, M., and Kostarelos, K. (2008) Carbon-nanotube shape and individualization critical for renal excretion, *Small*, **4**(8), 1130–1132.

176. Singh, R., Pantarotto, D., Lacerda, L., Pastorin, G., Klumpp, C., Prato, M., Bianco, A., and Kostarelos, K. (2006) Tissue biodistribution and blood clearance rates of intravenously administered carbon nanotube radiotracers, *Proc. Natl. Acad. Sci. USA*, **103**, 3357-3362.

177. Lacerda, L., Soundararajan, A., Singh, R., Pastorin, G., Al-Jamal, K. T., Turton, J., Frederik, P., Herrero, M. A., Li, S., Bao, A., Emfietzoglou, D., Mather, S., Phillips, W. T., Prato, M., Bianco, A., Goins, B., and Kostarelos, K. (2008) Dynamic imaging of functionalized multi-walled carbon nanotube systemic circulation and urinary excretion, *Adv. Mater.*, **20**(2), 225–230.

178. Deen, W. M. (2004) What determines glomerular capillary permeability? *J. Clin. Invest.*, **114**, 1412-1414.

179. Lacerda, L., Ali-Boucetta, H., Herrero, M. A., Pastorin, G., Bianco, A., Prato, M., and Kostarelos, K. (2008) Tissue histology and physiology following intravenous administration of different types of functionalized multiwalled carbon nanotubes, *Nanomedicine*, **3**(2), 149–161.

180. Kostarelos, K., Lacerda, L., Pastorin, G., Wu, W., Wieckowski, S., Luangsivilay, J., Godefroy, S., Pantarotto, D., Briand, J.-P., Muller, S., Prato, M., and Bianco, A. (2007) Cellular uptake of functionalized carbon nanotubes is independent of functional group and cell type, *Nat. Nanotechnol.*, **2**(2), 108–113.

181. Lacerda, L., Pastorin, G., Gathercole, D., Buddle, J., Prato, M., Bianco, A., and Kostarelos, K. (2007) Intracellular trafficking of carbon nanotubes by confocal laser scanning microscopy, *Adv. Mater.*, **19**(11), 1480–1484.

182. Kotov, N. A., Winter, J. O., Clements, I. P., Jan, E., Timko, B. P., Campidelli, S., Pathak, S., Mazzatenta, A., Lieber, C. M., Prato, M., Bellamkonda, R. V., Silva, G. A., Kam, N. W. S., Patolsky, F., and Ballerini, L. (2009) Nanomaterials for neural interfaces, *Adv. Mater.*, **21**(40), 3970–4004.

183. Mattson, M. P., Haddon, R. C., and Rao, A. M. (2000) Molecular functionalization of carbon nanotubes and use as substrates for neuronal growth, *J. Mol. Neurosci.*, **14**, 175–182.

184. Lovat, V., Pantarotto, D., Lagostena, L., Cacciari, B., Grandolfo, M., Righi, M., Spalluto, G., Prato, M., and Ballerini, L. (2005) Carbon nanotube substrates boost neuronal electrical signaling, *Nano Lett.*, **5**, 1107–1110.

185. Gaillard, C., Cellot, G., Li, S., Toma, F. M., Dumortier, H., Spalluto, G., Cacciari, B., Prato, M., Ballerini, L., and Bianco, A. (2009) Carbon nanotubes carrying cell-adhesion peptides do not interfere with neuronal functionality, *Adv. Mater.*, **21**, 2903–2908.

186. Dhoot, N. O., Tobias, C. A., Fischer, I., and Wheatley, M. A. (2004) Peptide-modified alginate surfaces as a growth permissive substrate for neurite outgrowth, *J. Biomed. Mater., Res. Part A*, **71**, 191–200.

187. Mazzatenta, A., Giugliano, M., Campidelli, S., Gambazzi, L., Businaro, L., Markram, H., Prato, M., and Ballerini, L. (2007) Interfacing neurons with carbon nanotubes: electrical signal transfer and synaptic stimulation in cultured brain circuits, *J. Neurosci.*, **27**(26), 6931–6936.

188. Ehli, C., Oelsner, C., Guldi, D. M., Mateo-Alonso, A., Prato, M., Schmidt, C., Backes, C., Hauke, F., and Hirsch, A. (2009) Manipulating single-wall carbon nanotubes by chemical doping and charge transfer with perylene dyes, *Nat. Chem.*, **1**(3), 243–249.

189. Angeles Herranz, M., Ehli, C., Campidelli, S., Gutierrez, M., Hug, G. L., Ohkubo, K., Fukuzumi, S., Prato, M., Martin, N., and Guldi, D. M. (2008) Spectroscopic characterization of photolytically generated radical ion pairs in single-wall carbon nanotubes bearing surface-immobilized tetrathiafulvalenes, *J. Am. Chem. Soc.*, **130**(1), 66–73.

190. Ballesteros, B., Campidelli, S., de la Torre, G., Ehli, C., Guldi, D. M., Prato, M., and Torres, T. (2007) Synthesis, characterization and photophysical properties of a SWNT-phthalocyanine hybrid, *Chem. Commun.*, **28**, 2950–2952.

191. Rahman, G. M. A., Troeger, A., Sgobba, V., Guldi, D. M., Jux, N., Tchoul, M. N., Ford, W. T., Mateo-Alonso, A., and Prato, M. (2008) Improving photocurrent generation: supramolecularly and covalently functionalized single-wall carbon nanotubes-polymer/porphyrin donor-acceptor nanohybrids, *Chem.—Eur. J.*, **14**(29), 8837–8846.

192. Campidelli, S., Ballesteros, B., Filoramo, A., Diaz, D. D., de la Torre, G., Torres, T., Rahman, G. M. A., Ehli, C., Kiessling, D., Werner, F., Sgobba, V., Guldi, D. M., Cioffi, C., Prato, M. and Bourgoin, J.-P. (2008) Facile decoration of functionalized single-wall carbon nanotubes with phthalocyanines via "click chemistry", *J. Am. Chem. Soc.*, **130**(34), 11503–11509.

193. Guryanov, I., Toma, F. M., Montellano Lopez, A., Carraro, M., Da Ros, T., Angelini, G., D'Aurizio, E., Fontana, A., Maggini, M., Prato, M., and Bonchio, M. (2009) Microwave-assisted functionalization of carbon nanostructures in ionic liquids. *Chem.—Eur. J.*, **15**(46), 12837–12845.

194. Quintana, M., and Prato, M. (2009) Supramolecular aggregation of functionalized carbon nanotubes. *Chem. Commun.*, **40**, 6005–6007.

195. Singh, P., Kumar, J., Toma, F. M., Raya, J., Prato, M., Fabre, B., Verma, S., and Bianco, A. (2009) Synthesis and characterization of nucleobase-carbon nanotube hybrids. *J. Am. Chem. Soc.*, **131**(37), 13555–13562.

196. Joung, S.-K., Okazaki, T., Kishi, N., Okada, S., Bandow, S., and Iijima, S. (2009) Effect of fullerene encapsulation on radial vibrational breathing-mode frequencies of single-wall carbon nanotubes, *Phys. Rev. Lett.*, **103**(2), 027403/1–027403/4.

197. R. B. Weisman and Bachilo, S. M., (2003) Dependence of optical transition energies on structure for single-walled carbon nanotubes in aqueous suspension: an empirical Kataura plot, *Nano Lett.*, **3**, 1235-1238.

198. Liu, Z., Joung, S.-K., Okazaki, T., Suenaga, K., Hagiwara, Y., Ohsuna, T., Kuroda, K., and Iijima, S. (2009) Self-assembled double ladder structure formed inside carbon nanotubes by encapsulation of $H_8Si_8O_{12}$, *ACS Nano*, **3**(5), 1160–1166.

199. Sato, Y., Suenaga, K., Bandow, S., and Iijima, S. (2008) Site-dependent migration behavior of individual cesium ions inside and outside C_{60} fullerene nanopeapods, *Small*, **4**(8), 1080–1083.

200. Guan, L., Suenaga, K., Shi, Z., Gu, Z., and Iijima, S. (2005) Direct imaging of the alkali metal site in K-doped fullerene peapods, *Phys. Rev. Lett.*, **94**, 045502.

201. Smith, B. W., Monthioux, M., and Luzzi, D. E. (1998) Encapsulated C_{60} in carbon nanotubes, *Nature*, **396**, 323–324.

202. Guan, L., Suenaga, K., Okazaki, T., Shi, Z., Gu, Z., and Iijima, S. (2007) Coalescence of C_{60} molecules assisted by doped iodine inside carbon nanotubes, *J. Am. Chem. Soc.*, **129**(29), 8954–8955.

203. Sato, Y., Suenaga, K., Okubo, S., Okazaki, T., and Iijima, S. (2007) Structures of D5d-C_{80} and Ih-Er3N@C_{80} fullerenes and their rotation inside carbon nanotubes demonstrated by aberration-corrected electron microscopy, *Nano Lett.*, **7**(12), 3704–3708.

204. Liu, Z., Yanagi, K., Suenaga, K., Kataura, H., and Iijima, S. (2007) Imaging the dynamic behaviour of individual retinal chromophores confined inside carbon nanotubes, *Nat. Nanotechnol.*, **2**(7), 422–425.

205. Guan, L., Suenaga, K., Okubo, S., Okazaki, T., and Iijima, S. (2008) Metallic wires of lanthanum atoms inside carbon nanotubes, *J. Am. Chem. Soc.*, **130**(7), 2162–2163.

206. Guan, L., Suenaga, K., Shi, Z., Gu, Z., and Iijima, S. (2007) Polymorphic structures of iodine and their phase transition in confined nanospace, *Nano Lett.*, **7**(6), 1532–1535.

207. Shukla B., Saito T., Yumura M., and Iijima S. (2009) An efficient carbon precursor for gas phase growth of SWCNTs, *Chem. Commun.*, **23**, 3422–3424.

208. Guan, L., Suenaga, K., and Iijima, S. (2008) Smallest carbon nanotube assigned with atomic resolution accuracy, *Nano Lett.*, **8**(2), 459–462.

209. Yamada, T., Maigne, A., Yudasaka, M., Mizuno, K., Futaba, D. N., Yumura, M., Iijima, S., and Hata, K. (2008) Revealing the secret of water-assisted carbon nanotube synthesis by microscopic observation of the interaction of water on the catalysts, *Nano Lett.*, **8**(12), 4288–4292.

210. Jin, C., Suenaga, K., and Iijima, S. (2008) How does a carbon nanotube grow? An in situ investigation on the cap evolution. *ACS Nano*, **2**(6), 1275–1279.

211. Okazaki, T., Okubo, S., Nakanishi, T., Joung, S.-K., Saito, T., Otani, M., Okada, S., Bandow, S., and Iijima, S. (2008) Optical band gap modification of single-walled carbon nanotubes by encapsulated fullerenes, *J. Am. Chem. Soc.*, **130**(12), 4122–4128.

212. Otani, M., Okada, S., and Oshiyama, A. (2003) Energetics and electronic structures of one-dimensional fullerene chains encapsulated in zigzag nanotubes, *Phys. Rev. B*, **68**, 125424.

213. Okada, S., Saito, S., and Oshiyama, A. (2001) Energetics and electronic structures of encapsulated C_{60} in a carbon nanotube, *Phys. Rev. Lett.*, **86**, 3835–3838.

214. Okada, S. (2007) Radial-breathing mode frequencies for nanotubes encapsulating fullerenes, *Chem. Phys. Lett.*, **438**, 59–62.

215. Smith, B. W., and Luzzi, D. E. (2001) Electron irradiation effects in single wall carbon nanotubes, *J. Appl. Phys.*, **90**, 3509–3515.

216. Suenaga, K., Sato, Y., Liu, Z., Kataura, H., Okazaki, T., Kimoto, K., Sawada, H., Sasaki, T., Omoto, K., Tomita, T., Kaneyama, T., and Kondo, Y. (2009) Visualizing and identifying single atoms using electron energy-loss spectroscopy with low accelerating voltage, *Nat. Chem.*, **1**(5), 415–418.

217. Mikami, F., Matsuda, K., Kataura, H., and Maniwa, Y. (2009) Dielectric properties of water inside single-walled carbon nanotubes, *ACS Nano*, **3**(5), 1279–1287.

218. Maniwa, Y., Matsuda, K., Kyakuno, H., Ogasawara, S., Hibi, T., Kadowaki, H., Suzuki, S., Achiba, Y., and Kataura, H. (2007) Water-filled single-wall carbon nanotubes as molecular nanovalves, *Nat. Mater.*, **6**(2), 135–141.

219. Rols, S., Cambedouzou, J., Chorro, M., Schober, H., Agafonov, V., Launois, P., Davydov, V., Rakhmanina, A. V., Kataura, H., and Sauvajol, J.-L. (2008) How confinement affects the dynamics of C_{60} in carbon nanopeapods, *Phys. Rev. Lett.*, **101**(6), 065507/1–065507/4.

220. Shiozawa, H., Pichler, T., Kramberger, C., Rummeli, M., Batchelor, D., Liu, Z., Suenaga, K., Kataura, H., and Silva, S. R. P. (2009) Screening the missing electron: nanochemistry in action, *Phys. Rev. Lett.*, **102**(4), 046804/1–046804/4.

221. Pfeiffer, R., Holzweber, M., Peterlik, H., Kuzmany, H., Liu, Z., Suenaga, K., and Kataura, H. (2007) Dynamics of carbon nanotube growth from fullerenes, *Nano Lett.*, **7**(8), 2428–2434.

222. Shiozawa, H., Pichler, T., Grueneis, A., Pfeiffer, R., Kuzmany, H., Liu, Z., Suenaga, K., and Kataura, H. (2008) A catalytic reaction inside a single-walled carbon nanotube. *Adv. Mater.*, **20**(8), 1443–1449.

223. Yanagi, K., Iakoubovskii, K., Matsui, H., Matsuzaki, H., Okamoto, H., Miyata, Y., Maniwa, Y., Kazaoui, S., Minami, N., and Kataura, H. (2007) Photosensitive function of encapsulated dye in carbon nanotubes. *J. Am. Chem. Soc.*, **129**(16), 4992–4997.

224. McIntosh, G. C., Tománek, D., and Park, Y. W. (2003) Energetics and electronic structure of a polyacetylene chain contained in a carbon nanotube. *Phys. Rev. B*, **67**, 125419–125424.

225. Tanaka, T., Jin, H., Miyata, Y., Fujii, S., Suga, H., Naitoh, Y., Minari, T., Miyadera, T., Tsukagoshi, K., and Kataura, H. (2009) Simple and scalable gel-based separation of metallic and semiconducting carbon nanotubes, *Nano Lett.*, **9**(4), 1497–1500.

226. Sato, Y., Yanagi, K., Miyata, Y., Suenaga, K., Kataura, H., and Iijima, S. (2008) Chiral-angle distribution for separated single-walled carbon nanotubes. *Nano Lett.*, **8**(10), 3151–3154.

Colour Insert

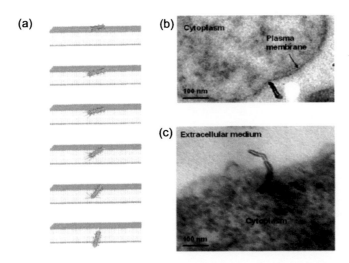

Figure 1.2 CNTs acting as nanoneedles. (a) A schematic of a CNT crossing the plasma membrane; (b) a TEM image of MWNT-NH$_3^+$ interacting with the plasma membrane of A549 cells; (c) a TEM image of MWNT-NH$_3^+$ crossing the plasma membrane of HeLa cells. Reproduced from Lacerda et al.[7] with permission.

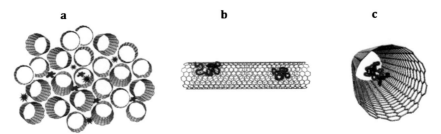

Figure 1.3 A schematic representation of how drugs can interact with CNTs. (a) A bundle of CNTs can act as a porous matrix encapsulating drug molecules between the grooves of individual CNTs. (b) Drug moieties can be attached to the exterior of a CNT either by covalent bonding to the CNT wall or by hydrophobic interaction. (c) The drug can be encapsulated within the internal nanochannel of a CNT. Reproduced from Foldvari and Bagonluri[10] with permission.

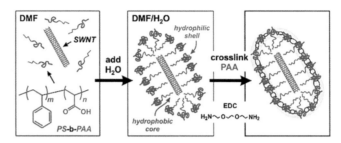

Figure 1.6 Schematic representation of the mechanism of encapsulation of SWCNTs into block copolymers. Reproduced from Kang *et al.*[29] with permission.

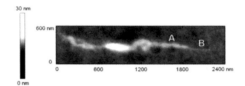

Figure 1.11 An AFM image of SWNT treated with octyl chitosan (0.5 mg/mL) after centrifugation. A = coated region (height = 6.6 ± 1.2 nm, n = 10) and B = uncoated region (height = 1.9 ± 0.4 nm, n = 10).

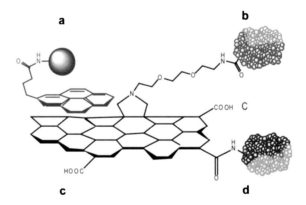

Scheme 2.1 Examples of functionalisations on CNTs' sidewalls and tips for drug delivery purposes: (a) non-covalent, (b) covalent (1,3-dipolar cycloaddition), (c) "defect" and (d) covalent (via oxidation).

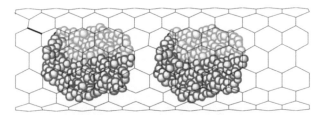

Scheme 2.2 Encapsulation of bioactive molecules in the inner cavity of CNTs.

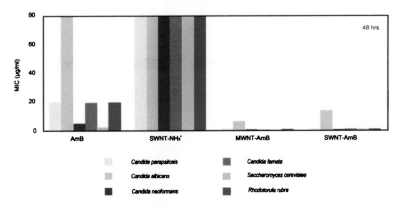

Scheme 2.7 MIC of AmB-FITC-functionalised carbon nanotubes on fungi and yeast after 48 hours of incubation. AmB alone (**AmB**), CNTs alone (**SWNT-NH$_3$$^+$**) and CNTs with AmB covalently bound (**MWNT-AmB** and **SWNT-AmB**).

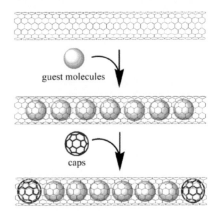

Scheme 2.8 Anticancer drug hexamethylmelamine (HMM) encapsulated inside SWCNTs and capped with fullerenes (C$_{60}$). Reproduced from Ren and Pastorin[88] with permission.

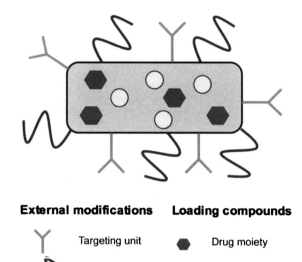

External modifications

Y Targeting unit

∿ Solubilising unit

Loading compounds

⬡ Drug moiety

○ Contrast agent

Figure 3.1 Schematic representation of nanovectors.

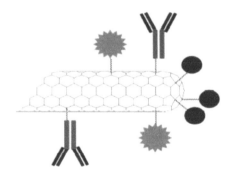

Figure 3.2 An ideal example of multifuncionalised CNTs.

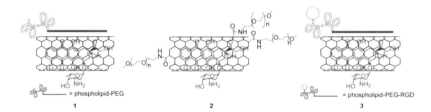

Figure 3.3 Structures of compounds **1–3**.

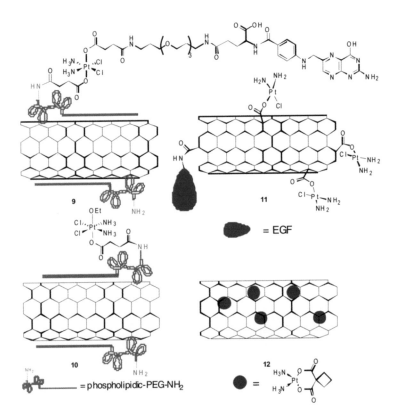

Figure 3.6 Structures of compounds **9–12**.

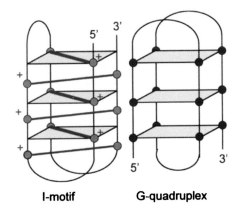

Figure 3.7 I-motif and G-quadruplex structures.

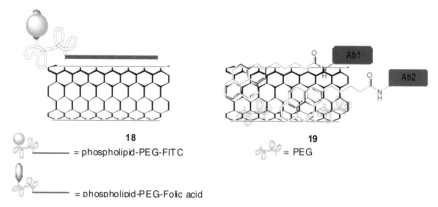

Figure 3.12 Structures of compounds **18** and **19**.

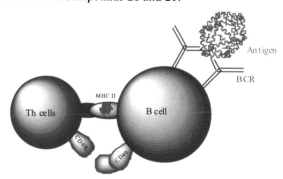

Figure 4.1 Recognition and immobilisation of an antigen at the surface of B cells through a B-cell receptor (BCR), as well as subsequent digestion and presentation of the antigen in the form of processed peptide to T cells (helper), which induce B cells to produce antibodies against the antigen.

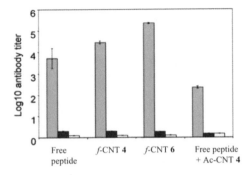

Figure 4.3 Anti-peptide antibody responses following immunisation with peptide and peptide-CNT conjugates. Serum samples were screened by ELISA for the presence of antibodies using FMDV 141–159 peptide conjugated to BSA (cyan bar), control peptide conjugated to BSA (magenta bar) or CNTs 1 functionalised with a maleimido group without peptide (white bar) as solid-phase antigens. Reproduced from Davide Pantarotto *et al.*[7] with permission.

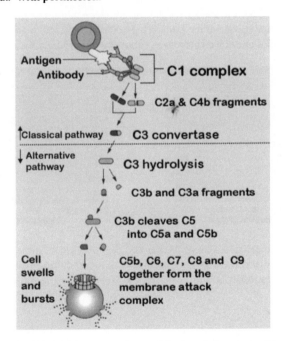

Figure 4.7 Complement activation via classical (up) and alternative (down) pathways. Reproduced from Wikimedia Commons (http://en.wikipedia.org/wiki/Alternative_complement_pathway).

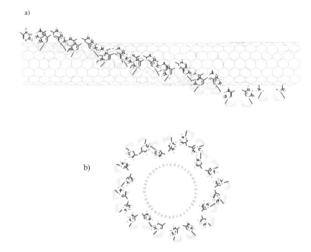

Figure 5.1 Binding model of a carbon nanotube wrapped by a poly(T) sequence. (a) The bases (red) orient to stack with the surface of the nanotube and extend away from the sugar–phosphate backbone (yellow). (b) The DNA wraps to provide a tube within which the carbon nanotube can reside, thus converting it into a water-soluble object.

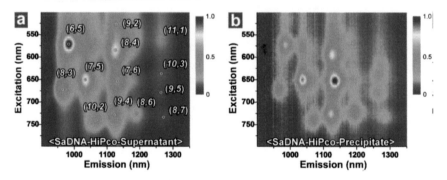

Figure 5.5 Normalised PLE emission contour plot (a, b) from SaDNA dispersed SWCNTs in the ultracentrifuged supernatant (a) and the redispersed-precipitate fractions (b), respectively. Reproduced from Kim *et al.*[18] with permission.

Figure 5.6 Simulation of double-stranded DNA onto uncharged CNTs. Reproduced from Zhao and Johnson[21] with permission.

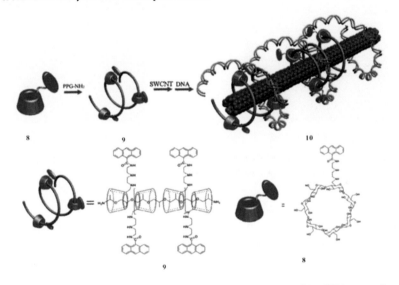

Figure 5.9 Preparation of a SWCNT–ACD-PPR conjugate and its DNA wrapping.

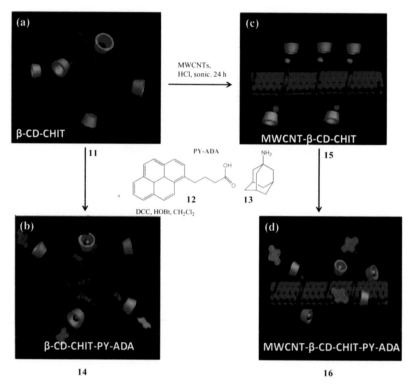

Figure 5.10 Preparation of the supramolecular architecture MWCNT-β-CD-CHIT-PY-ADA. Modified from Liu *et al.*[37] with permission.

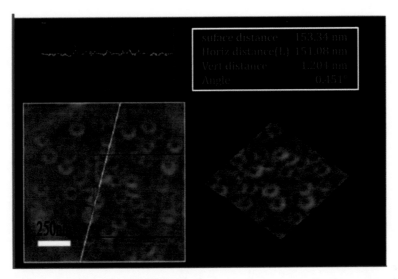

Figure 5.11 AFM images of DNA in the presence of the supramolecular assemblies. Reproduced from Liu *et al.*[37] with permission.

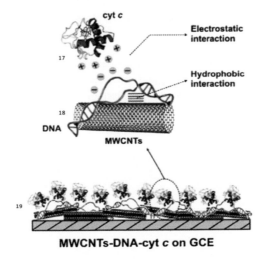

Figure 5.12 Possible interaction between MWCNTs, DNA and cyt *c* for the formation of MWCNT–DNA–cyt *c* biocomposite film modified electrodes. Reproduced from Shie *et al.*[38] with permission.

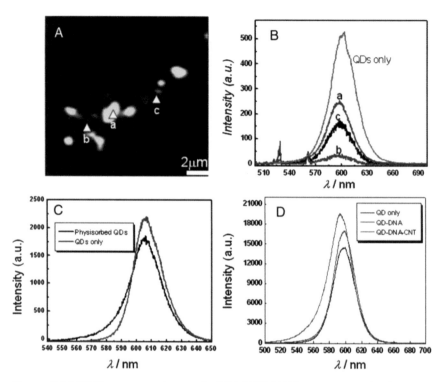

Figure 5.14 (A) Fluorescence image of a DNA-SWCNT/QD conjugate and (B) spectra from different sample locations (a–c). (C) Fluorescence spectra from unconjugated QDs and QDs physisorbed onto DNA-SWCNTs. (D) Fluorescence spectra from unconjugated QDs, DNA-conjugated QDs in the absence of SWCNTs, and DNA-SWCNT/QD conjugates. Reproduced from Zhou *et al.*[41] with permission.

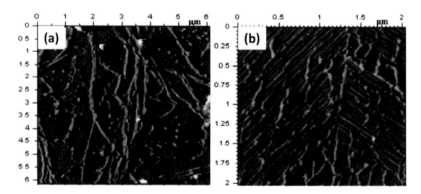

Figure 5.16 AFM images of SWCNT–adenine hybrids without (*f*-SWCNT 2a, left) and with (*f*-SWCNT 2b, right) the tri-ethylene glycol chain. Reproduced from Singh *et al.*[42] with permission.

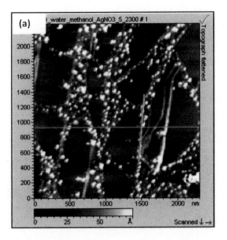

Figure 5.17 AFM image of *f*-SWCNT **2a** on HOPG showing the attachment of Ag(I) nanoparticles all over the nanotube network. Reproduced from Singh *et al.*[42] with permission.

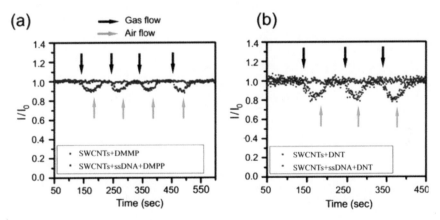

Figure 5.19 (a) Change in the device current when sarin simulant DMMP is applied to SWCNT-FETs before and after ssDNA functionalisation. (b) Sensor response to DNT. Reproduced from Staii *et al.*[70] with permission.

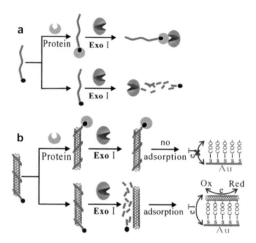

Figure 5.20 Terminal protection assay of small-molecule-linked ssDNA. Small-molecule-linked ssDNA is hydrolysed successively into mononucleotides from the 3′ end by Exo I, while protected from the hydrolysis when the small molecule moiety is bound to its protein target (A). SWNT-wrapping ssDNA terminally tethered to the small molecule is degraded by Exo I, rendering SWNTs assembled on MHA-SAM, which mediates electron transfer between electroactive species and the electrode. Protein binding of small-molecule-linked ssDNA prevents digestion of ssDNA, precluding adsorption of DNA-wrapped SWNTs on MHA-SAM with no redox current generated (B). Reproduced from Wu *et al.*[71] with permission.

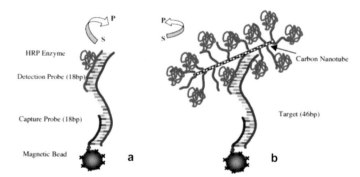

Figure 5.21 Schematic representation of (a) conventional probes and (b) multiple HRP and DP-conjugated CNT-based labels. Reproduced from Lee *et al.*[78] with permission.

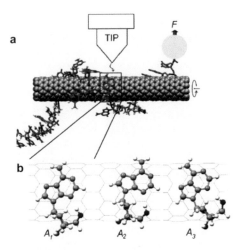

Figure 5.22. (a) Proposed experimental setup for single base measurement: an ssDNA fragment is in partial contact with the CNT and is being pulled at one end. (b) Representative optimal structures of adenine on the (10,0) CNT. The gray, blue, red and white balls represent C, N, O and H atoms, respectively. Reproduced from Meng *et al.*[82] with permission.

Figure 5.23 Isodensity surface of the charge density difference for adenine–CNT. The charge density difference is calculated by subtracting the charge density of the individual adenine (A) and CNT systems, each fixed at its respective position when it is part of the A-CNT complex. Electron accumulation–depletion regions are shown in blue (+) and red (–). Reproduced from Meng *et al.*[82] with permission.

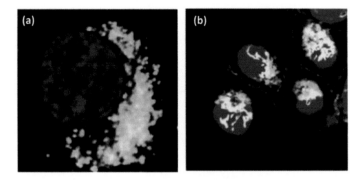

Figure 5.28 Confocal image of *in vitro* HeLa cells after a 12 h incubation in a fluorescent-DNA-SWCNT solution. (a) Dual detection of fluorescent-DNA-SWCNTs (green) internalised into a HeLa cell with the nucleus stained by DRAQ5 (red). (b) Co-localisation (yellow) of fluorescent DNA (green) in cell nucleus (red), after NIR irradiation of 2.5–5 mg/L of SWCNT-DNA, indicating translocation of DNA to the nucleus. Reproduced from Kam *et al.*[92] with permission.

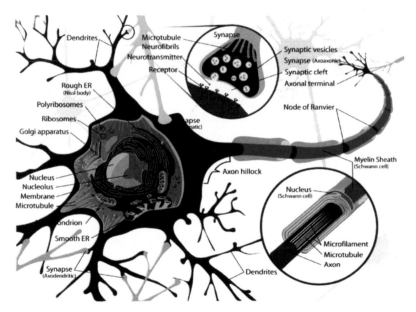

Figure 6.1 Structure of a typical myelinated vertebrate motoneuron.

Figure 7.7 Poly(9,9-dioctyfluorenyl-2,7-diyl)-wrapped CNTs (7,5) showed fluorescence emission. Reproduced from Chen *et a/*.47.with permission.

Figure 7.9 SWNT-Rituxan targeted B cell and labelled it with NIR fluorescence. T cell, without CD20 receptor, was not labelled. Reproduced from Leeuw *et a/*.56 with permission.

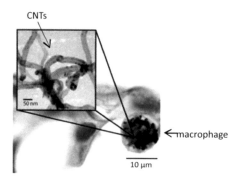

Figure 8.4 Inhaled carbon nanotubes accumulate within cells at the pleural lining of the lung as visualised by light microscopy. Reproduced from www.nanotech-now.com/news_images/35119.jpg with permission of Dr James Bonner, North Carolina State University.

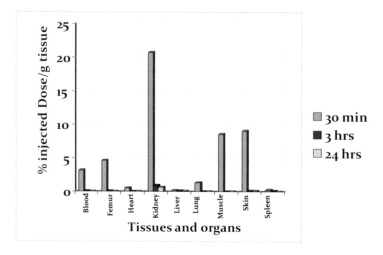

Figure 8.5 Biodistribution per collected gram of tissue of [^{111}In]DTPA–SWCNTs after intravenous administration of 60 µg of SWCNTs in 200 µL of phosphate buffered saline (PBS). Reproduced from Singh *et al.*[56] with permission.

Figure 9.3 Schematic representation of the exciton decay mechanisms: [a) multiphonon decay (MPD) and (b) phonon-assisted indirect exciton ionisation (PAIEI) in CNTs. Figure redrawn from Perebeinos and Avouris.32

Figure 9.7 Schematic representation of non-covalent functionalisations onto gold nanoparticles (NPs), single-walled carbon nanotubes (SWCNTs) and gold nanorods (NRs). Figure modified from Prencipe *etal.*35with permission.

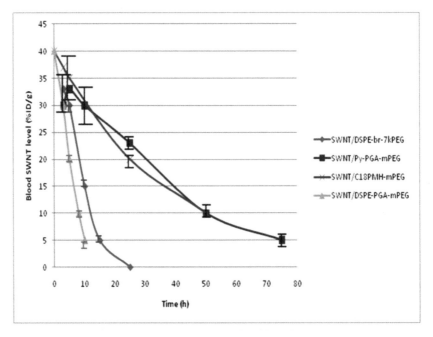

Figure 9.8 Blood circulation data of SWCNTs with different functionalisations in BALB/c mice. Blood circulation curves of **1-3** coated nanotubes compared with DSPE-branched-7kPEG coated nanotubes. The latter one was previously reported by the same research group. Graph modified from Prencipe *et al.*[55] with permission.

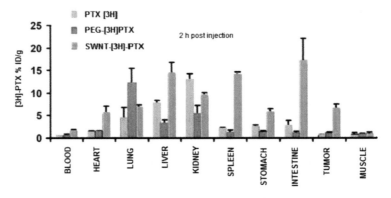

Figure 9.10 ^{3}H-PTX biodistribution in 4T1 tumour-bearing mice injected with ^{3}H-labelled Taxol, PEG-PTX and SWCNT-PTX at 2 hours after injection. Reproduced from Liu *et al.*[60] with permission.

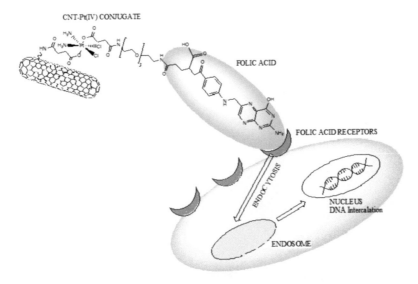

Figure 9.11 Cancer cell uptake of SWCNT-Pt(IV)-FA complexes and subsequent reduction and release of Pt(II) derivative into the nuclear DNA. Figure modified from Dhar *et al.*[68] with permission.

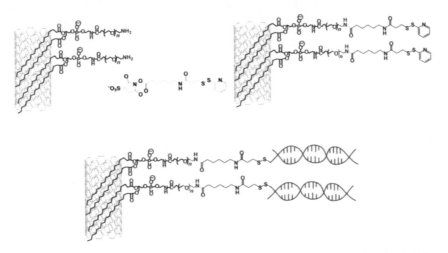

Figure 9.12 A scheme showing siRNA conjugation to SWCNTs through a disulphide bond. PL-PEG2000-amine-functionalised SWCNTs are activated by the sulpho-LC-SPDP bifunctional linker. The pyridyl disulphide group can then be coupled to thiolated siRNA to create a disulphide linkage through a thiol exchange reaction. Reproduced from Liu *et al.*[70] with permission.

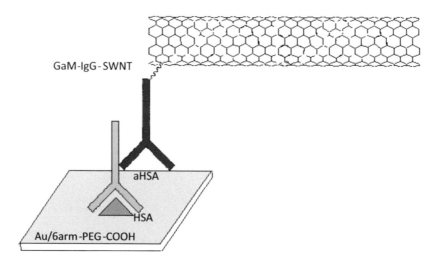

GaM-IgG-SWNT

aHSA

HSA

Au/6arm-PEG-COOH

Figure 9.13 Sandwich assay scheme. Immobilised proteins in a surface spot were used to capture an analyte (antibody) from a serum sample. Detection of the analyte by Raman scattering measurement was carried out after incubation of SWCNTs conjugated to goat anti-mouse antibody (GaM-IgG–SWCNTs), specific to the captured analyte. Figure redrawn from Chen *et al.*[71]

Figure 9.14 SWCNTs with different Raman colours, (a) Schematic of SWCNTs conjugated with different targeting ligands, *(b)* Solution-phase Raman spectra of the three SWCNT conjugates under 785 nm laser excitation. Figure partially modified from Liu *et al.* 72 with permission.

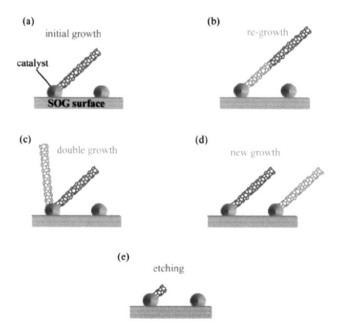

Figure 9.15 Structure of SWCNT-RGD. Figure redrawn from De La Zerda *et al.*[67]

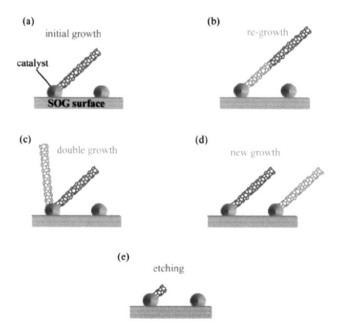

Figure 9.17 Schematic representations of (a) initial growth, (b) regrowth (amplification), (c) double growth, (d) new growth (nucleation from initially inactive catalysts) and (e) CNT etching. The red CNT represents material grown during the first growth run, whereas the green CNT represents new material formed during subsequent growth runs. Reproduced from Crouse *et al.*[112] with permission.

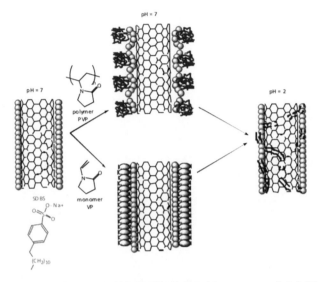

Figure 9.21 Scheme of (Left) SDBS-SWCNT at pH 7. (Centre-up and right) PVP-SDBS-SWCNT at pH 7 and 2. (Centre-below and right) VP-SDBS-SWCNT at pH 7 and 2 for which cationic polymerisation occurred. Figure modified from Duque *et al.*[131] with permission.

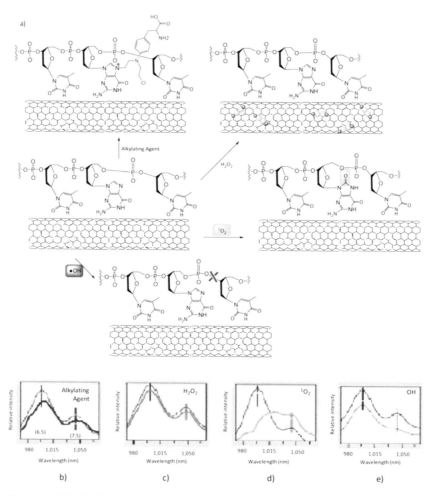

Figure 9.22 Multimodal detection of four reaction pathways. (a) Scheme of interactions on the SWCNT-DNA complex – an alkylating agent reaction with guanine, H_2O_2 adsorption on the nanotube sidewall, a singlet oxygen ($^1O_2$) reaction with DNA, and hydroxyl radical (•OH) damage to DNA. (b–e) SWCNT-DNA photoluminescence spectra before (grey) and after introducing the alkylating agent (red) (b), H_2O_2 (blue) (c), singlet oxygen (yellow) (d) and hydroxyl radicals (green) (e). Figure redrawn from Heller *et al.*[137]

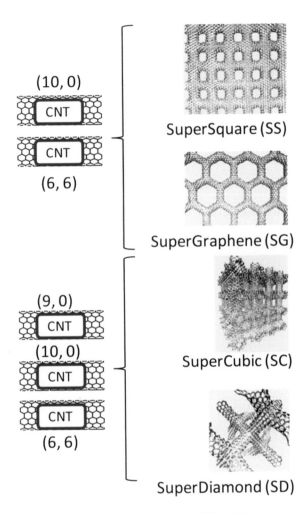

Figure 9.26 Ordered networks based on CNTs (1D nanostructures). (a, b) Super-square and super-graphene correspond to 2D networks, whereas (c, d) super-cubic and super-diamond represent 3D network examples. The four families are constructed from either armchair or zigzag CNTs in order to study the chirality effects. The red rings point out the non-hexagonal carbon rings in each node. Figure modified from Romo-Herrera *et al.*[164] with permission.

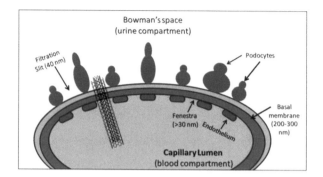

Figure 9.28 Schematic representation of renal clearance of MWCNTs. MWCNT (in black, with diameter ≤40 nm) a few minutes after intravenous (tail vein) injection. The tubes can crossthe renal filtration membrane after orienting perpendicularly.

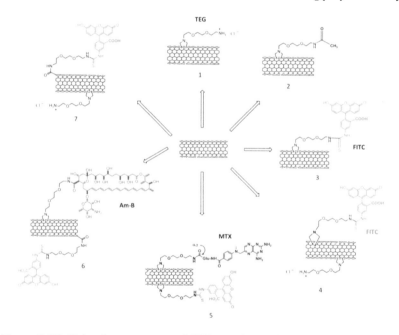

Figure 9.29 Molecular structures of CNTs covalently functionalised with different types of small molecules. (1) Ammonium-functionalised CNTs, (2) acetamido-functionalised CNTs, (3) CNTs functionalised with fluorescein isothiocyanate (FITC), (4) CNTs bifunctionalised with ammonium groups and FITC, (5) CNTs bifunctionalised with methotrexate (MTX) and FITC, (6) shortened CNTs bifunctionalised with amphotericin B (AmB) and FITC, and (7) shortened CNTs bifunctionalised with ammonium groups (via 1,3-dipolar cycloaddition) and FITC (through an amide linkage). Figure redrawn from Kostarelos *et al.*[180]

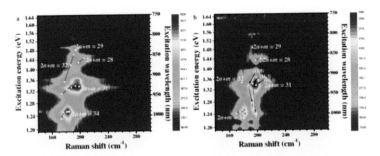

Figure 9.31 2D Raman intensity maps of RBM phonon regions of (a) SWCNT control samples and (b) C_{60} nanopeapods in SDBS-D_2O solution. Reproduced with permission from Joung *et al.*[196] Copyright (2009) by the American Physical Society.

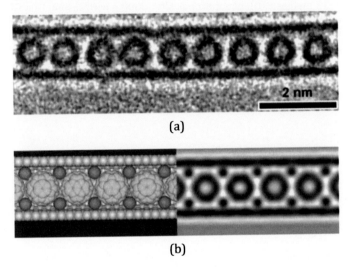

(a)

(b)

Figure 9.32 (a) TEM images of C_{60} nanopeapods after doping with cesium INSIDE. (b) Schematic illustration of the ordered Cs_2C_{60} structure inside a SWCNT (*right*) and its simulated TEM image (*left*). Reproduced from Sato *et al.*[199] with permission.

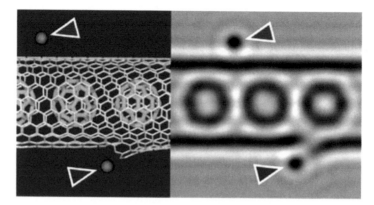

Figure 9.33 Images of C_{60} nanopeapods after doping with cesium OUTSIDE. Cs⁺ ions trapped at the defects in the nanotubes are indicated by blue triangles, and that of the defect-free part of the nanotube wall is indicated by a red triangle. Reproduced with permission from Sato *et al.*[199]

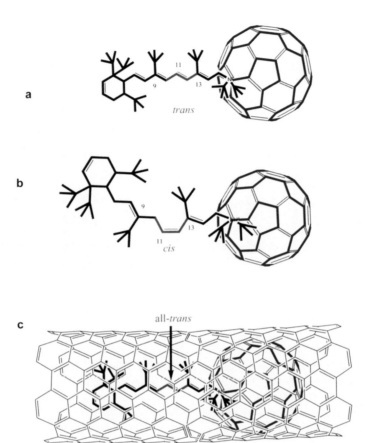

Figure 9.36 (a, b) Two isomeric forms of retinal attached onto C$_{60}$ molecules: *trans* (a) and *cis* (b). (c) Best-fit model of a retinal-C$_{60}$ molecule inside an SWCNT, suggesting that the image in C is of the all-*trans* isomer. Figure redrawn from Liu *et al.*[204]

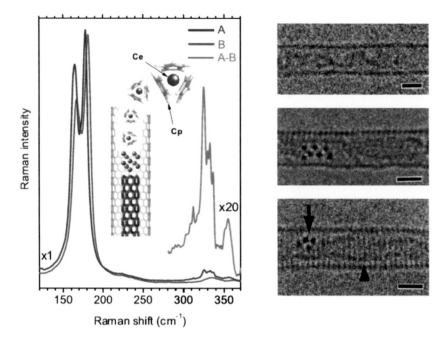

Figure 9.38 *Left*: Raman spectra for Ce@DWCNTs (A), CeCp$_3$@SWCNTs (B), and their difference (A–B) at the radial breathing mode region. *Right*: TEM micrographs of CeCp$_3$@SWCNT (*top*) and Ce@DWCNT (*middle* and *bottom*). Scale bar = 1 nm. Reprinted with permission from Shiozawa *et al.*[220] Copyright (2009) by the American Physical Society.

Index